Bio Mat
£35⁰⁰

D0710305

Applied Mixed Models in Medicine

Applied Mixed Models in Medicine

Helen Brown

and

Robin Prescott

Medical Statistics Unit
Department of Community Health Sciences
University of Edinburgh
UK

JOHN WILEY & SONS, LTD

Chichester • New York • Weinheim • Brisbane • Singapore • Toronto

Copyright © 1999 by John Wiley & Sons Ltd
Baffins Lane, Chichester,
West Sussex, PO19 1UD, England

National　　01243 779777
International　(+44) 1243 779777

e-mail (for orders and customer service enquiries): cs-books@wiley.co.uk

Visit our Home Page on http://www.wiley.co.uk or http://www.wiley.com

Reprinted March 2001, December 2001

All Rights Reserved. No part of this publication may be reproduced, stored in a retrieval system,
or transmitted, in any form or by any means, electronic, mechanical, photocopying, recording,
scanning or otherwise, except under the terms of the Copyright, Designs and Patents Act 1988
or under the terms of a licence issued by the Copyright Licensing Agency, 90 Tottenham Court
Road, London, W1P 9HE, UK, without the permission in writing of the Publisher.

Other Wiley Editorial Offices

John Wiley & Sons, Inc., 605 Third Avenue,
New York, NY 10158-0012, USA

Wiley-VCH Verlag GmbH, Pappelallee 3,
D-69469 Weinheim, Germany

Jacaranda Wiley Ltd, 33 Park Road, Milton,
Queensland 4064, Australia

John Wiley & Sons (Asia) Pte Ltd, 2 Clementi Loop #02-01,
Jin Xing Distripark, Singapore 129809

John Wiley & Sons (Canada) Ltd, 22 Worcester Road,
Rexdale, Ontario, M9W 1L1, Canada

Library of Congress Cataloging-in-Publication Data

Brown, Helen.
　　Applied mixed models in medicine / Helen Brown and Robin Prescott.
　　　　p.　cm. — (Statistics in practice)
　　Includes bibliographical references and index.
　　ISBN 0-471-96554-5 (alk. paper)
　　　1. Medicine — Research — Statistical methods.　2. Medicine-
　-Mathematical models.　3. Statistics.　I. Prescott, Robin.
　II. Title.　III. Series: Statistics in practice (Chichester,
　England)
　　　[DNLM: 1. Medicine.　2. Statistics — methods.　3. Models,
　Statistical.　QH 323.5 B878a 1999]
　R853.S7B76　1999
　610'.72'7 — dc21
　DNLM/DLC
　for Library of Congress　　　　　　　　　　　　　99-14216
　　　　　　　　　　　　　　　　　　　　　　　　　　CIP

British Library Cataloguing in Publication Data

A catalogue record for this book is available from the British Library

ISBN 0-471-96554-5

Typeset in 10/12pt Photina by Laser Words, Madras, India
Printed and bound in Great Britain by Antony Rowe Ltd, Chippenham, Wiltshire
This book is printed on acid-free paper responsibly manufactured from sustainable forestry
in which at least two trees are planted for each one used for paper production.

Contents

Preface

Analysis of variance and regression analysis have for many years been the mainstay of statistical modelling. These techniques usually have as a basic assumption that the residual or error terms are independently and identically distributed. Mixed models are an important new approach to modelling which allows us to relax the independence assumption and take into account more complicated data structures in a flexible way. Sometimes this interdependence of observations is modelled directly in a mixed model. For example, if a number of repeated measurements are made on a patient, mixed models allow us to specify a pattern for the correlation between these measurements. In other contexts, such as the cross-over clinical trial, specifying that patient effects are normally distributed, rather than fixed as in the classical approach, induces observations on the same patient to be correlated.

There are many benefits to be gained from using mixed models. In some situations the benefit will be an increase in the precision of our estimates. In others, we will be able to make wider inferences. We will sometimes be able to use a more appropriate model which will give us greater insight into what underpins the structure of the data. However, it is only the relatively recent availability of software in versatile packages such as SAS® which has made these techniques widely accessible. It is now important that suitable information on their use becomes available so that they may be applied confidently on a routine basis.

Our intention in this book is to put all types of mixed model into a general framework and to consider the practical implications of their use. We aim to do this at a level that can be understood by applied statisticians and numerate scientists. Greatest emphasis is placed on skills required for their application and interpretation. An in-depth understanding of the mathematical theory underlying mixed models is not essential to gain these, but an awareness of the practical consequences of fitting different types of mixed models is necessary. While many publications are available on various aspects of mixed models, these generally relate to specific types of model and often differ in their use of terminology. Such publications are not always readily comprehensible to the applied statisticians who will be the most frequent users of the methods. An objective of this book is to help overcome this deficit.

Examples given will primarily relate to the medical field. However, the general concepts of mixed models apply equally to many other areas of application, e.g. social sciences, agriculture, official statistics. (In the social sciences mixed models are often referred to as 'multi-level' models.) Data are becoming easier to collect with the consequence that datasets are now often large and complex. We believe that mixed models provide useful tools for modelling the complex structures that occur in such data.

Chapter 1 provides an introduction to the capabilities of mixed models, defines general concepts and gives their basic statistical properties. Chapter 2 defines models and fitting methods for normally distributed data. Chapter 3 first introduces generalised linear models which can be used for the analysis of data which are binomial or Poisson or from any other member of the exponential family of distributions. These methods are then extended to incorporate mixed model concepts under the heading of generalised linear mixed models. The fourth chapter then examines how mixed models can be applied when the variable to be analysed is categorical. The main emphasis in these chapters, and indeed in the whole book, is on classical statistical approaches to inference, based on significance tests and confidence intervals. However, the Bayesian approach is also introduced, since it has several potential advantages and its use is becoming more widespread. Although the overall emphasis of the book is on the application of mixed model techniques, these chapters can also be used as a reference guide to the underlying theory of mixed models.

Chapters 5–7 consider the practical implications of using mixed models for particular designs. Each design illustrates a different feature of mixed models.

Multi-centre trials and meta-analyses are considered in Chapter 5. These are examples of hierarchical data structures and the use of a mixed model allows for any additional variation in treatment effects occurring between centres (or trials) and hence makes results more generalisable. The methods shown can be applied equally to any type of hierarchical data.

In Chapter 6 the uses of covariance pattern models and random coefficients models are described using the repeated measures design. These approaches take into account the correlated nature of the repeated observations and give more appropriate treatment effect estimates and standard errors. The material in this chapter will apply equally to any situation where repeated observations are made on the same units.

Chapter 7 considers cross-over designs where each patient may receive several treatments. In this design more accurate treatment estimates are often achieved by fitting patient effects as random. This improvement in efficiency can occur for any dataset where a fixed effect is 'crossed' with a random effect.

In Chapter 8 a variety of other designs and data structures are considered. These either incorporate several of the design aspects covered in Chapters 5–7 or have structures that have arisen in a more unplanned fashion. They help to illustrate the broad scope of application of mixed models.

Chapter 9 gives information on software available for fitting mixed models. Most of the analyses in the book are carried out using PROC MIXED in SAS and we give basic details on its usage. This information should be sufficient for fitting most of the normal mixed model analyses described. However, the full SAS documentation on PROC MIXED should be used by those who wish to use more complex features. The SAS code used for most of the examples is supplied within the text. Additionally, the example datasets and SAS code may be obtained electronically from www.med.ed.ac.uk/phs/mixed.

This book has been written to provide the reader with a thorough understanding of the concepts of mixed models and we trust it will serve well for this purpose. However, readers wishing to take a short cut to the fitting of normal mixed models should read Chapter 1 for an introduction, Section 2.4 for practical details, and the chapter relevant to their design. To fit non-normal or categorical mixed models, Section 3.3 or Section 4.4 should be read in addition to Section 2.4. In an attempt to make this book easier to use, we have presented at the beginning of the text a summary of the notation we have used, while at the end we list some key definitions in a glossary.

Our writing of this book has been aided in many ways. It has evolved from a constantly changing set of course notes that has accompanied a three-day course on the subject, run regularly over the past six years. We are very grateful to participants who have contributed to discussions and have thereby helped to shape our views. We are also grateful to other colleagues who have read and commented on various sections of the manuscript. We hope that readers will find the resulting book a useful reference in an interesting and expanding area of statistics.

Helen Brown
Robin Prescott
Edinburgh
1999

Series Preface

Statistics in Practice is an important international series of texts which provide detailed coverage of statistical concepts, methods and worked case studies in specific fields of investigation and study.

With sound motivation and many worked practical examples, the books show in down-to-earth terms how to select and use an appropriate range of statistical techniques in a particular practical field within each title's special topic area.

The books meet the need for statistical support required by professionals and research workers across a range of employment fields and research environments. The series covers a variety of subject areas: in medicine and pharmaceutics (e.g. in laboratory testing or clinical trials analysis): in industry, finance and commerce (e.g. for design or forecasting): in the public services (e.g. in forensic science): in the earth and environmental sciences, and so on.

But the books in the series have an even wider relevance than this. Increasingly, statistical departments in universities and colleges are realizing the need to provide at least a proportion of their course-work in highly directed areas of study to equip their graduates for the work environment. For example, it is common for courses to be given on statistics applied to medicine, industry, social and administrative affairs etc., and the books in this series provide support for such courses.

It is our aim to present judiciously chosen and well-written workbooks to meet everyday practical needs. Feedback of views from readers will be most valuable to monitor the success of this aim.

Vic Barnett
Series Editor
1999

Mixed Models Notation

The notation below is provided for quick reference. Models are defined more fully in Sections 2.1, 3.1 and 4.1.

Normal mixed model

$$\mathbf{y} = \mathbf{X}\boldsymbol{\alpha} + \mathbf{Z}\boldsymbol{\beta} + \mathbf{e},$$
$$\boldsymbol{\beta} \sim N(\mathbf{0}, \mathbf{G}),$$
$$\text{var}(\mathbf{e}) = \mathbf{R},$$
$$\text{var}(\mathbf{y}) = \mathbf{V} = \mathbf{Z}\mathbf{G}\mathbf{Z}' + \mathbf{R}.$$

Generalised linear mixed model

$$\mathbf{y} = \boldsymbol{\mu} + \mathbf{e},$$
$$g(\boldsymbol{\mu}) = \mathbf{X}\boldsymbol{\alpha} + \mathbf{Z}\boldsymbol{\beta},$$
$$\boldsymbol{\beta} \sim N(\mathbf{0}, \mathbf{G}),$$
$$\text{var}(\mathbf{e}) = \mathbf{R},$$
$$\text{var}(\mathbf{y}) = \mathbf{V} = \text{var}(\boldsymbol{\mu}) + \mathbf{R},$$
$$\approx \mathbf{B}\mathbf{Z}\mathbf{G}\mathbf{Z}'\mathbf{B} + \mathbf{R}, \quad \text{(a first-order approximation)}$$

where
$\mathbf{y}$ = dependent variable,
$\mathbf{e}$ = residual error,
$\mathbf{X}$ = design matrix for fixed effects,
$\mathbf{Z}$ = design matrix for random effects,
$\boldsymbol{\alpha}$ = fixed effects parameters,
$\boldsymbol{\beta}$ = random effects parameters,
$\mathbf{R}$ = residual variance matrix,
$\mathbf{G}$ = matrix of covariance parameters,
$\mathbf{V}$ = var($\mathbf{y}$) variance matrix,
$\boldsymbol{\mu}$ = expected values,
$\mathbf{g}$ = link function,
$\mathbf{B}$ = diagonal matrix of variance terms (e.g. $\mathbf{B} = \text{diag}\{\mu_i(1 - \mu_i)\}$ for binary data)

Ordered categorical mixed model

$\mathbf{y} = \boldsymbol{\mu} + \mathbf{e}$,

$\text{logit}(\boldsymbol{\mu}^{[c]}) = \mathbf{X}\boldsymbol{\alpha} + \mathbf{Z}\boldsymbol{\beta}$,

$\boldsymbol{\beta} \sim N(\mathbf{0}, \mathbf{G})$,

$\text{var}(\mathbf{y})$ is defined as in the GLMM,

where

$\boldsymbol{\mu} = (\mu_{11}, \mu_{12}, \mu_{13}, \ \mu_{21}, \mu_{22}, \mu_{23}, \ \dots, \ \mu_{n1}, \mu_{n2}, \mu_{n3})'$,

$\mu_{ij} = $ probability observation i is in category j,

$\boldsymbol{\mu}^{[c]} = (\mu_{11}^{[c]}, \mu_{12}^{[c]}, \mu_{13}^{[c]}, \ \mu_{21}^{[c]}, \mu_{22}^{[c]}, \mu_{23}^{[c]}, \ \dots \ \mu_{n1}^{[c]}, \mu_{n2}^{[c]}, \mu_{n3}^{[c]})'$,

$\mu_{ij}^{[c]} = $ probability $(y_i <= j) = \sum_{k=1}^{j} \mu_{ik}$.

1

Introduction

At the start of each chapter we will 'set the scene' by outlining its content. In this introductory chapter we start in Section 1.1 by describing some situations where a mixed model analysis will be particularly helpful. In Section 1.2 we describe a simplified example and use it to illustrate the idea of a statistical model. We then introduce and compare fixed effects and random effects models. In the next section we consider a more complex 'real life' multi-centre trial and look at some of the variety of models which could be fitted (Section 1.3). This example will be used for several illustrative examples throughout the book. In Section 1.4 the use of mixed models to analyse a series of observations (repeated measures) is considered. Section 1.5 broadens the discussion on mixed models, and looks at mixed models with a historical perspective of their use. In Section 1.6 we introduce some technical concepts: containment, balance and error strata.

We will assume in our presentation that the reader is already familiar with some of the basic statistical concepts as found in elementary statistical textbooks.

1.1   THE USE OF MIXED MODELS

In the course of this book we will encounter many situations in which a mixed model approach has advantages over the conventional type of analysis which would be accessible via introductory texts on statistical analysis. Some of them are introduced in outline here, and will be dealt with in detail later on.

Example 1: Utilisation of incomplete information in a cross-over trial   Cross-over trials are often utilised to assess treatment efficacy in chronic conditions, such as asthma. In such conditions an individual patient can be tested for response to a succession of two or more treatments, giving the benefit of a 'within-patient' comparison. In the most commonly used cross-over design, just two treatments are compared. If, for generality, we call these treatments A and B, then patients will be assessed either on their response to treatment A, followed by their response to treatment B, or vice versa. If all patients complete the trial, and both treatments are assessed, then the analysis is fairly straightforward. However, commonly, patients drop out during the trial, and may have a valid observation from only the first treatment period. These incomplete observations cannot be utilised in a conventional analysis. In contrast, the use of a mixed model will allow all of the observations to be analysed, resulting in more accurate comparisons of the efficacy of treatment. This benefit, of more efficient use of the data, applies to all types of cross-over trial where there are missing data.

Example 2: Cross-over trials with fewer treatment periods than treatments   In cross-over trials, for logistical reasons, it may be impractical to ask a patient to evaluate more than two treatments (for example, if the treatment has to be given for several weeks). Nevertheless, there may be the need to evaluate three or more treatments. Special types of cross-over design can be used in this situation, but a simple analysis will be very inefficient. Mixed models provide a straightforward method of analysis which fully uses the data, resulting again in more precise estimates of the effect of the treatments.

Example 3: A surgical audit   A surgical audit is to be carried out to investigate how different hospitals compare in their rates of post-operative complications following a particular operation. As some hospitals carry out the operation commonly, while other hospitals perform the operation rarely, the accuracy with which the complication rates are estimated will vary considerably from hospital to hospital. Consequently, if the hospitals are ordered according to their complication rates, some may appear to be outliers, compared with other hospitals, purely due to chance variation. When mixed models are used to analyse data of this type, the estimates of the complication rates are adjusted to allow for the number of operations, and rates based on small numbers become less extreme.

Example 4: Analysis of a multi-centre trial   Many clinical trials are organised on a multi-centre basis, usually because there is an inadequate number of suitable patients in any single centre. The analysis of multi-centre trials often ignores the centres from which the data were obtained, making the implicit assumption that all centres are identical to one another. This assumption may sometimes be dangerously misleading. For example, a multi-centre trial comparing two surgical treatments for a condition could be expected to show major differences between centres. There could be two types of differences. Firstly, the centres may differ in the overall success, averaged over the two surgical treatments. More importantly, there may be substantial differences in the relative benefit of the two treatments across different centres. Surgeons who have had more experience with one operation (A) may produce better outcomes with A, while surgeons with more experience with the alternative operation (B) may obtain better results with B. Mixed models can provide an insightful analysis of such a trial by allowing for the extent to which treatment effects differ from centre to centre. Even when the difference between treatments can be assumed to be identical in all centres, a mixed model can improve the precision of the treatment estimates by taking appropriate account of the centres in the analysis.

Example 5: Repeated measurements over time   In a clinical trial, the response to treatment is often assessed as a series of observations over time. For example, in a trial to assess the effect of a drug in reducing blood pressure, measurements might be taken at two, four, six and eight weeks after starting treatment. The analysis will usually be complicated by a number of patients failing to appear for some assessments, or withdrawing from the study before it is complete. This complication can cause considerable difficulty in a conventional analysis. A mixed model analysis of such a study does not require complete data from all subjects. This results in more appropriate estimates of the effect of treatment and their standard errors. The mixed model also gives great flexibility in analysis, in that it can allow for a wide variety of ways in which the successive observations are correlated with one another.

1.2 INTRODUCTORY EXAMPLE

We consider a very simple cross-over trial using artificial data. In this trial each patient receives each of treatments A and B for a fixed period of time. At the end of each treatment period, a measurement is taken to assess the response to that treatment. In the analysis of such a trial we commonly refer to treatments being *crossed* with patients, meaning that the categories of 'treatments' occur in combination with the categories of 'patients'. For the purpose of this illustration we will suppose that the response to each treatment is unaffected by whether it is received in the first or second period. The table shows the results from the six patients in this trial.

Patient	Treatment A	B	Difference A − B	Patient mean
1	20	12	8	16.0
2	26	24	2	25.0
3	16	17	−1	16.5
4	29	21	8	25.0
5	22	21	1	21.5
6	24	17	7	20.5
Mean	22.83	18.67	4.17	20.75

1.2.1 Simple model to assess the effects of treatment (Model A)

We introduce here a very simple example of a statistical model using the above data. A model can be thought of as an attempt to describe quantitatively the effect of a number of factors on each observation. Any model we describe is likely to be a gross oversimplification of reality. In developing models we are seeking ones which are as simple as possible, but which contain enough truth to ask questions of interest. In this first simple model we will deliberately be oversimplistic in order to introduce our notation. We just describe the effect of the two treatments. The model may be expressed as

$$y_{ij} = \mu + t_j + e_{ij},$$

where

$j =$ A or B,
$y_{ij} =$ observation for treatment j on the ith patient,
$\mu =$ overall mean,
$t_j =$ effect of treatment j,
$e_{ij} =$ error for treatment j on the ith patient.

The constant μ represents the overall mean of the observations. $\mu + t_A$ corresponds to the mean in the treatment group A, while $\mu + t_B$ corresponds to the mean in the treatment group B. The constants μ, t_A and t_B can thus be estimated from the data. In our example we can estimate the value of μ to be 20.75, the overall mean value. From the mean value in the first treatment group we can estimate $\mu + t_A$ as 22.83 and hence our estimate of t_A is $22.83 - 20.75 = 2.08$. Similarly, from the mean of the second treatment group we estimate t_B as -2.08. The term t_j can therefore be thought of as a measure of the relative effect that treatment j has had on our outcome variable.

The error term, e_{ij}, or *residual* is what remains for each patient in each period when $\mu + t_j$ is deducted from their observed measurement. This represents random variation about the mean value for each treatment. As such, the residuals can be regarded as the result of drawing random samples from a distribution. We will assume that the distribution is a Gaussian or normal distribution, with standard deviation σ, and that the samples drawn from the distribution are independent of each other. The mean of the distribution can be taken as zero, since any other value would simply cause a corresponding change in the value of μ. Thus, we will write this as

$$e_{ij} \sim N(0, \sigma^2),$$

where σ^2 is the variance of the residuals. In practice, checks should be made to determine whether this assumption of normally distributed residuals is reasonable. Suitable checking methods will be considered in Section 2.4.6. As individual observations are modelled as the sum of $\mu + t_j$, which are both constants, plus the residual term, it follows that the variance of individual observations equals the residual variance:

$$\text{var}(y_{ij}) = \sigma^2.$$

The covariance of any two separate observations y_{ij} and $y_{i'j'}$ can be written

$$\text{cov}(y_{ij}, y_{i'j'}) = \text{cov}(\mu + t_i + e_{ij}, \mu + t_{i'} + e_{i'j'})$$

$$= \text{cov}(e_{ij}, e_{i'j'}) \text{ (since other terms are constants)}.$$

Since all the residuals are assumed independent (i.e. uncorrelated), it follows that

$$\text{cov}(y_{ij}, y_{i'j'}) = 0.$$

The residual variance, σ^2, can be estimated using a standard technique known as analysis of variance (ANOVA). The essence of the method is that the total variation in the data is decomposed into components which are associated with possible causes of this variation, e.g. that one treatment may be associated with higher observations, with the other being associated with lower observations. For this first model, using this technique we obtain the following ANOVA table:

Source of variation	Degrees of freedom	Sums of squares	Mean square	F	p
Treatments	1	52.08	52.08	2.68	0.13
Residual	10	194.17	19.42		

Note: F = value for the F test (ratio of mean square for treatments to mean square for residual).
p = significance level corresponding to the F test.

The residual mean square of 19.42 is our estimate of the residual variance, σ^2, for this model. The key question often arising from this type of study is: 'Do the treatment effects differ significantly from each other?' This can be assessed by the F test, which assesses the null hypothesis of no mean difference between the treatments (the larger the treatment difference, the larger the treatment mean square, and the higher the value of F). The p-value of 0.13 is greater than the conventionally used cutoff point for statistical significance of 0.05. Therefore, we cannot conclude here that the treatment effects are significantly different. The difference between the treatment effects and the standard error (SE) of this difference provides a measure of the size of the treatment difference, and the accuracy with which it is estimated:

$$\text{difference} = t_A - t_B = 2.08 + 2.08 = 4.16.$$

The standard error of the difference is given by the formula

$$\text{SE}(t_A - t_B) = \sqrt{\sigma^2(1/n_A + 1/n_B)}$$
$$= \sqrt{(2 \times \sigma^2/6)} = \sqrt{6.47} = 2.54.$$

Note that a t test can also be constructed from this difference and standard error, giving $t = 4.16/2.54 = 1.63$. This is the square root of our F statistic of 2.68 and gives an identical t test p-value of 0.13.

1.2.2   A model taking patient effects into account (Model B)

Model A above did not utilise the fact that pairs of observations were taken on the same patients. It is possible, and indeed likely, that some patients will tend to have systematically higher measurements than others and we may be able to improve the model by making allowance for this. This can be done by additionally including patient effects into the model:

$$y_{ij} = \mu + p_i + t_j + e_{ij},$$

where p_i are constants representing the ith patient effect.

The ANOVA table arising from this model is

Source of variation	Degrees of freedom	Sums of squares	Mean square	F	p
Patients	5	154.75	30.95	3.93	0.08
Treatments	1	52.08	52.08	6.61	0.05
Residual	5	39.42	7.88		

The estimate of the residual variance, σ^2, is now 7.88. It is lower than in Model A because it represents the *within-patient* variation as we have taken account of patient effects. The F test p-value of 0.05 indicates here that the treatment effects are now significantly different. The difference between the treatment effects is the same as in Model A, 4.16, but its standard error is now

$$\mathrm{SE}(t_\mathrm{A} - t_\mathrm{B}) = \sqrt{(2 \times \sigma^2/6)} = \sqrt{2.63} = 1.62.$$

(Note that the standard error of the treatment difference could alternatively have been obtained directly from the differences in patient observations.)

Model B is perhaps the 'obvious' one to think of for the above dataset. However, even in this simple case, by comparison with Model A we can see that the statistical modeller has some flexibility in his/her choice of model. In most situations there is no single 'correct' model and in fact models are rarely completely adequate. The job of the statistical modeller is to choose that model which most closely achieves the objectives of the study.

1.2.3   Random effects model (Model C)

In Models A and B the only assumption we made about variation was that the residuals were normally distributed. We did not assume that patient or treatment effects arose from a distribution. They were assumed to take constant values. These models can be described as *fixed effects models* and all effects fitted within them are *fixed effects*.

An alternative approach available to us is to assume that some of the terms in the model, instead of taking constant values, are realisations of values from a probability distribution. If we assumed that patient effects also arose from independent samples from a normal distribution, then the model could be expressed as

$$y_{ij} = \mu + p_i + t_j + e_{ij},$$
$$e_{ij} \sim N(0, \sigma^2)$$

and

$$p_i \sim N(0, \sigma_p^2).$$

The p_i are now referred to as *random effects*. Such models, which contain a mixture of fixed effects and random effects, provide an example of a *mixed model*. In this book we will meet several different types of mixed model and we describe in Section 1.5 the common feature which distinguishes them from fixed effects models. To distinguish the class of models we have just met from those we will meet later, we will refer to this type of model as a *random effects model*.

Each random effect in the model gives rise to what is known as a *variance component*. This is a model parameter that quantifies random variation due to that effect only. Here, the patient variance component is σ_p^2. We can describe variation at this level (between patients) as occurring within the patient *error stratum* (see Section 1.6 for a full description of the error stratum). This random variation occurs in addition to the residual variation. (The residual variance can also be defined as a variance component.)

Defining the model in this way causes some differences in its statistical properties, compared with the fixed effects model met earlier.

The variance of individual observations in a random effects model is the sum of all the variance components. Thus,

$$\text{var}(y_{ij}) = \sigma_p^2 + \sigma^2.$$

This contrasts with the fixed effects models where we had

$$\text{var}(y_{ij}) = \sigma^2.$$

The effect on the covariance of pairs of observations in the random effects model is interesting, and perhaps surprising. Since $y_{ij} = \mu + p_i + t_j + e_{ij}$, we can write

$$\text{cov}(y_{ij}, y_{i'j'}) = \text{cov}(\mu + p_i + t_j + e_{ij}, \quad \mu + p_{i'} + t_{j'} + e_{i'j'})$$
$$= \text{cov}(p_i + e_{ij}, p_{i'} + e_{i'j'}).$$

When observations from different patients are being considered (i.e. $i \neq i'$), because of the independence of the observations, $\text{cov}(y_{ij}, y_{i'j'}) = 0$. However, when two samples from the same patient are considered (i.e. $i = i'$), then

$$\text{cov}(y_{ij}, y_{i'j'}) = \text{cov}(p_i + e_{ij}, p_i + e_{ij'})$$
$$= \text{cov}(p_i, p_i) = \sigma_p^2.$$

Thus, observations on the same patient are correlated and have covariance equal to the patient variance component, while observations on different patients are uncorrelated. This contrasts with the fixed effects models where the covariance of any pair of observations is zero.

The ANOVA table for the random effects model is identical to that for the fixed effects model. However, we can now use it to calculate the patient variance component using results from the statistical theory which underpins the ANOVA method. The theory tells us the expected values for each of the mean square terms in the ANOVA table, in terms of σ^2, σ_p^2 and the treatment effects. These are tabulated below. We can now equate the expected value for the mean squares expressed in terms of the variance components to the observed values of the mean squares to obtain estimates of σ^2 and σ_p^2.

Source of variation	Degrees of freedom	Sums of squares	Mean square	E(MS)
Patients	5	154.75	30.95	$2\sigma_p^2 + \sigma^2$
Treatments	1	52.08	52.08	$\sigma^2 + 6\Sigma t_i^2$
Residual	5	39.42	7.88	σ^2

Note: E(MS) = expected mean square.

Thus, from the residual line in the ANOVA table, $\hat{\sigma}^2 = 7.88$. Also, by subtracting the third line of the table from the first:

$$2\hat{\sigma}_p^2 = (30.95 - 7.88), \text{ and } \hat{\sigma}_p^2 = 11.54$$

(We are introducing the notation $\hat{\sigma}_p^2$ to denote that this is an estimate of the unknown σ_p^2, and $\hat{\sigma}^2$ is an estimate of σ^2.)

In this example we obtain identical treatment effect results to those for the fixed effects model (Model B). This occurs because we are, in effect, only using within-patient information to estimate the treatment effect (since all information on treatment occurs in the within-patient residual error stratum). Again, we obtain the treatment difference as -4.16 with a standard error of 1.62. Thus, here it makes no difference at all to our conclusions about treatments whether we fit patient effects as fixed or random. However, had any of the values in the dataset been missing this would not have been the case. We now consider this situation.

Dataset with missing values

We will now consider analysing the dataset with two of the observations set to missing.

Patient	Treatment A	Treatment B	Difference A − B	Patient mean
1	20	12	8	16.0
2	26	24	2	25.0
3	16	17	−1	16.5
4	29	21	8	25.0
5	22	—	—	22.0
6	—	17	—	17.0
Mean			4.25	

As before, there are two ways we can analyse the data. We can base our analysis on a model where the patient effects are regarded as fixed (Model B) or we can regard patient effects as random (Model C).

The fixed effects model   For this analysis we apply ANOVA in the standard way, and the result of that analysis is summarised below.

Source of variation	Degrees of freedom	Sums of squares	Mean square	F	p
Patients	5	167.90	33.58	3.32	0.18
Treatments	1	36.13	36.13	3.57	0.16
Residual	3	30.38	10.12		

In the fitting of Model B it is interesting to look at the contribution which the data from patient 5 is making to the analysis. The value of 22 gives us information which will allow us to estimate the level in that patient, but it tells us nothing at all about the difference between the two treatments. Nor does it even tell us anything about the effect of treatment A, which was received, because all the information in the observed value of 22 is used up in estimating the patient effect. The same comment applies to the data from patient 6.

Thus, in this fixed effects model the estimate of the mean treatment difference, $\hat{t}_{FE}$, will be calculated only from the treatment differences for patients 1–4, who have complete data:

$$\hat{t}_{FE} = 4.25.$$

The variance of $\hat{t}_{FE}$ can be calculated from the residual, $\hat{\sigma}^2 = 10.12$, as

$$\mathrm{var}(\hat{t}_{FE}) = \hat{\sigma}^2(1/n_p + 1/n_p) = 10.12 \times (1/4 + 1/4) = 5.06,$$

where n_p is the number of observations with data on treatments A and B. The standard error of the treatment difference is $\sqrt{5.06} = 2.25$.

The random effects model   When patient effects are fitted as random, the variance components cannot be derived in a straightforward way from an ANOVA table since the data are unbalanced. They are found computationally (using PROC MIXED, a SAS procedure, which is described more fully in Chapter 9) as

$$\hat{\sigma}_p^2 = 12.63,$$

$$\hat{\sigma}^2 = 8.90.$$

The treatment difference is estimated from the model to be 4.32 with a standard error of 2.01. Thus, the standard error is smaller than the standard error of 2.25 obtained in the fixed effects model. This is due partly to a fortuitously lower estimate of σ^2, but is also due to the fact that the random effects model utilises information on treatment from both the patient error stratum (between patients) and from the residual stratum (within patients). As noted above, the standard error of the estimates is less than that in the fixed effects model, which only uses information from within patients. The use of this extra information compared with the fixed effects model can be referred to as the *recovery* of between-patient information.

In practice, we would recommend that random effects models are always fitted computationally using a procedure such as PROC MIXED. However, in our simple example here it may be of help to the understanding of the concept of recovery of information if we illustrate how the treatment estimates can be obtained manually.

Manual calculation   In this example the estimate of the treatment difference for the random effects model may be obtained by combining estimates from the between-patient and within-patient (residual) error strata. It is calculated by a weighted average of the two estimates with the inverses of the variances of the estimates used as weights. The within-patient estimate, $\hat{t}_W$, is obtained as in the fixed effects model from patients 1–4 as 4.25. However, its variance is now calculated from the new estimate of σ^2 as

$$\text{var}(\hat{t}_W) = \sigma^2(1/n_p + 1/n_p) = 8.90 \times (1/4 + 1/4) = 4.45.$$

The between-patient estimate, $\hat{t}_B$, here is simply the difference between the single values for patients 5 and 6:

$$\hat{t}_B = 22 - 17 = 5$$

and has variance

$$\text{var}(\hat{t}_p) = (\sigma^2 + \sigma_p^2) \times (1/1 + 1/1) = (8.90 + 12.63) \times 2 = 43.06.$$

The combined random effects model estimate, $\hat{t}_{RE}$, is obtained as a weighted average of $\hat{t}_W$ and $\hat{t}_B$:

$$\hat{t}_{RE} = K \times (\hat{t}_W/\text{var}(\hat{t}_W) + \hat{t}_p/\text{var}(\hat{t}_p)),$$

where
$$K = 1/(1/\operatorname{var}(\hat{t}_W) + 1/\operatorname{var}(\hat{t}_B)).$$

For our data,
$$K = (1/(1/4.45 + 1/43.06)) = 4.03,$$

giving,
$$\hat{t}_{RE} = 4.03 \times (4.25/4.45 + 5/43.06) = 4.03 \times 1.07 = 4.32.$$

To calculate $\operatorname{var}(\hat{t}_{RE})$ we use the property, $\operatorname{var}(nx) = n^2 \operatorname{var}(x)$, so that
$$\operatorname{var}(\hat{t}_{RE}) = K^2 \times \{\operatorname{var}(\hat{t}_W)/(\operatorname{var}(\hat{t}_W))^2 + \operatorname{var}(\hat{t}_B)/(\operatorname{var}(\hat{t}_B))^2\},$$

giving
$$\operatorname{var}(\hat{t}_{RE}) = K^2 \times (1/\operatorname{var}(\hat{t}_W) + 1/\operatorname{var}(\hat{t}_B))$$
$$= K.$$

Thus, for our data:
$$\operatorname{var}(\hat{t}_{RE}) = 4.03,$$

and
$$\operatorname{SE}(\hat{t}_{RE}) = 2.01.$$

These results are identical to those we obtained initially using PROC MIXED. However, it is not usually quite so simple to combine estimates manually from different error strata. A general formula for calculating the mixed model, fixed effect estimates for all types of mixed model will be given in Section 2.2.2.

The point which we hope has been made clear by the example is the way in which the random effects model has used the information from patients 5 and 6, which would have been lost in a fixed effects analysis.

1.2.4 Estimation (or prediction) of random effects

In the previous model the patient terms were regarded as random effects. That is, they were defined as realisations of samples from a normal distribution with mean equal to zero, and with variance σ_p^2. Thus, their expected values are zero. We know, however, that patients may differ from one another, and the idea that all have the same expected value is counter-intuitive. We resolve this paradox by attempting to determine for each individual patient a *prediction* of the location within the normal distribution from which that patient's observations have arisen. This prediction will be affected by the prediction for all other patients, and will differ from the corresponding estimate in the fixed effects model. The predictions will be less widely spread than the fixed effects estimates and because of this they are described as *'shrunken'*. The extent of this shrinkage depends on the relative sizes of the patient and residual variance components. In the extreme case where the estimate of the patient variance component is zero, all patients will have equal predictions. Shrinkage will also be relatively greater when there are fewer observations per patient. Shrinkage occurs for both balanced and unbalanced data and the relevant formula is given in Section 2.2.3. Although, on technical grounds, it is more accurate to refer to *predictions* of *random effect categories* (e.g. of individual patients), in this book we will use the more colloquial form of expression, and refer to estimates of patient effects.

In our example, using the complete trial data, the random effect estimates can be obtained computationally using PROC MIXED. They are listed below along with the fixed effects patient means.

Patient number	1	2	3	4	5	6
Fixed patients	16.0	25.0	16.5	25.0	21.5	20.5
Random patients	17.2	23.9	17.6	23.9	21.3	20.6

We observe that the mean estimates are indeed 'shrunken' towards the grand mean of 20.8. Shrinkage has occurred because patients are treated as a sample from the overall patient population.

1.3   A MULTI-CENTRE HYPERTENSION TRIAL

We now introduce a more complex 'real life' clinical trial. Measurements from this trial will be used to provide data for several examples in future chapters. Although it is by no means the only example we will be presenting, by the repeated use of this trial we hope that the reader will identify more readily with the analyses.

The trial was a randomised double blind comparison of three treatments for hypertension and has been reported by Hall *et al.* (1991). One treatment was a new drug (A) and the other two (B and C) were standard drugs for controlling hypertension (A = Carvedilol, B = Nifedipine, C = Atenolol). Twenty-nine centres participated in the trial and patients were randomised in order of entry. Two pre-treatment and four post-treatment visits were made as follows:

- Visit 1 (week 0): Measurements were made to determine whether patients met the eligibility criteria for the trial. Patients who did so received a placebo treatment for one week, after which they returned for a second visit.

- Visit 2 (week 1): Measurements were repeated and patients who still satisfied the eligibility criteria were entered into the study and randomised to receive one of the three treatments.

- Visits 3–6 (weeks 3,5,7,9): Measurements were repeated at four post-treatment visits, which occurred at two-weekly intervals.

- Three hundred and eleven patients were assessed for entry into the study. Of these, 288 patients were suitable and were randomised to receive one of the three treatments. Thirty patients dropped out of the study prior to Visit 6.
- Measurements on cardiac function, laboratory values and adverse events were recorded at each visit. Diastolic blood pressure (DBP) was the primary endpoint and we will consider its analysis here.
- The frequencies of patients attending at least one post-treatment visit at each of the 29 centres are shown in Table 1.1.

Table 1.1 Number of patients included in analyses of final visits by treatment and centre.

Centre	Treatment			Total
	A	**B**	**C**	
1	13	14	12	39
2	3	4	3	10
3	3	3	2	8
4	4	4	4	12
5	4	5	2	11
6	2	1	2	5
7	6	6	6	18
8	2	2	2	6
9	0	0	1	1
11	4	4	4	12
12	4	3	4	11
13	1	1	2	4
14	8	8	8	24
15	4	4	3	11
18	2	2	2	6
23	1	0	2	3
24	0	0	1	1
25	3	2	2	7
26	3	4	3	10
27	0	1	1	2
29	1	0	2	3
30	1	2	2	5
31	12	12	12	36
32	2	1	1	4
35	2	1	1	4
36	9	6	8	23
37	3	1	2	6
40	1	1	0	2
41	2	1	1	4
Total	100	91	94	288

Note: Several additional centres were numbered but did not eventually participate in the study.

1.3.1   Modelling the data

The main purpose of this trial was to assess the effect of the three treatments on the primary endpoint, DBP recorded at the final visit. As in the previous example we can do this by forming a statistical model. We will now describe several possible models. A simple model (Model A) to assess just the effects of treatment could be expressed as

$$DBP_i = \mu + t_k + e_i,$$

where

DBP_i = diastolic blood pressure at final visit for patient i,
μ = intercept,
t_k = kth treatment effect (where patient i has received treatment k),
e_i = error term (residual) for the ith patient.

Before the model is fitted we should be certain that we have the most relevant dataset for our objectives. In this trial 30 patients dropped out of the study before their final visit. If treatments have influenced whether patients dropped out, omitting these patients from the analysis could give rise to biased estimates of treatment effects. We therefore adopt a 'last value carried forward' approach and substitute the last recorded value for the final visit values in these patients. (The issue of how to deal with missing data will be considered again in Section 2.4.7.)

1.3.2   Including a baseline covariate (Model B)

Model A was a very simple model for assessing the effect of treatment on DBP. It is usually reasonable to assume that there may be some relationship between pre- and post-treatment values on individual patients. Patients with relatively high DBP before treatment are likely to have higher values after treatment, and likewise for patients with relatively low DBPs. We can utilise this information in the model by fitting the baseline (pre-treatment) DBP as an additional effect in Model A:

$$DBP_i = \mu + b \cdot pre + t_k + e_i,$$

where

b = baseline covariate effect,
pre = baseline (pre-treatment) DBP.

Here, we will take the values recorded at visit 2 as the baseline values. We could, of course, have considered using either the visit 1 value, or the average of the visit 1 and visit 2 values, instead. The visit 2 value was chosen because it measured the DBP immediately prior to randomisation, after one week during which all patients received the same placebo medication. The baseline DBP is measured on a quantitative scale (unlike treatments). Such quantitative variables are commonly described as *covariate effects* and an analysis based on the above model is often referred to as *analysis of covariance*. The term b is a constant which has to be estimated from our data. There is an implicit assumption in our model that the relationship between the final DBP and the baseline value is linear. Also, that within each treatment group, an increase of 1 unit in the baseline DBP is associated with an average increase of b units in the final DBP. Figure 1.1 shows the results from fitting this model to the data (only a sample of data points is shown, for clarity).

This demonstrates that performing an analysis of covariance is equivalent to fitting separate parallel lines for each treatment to the relationship between post-treatment DBP and baseline DBP. The separation between the lines represents the magnitude of the treatment effects. The analysis will be considered in much greater detail in Section 2.5, but we note for now that two of the treatments appear to be similar to one another, while the lowest post-treatment blood pressures occur with treatment C.

The use of a baseline covariate will usually improve the precision of the estimates of the treatment effects. It will also compensate for any differences between the mean levels of the covariate in the treatment groups prior to treatment being received. Of course, our assumption that there is a linear relationship between pre- and post-treatment values may not be true. If this were the case, fitting a baseline

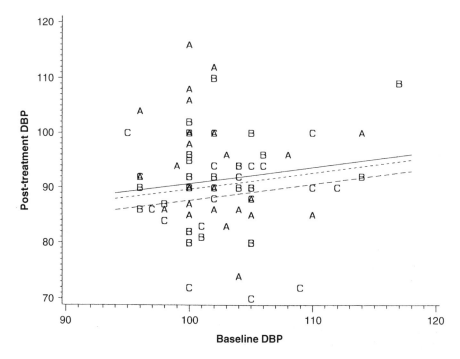

Figure 1.1   Plot to illustrate the analysis of covariance.
Treatment: ———A; - - - - -B; — — — —C.

covariate could lead to less precise results. However, in practice the assumption is very frequently justified in medicine and it has become almost standard to take baseline values into account in the model if they are available.

An alternative way of using baseline values (which we do not recommend) is to analyse the differences between pre- and post-treatment values. However, this generally leads to less accurate results than the 'covariate' approach, particularly when the relationship between pre- and post-treatment values is weak.

1.3.3 Modelling centre effects (Model C)

So far, the model has taken no account of the fact that the data are recorded at different centres. It is possible that values in some centres may tend to be higher than those in other centres. Such differences could be due, for example, to differences in the techniques of personnel across centres. It is also possible that some centres/clinics may recruit patients with differing degrees of severity of hypertension (within the bounds of the study entry criteria) who could, on average, have higher or lower values of DBP. We can allow for these possibilities by adding centre effects to Model B:

$$DBP_i = \mu + b \cdot pre + t_k + c_j + e_i,$$

where

$$c_j = \text{the jth centre effect.}$$

Thus, part of the residual term in Model B may now be explained by the centre effects, c_j. If there are differences between the centres, this model will have a smaller residual variance than Model B (i.e. a smaller σ^2). This in turn allows treatment effects to be calculated with greater accuracy.

1.3.4 Including centre-by-treatment interaction effects (Model D)

In Model C we took account of the fact that there may be an underlying difference in DBP between the centres. We did so in such a way that the effect of a patient being in a particular centre would be additive to the effect of treatment. Another possibility is that the response of patients to treatments may vary between the centres. That is, the effects of centre and treatment are non-additive, or that there is an *interaction*. For example, in any multi-centre trial, if some centres tended to have more severely ill patients, it is plausible that the reaction of these patients to the treatments would differ from that of patients at other centres who are less severely ill. We can take this possibility into account in the model by allowing the treatment effects to vary between the centres. This is achieved by adding a centre·treatment *interaction* to Model C. It causes a separate set of treatment effects to be fitted for each centre.

$$\mathrm{DBP}_i = \mu + b \cdot pre + t_k + c_j + (ct)_{jk} + e_i,$$

where

$(ct)_{jk} =$ the kth treatment effect at the jth centre.

Throughout this book we will refer to such interactions using the notation 'centre·treatment'. When Model D is fitted, the first question of interest is whether the centre·treatment effect is statistically significant. If the interaction term is significant, then we have evidence that the treatment effect differs between the centres. It will then be inadvisable to report the overall treatment effect across the centres. Results will need to be reported for each centre. If the interaction is not significant, centre·treatment may be removed from the model and the results from Model C reported. Further discussion on centre·treatment interactions appears in Chapter 5.

As we will see in more detail in Section 2.5, the centre·treatment effect is non-significant for our data ($p = 0.19$) and the results of Model C can be presented. Centre effects are statistically significant in Model C ($p = 0.004$), so this model will be preferred to Model B.

From our data, b is estimated as 0.22 with a standard error of 0.11. Thus, if the baseline DBPs of two patients receiving the same treatment differ by 10 mmHg, we can expect that their final DBPs will differ by only 2.2 mmHg (0.22×10), as illustrated in Figure 1.1. The relationship is therefore weak and hence we can anticipate that the analysis of covariance approach will be preferable to a simple analysis of change in DBP. In fact, the statistical significance of the treatment differences is $p = 0.054$ using the analysis of covariance, compared with $p = 0.072$ for the analysis of change.

1.3.5 Modelling centre and centre·treatment effects as random (Model E)

Models A−D can all be described as fixed effects models and only the residual term is assumed to have a distribution. Alternatively, we could assume that the centre and centre·treatment effects also arose from a distribution. We again write the model:

$$\mathrm{DBP}_i = \mu + b + t_k + c_j + (ct)_{jk} + e_i,$$

but now we assume that the residual, centre and centre·treatment effects are all realisations of separate distributions, all with zero means:

$$e_i \sim \mathrm{N}(0, \sigma^2),$$

$$c_j \sim \mathrm{N}(0, \sigma_c^2),$$

$$(ct)_{jk} \sim \mathrm{N}(0, \sigma_{ct}^2).$$

Hence, c_j and $(ct)_{jk}$ are now random effects, and b and t_k are fixed effects. This random effects model can be described as *hierarchical* since treatment effects are *contained* within the random centre·treatment effects. The concept of containment will be picked up again in Section 1.6.

Since we have assumed that centre·treatment effects have a distribution, i.e. that differences between treatments vary randomly across the centres, we can relate our results to the population of potential centres. This is in contrast to Model D, where treatment effects are assumed to be specific to the centres observed.

There are no hard and fast rules about whether effects should be modelled as fixed or random (or indeed whether some effects should be fitted at all). Here, various approaches are acceptable but they offer us different interpretations of the results. These various approaches will be discussed in much greater detail in Section 2.5, but for now we pick up on just one point: the precision with which treatment effects are estimated. We have seen above that fitting centre and centre·treatment effects as random enables our inferences to apply to a 'population' of centres. There is a price to be paid, however. The standard errors of the treatment estimates will be inflated, because we allow the treatment effects to vary randomly across centres. Thus, the mean difference in final DBP between treatments A and C is estimated as 2.92 mmHg with a standard error of 1.37 mmHg. In contrast, using Model C, the corresponding estimate is 2.99 mmHg with a smaller standard error of 1.23 mmHg. Arguments in favour of the random effects model are the wider scope of the inferences, and perhaps a more appropriate modelling of the data. In some circumstances, however, it is adequate to establish treatment differences in a specific set of centres. Statisticians in the pharmaceutical industry, for example, may prefer to avoid the penalty of less precise treatment estimates, with a corresponding reduction in *power* (the probability of obtaining statistically significant treatment differences when treatments do differ in their effect), and will often use a fixed effects model. This discussion point will be taken up again in Chapter 5.

1.4   REPEATED MEASURES DATA

There were four post-treatment visits in the multi-centre hypertension trial introduced in the previous section. However, so far in this chapter we have chosen only to model measurements made at the final visit, which were of primary interest. An alternative strategy would be to include measurements from all four post-treatment visits in the model. Since measurements are made repeatedly on the same patients, we can describe this type of data as *repeated measures* data. For illustrative purposes we now assume that the centre has no effect at all on the results, and consider which models are appropriate for analysing repeated measures data. The mean levels for the three treatments at all time points are shown in Figure 1.2.

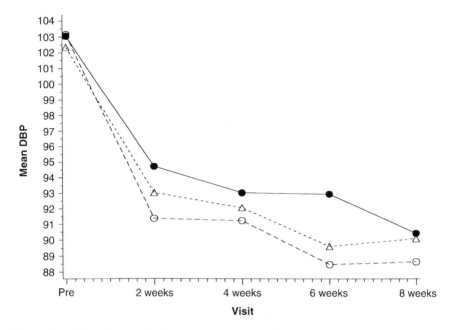

Figure 1.2   Plot of mean DBP by treatment group and visit.
Treatment: ————A; - - - - -B; — — — —C.

1.4.1   Covariance pattern models

Again, our primary objective is to assess the effect of the treatments on DBP and we might again consider models which fit treatment and baseline DBP as in Model B in Section 1.3. The models will, of necessity, be more complicated as we now have four observations per patient. Additionally, it is possible that there is an underlying change in DBP over the four post-randomisation visits and we can allow for this in the model by including a time effect, which we will denote by m. It is also possible that treatment effects may differ across time points and to allow for this we can also include a treatment-by-time interaction, (tm). Thus, the jth observation on patient i can be modelled as

$$DBP_{ij} = \mu + b \cdot pre + t_k + m_j + (tm)_{jk} + e_{ij},$$

where

m_j = time effect at the jth post-treatment visit,

$(tm)_{jk}$ = the kth treatment effect at the jth post-treatment visit,

e_{ij} = residual term for the ith patient at the jth post-treatment visit.

So far in developing this model we have taken no account of the fact that post-treatment measurements taken on the same patient may not be independent of one another. A straightforward way to do this would be to assume that there is a constant correlation for all pairs of measurements on the same patient. Then we could write the correlation between the residuals as

$$\text{corr}(e_{ij}, e_{ij'}) = \rho, \quad j \neq j'.$$

Alternatively, it is possible that the correlation between pairs of measurements decays as they become more widely separated in time. We could then write

$$\text{corr}(e_{ij}, e_{ij'}) = \rho^{|j'-j|}, \quad j \neq j'.$$

In the extreme, we can set a separate correlation for each pair of visits and may write

$$\text{corr}(e_{ij}, e_{ij'}) = \rho_{j,j'}, \quad j \neq j'.$$

A *covariance pattern model* can be used to fit any of these covariance (or correlation) patterns. This type of model forms another class of mixed models. Fitting covariance patterns leads to a more appropriate analysis than occurs when the fact that the repeated observations are correlated is ignored. The covariance parameter estimates may also uncover additional information about the data. They are considered in more detail in Section 6.2 and the analysis of this example is presented in Section 6.3.

1.4.2   Random coefficients models

In the previous section the pattern of covariance between the repeated observations was modelled. An alternative approach to modelling repeated measures data would be to devise a model that explained arithmetically the relationship between DBP and time. A very simple way to do this would be to include a quantitative time effect (e.g. in measured weeks) as a covariate in the model.

$$\text{DBP}_{ij} = \mu + b \cdot pre + t_k + m \cdot time_{ij} + e_{ij},$$

where

$time_{ij} =$ time of observation j for patient i (weeks),
$m =$ constant representing the change in DBP for unit time (week).

Thus, we obtain a time slope with gradient m which defines a linear relationship between DBP and time. It is also possible (and indeed likely) that the relationship between DBP and time would vary between patients. To allow for this we could model a separate regression of DBP on time for each patient. To do this we fit patient effects to provide the intercept terms for each patient, and a patient.time interaction to provide the slopes for each patient.

$$\text{DBP}_{ij} = \mu + b \cdot pre + t_k + p_i + m \cdot time_{ij} + (pm)_i \cdot time_{ij} + e_{ij},$$

where

$(pm)_i$ = difference in slope for the ith patient, from the average slope,

p_i = difference from average in the intercept term for the ith patient.

It would seem reasonable to regard the values of patient effects and their slopes against time as arising from a distribution. Thus, patient and patient·time effects can both be fitted as random effects. However, the statistical properties of a model where some of the random effects involve covariate terms (time in this example) differ from ordinary random effects models (where the random effects do not involve any covariates). For this reason we distinguish these models from ordinary random effects models and refer to them as *random coefficients models*. They form a third class of mixed models.

The statistical properties of random coefficients models are similar in many respects to random effects models. The residuals again are assumed to be independent and to have a normal distribution, with zero mean:

$$\text{var}(e_{ij}) = \sigma^2.$$

The main statistical difference from ordinary random effects models arises from the fact that when we fit a straight line, the estimates of the slope and the intercept are not independent. Thus, the patient effects (intercepts) and patient·time effects (slopes) are correlated within each patient. We therefore need to extend the approach met earlier, where separate normal distributions were used for each random effect. We do this by use of the bivariate normal distribution. As well as terms for the means of both effects (which, as usual, are zero) and the variance components σ_p^2 and σ_{pm}^2 for patients and patient·time, this incorporates a covariance parameter $\sigma_{p,pm}$ We denote the bivariate normal distribution as

$$\begin{pmatrix} p_i \\ pm_i \end{pmatrix} \sim N(\mathbf{0}, \mathbf{G}),$$

where

$$\mathbf{G} = \begin{pmatrix} \sigma_p^2 & \sigma_{p,pm} \\ \sigma_{p,pm^2} & \sigma_{pm}^2 \end{pmatrix}$$

Thus, repeated measures data can be modelled using two alternative types of mixed model. Either the pattern of covariance between the repeated observations is modelled using a covariance pattern model, or the relationship with time can be modelled using a random coefficients model. The latter approach is usually more appropriate if the repeated measurements do not occur at fixed intervals, or when the relationship with time is of particular interest.

1.5 MORE ABOUT MIXED MODELS

In Sections 1.2–1.4 we used examples to introduce various concepts and types of mixed models. In this section we pull together some of the ideas introduced earlier and define them more concisely. We also discuss some general points about mixed models. Finally, we present a perspective of mixed models, giving an outline of the history of their development.

1.5.1   What is a mixed model?

We have already met a number of models which have been described as mixed models, but it may not be clear what unites them. The key distinguishing feature of mixed models compared with fixed effects models is that they are able to model data in which the observations are not independent. To express this more positively, we say that mixed models are able to model the covariance structure of the data.

A simple type of mixed model is the *random effects model* which was introduced in Sections 1.2 and 1.3. Here, certain effects in the model are assumed to have arisen from a distribution and thus give rise to another source of random variation in addition to the residual variation. These effects are referred to as *random effects*. For example, when patient effects were fitted in the trial introduced in Section 1.2, random variation occurred both between patients and as residual variation. Any number of random effects can be specified in a model; for example, in a multi-centre trial (as in Section 1.3) both centre and centre-treatment effects can be fitted as random, giving rise to two additional sources of variation.

In *random coefficients models* a covariate effect is allowed to vary randomly. For example, in the repeated measures hypertension data considered in Section 1.4, interest might centre on the rate of change of DBP measured over the four treatment visits in the three arms of the trial. The random coefficients model allows this rate of change (or slope) to vary randomly between patients. This is achieved technically by fitting patients and the patient-slope interaction as random and these effects are referred to as *random coefficients*.

The *covariance pattern model*, introduced in Section 1.4, is a third type of mixed model which directly models a pattern of correlations between observations. For example, in repeated measures trials interest is focused on several observations of the response variable made over a period of time and we can allow for the correlations (or, equivalently, covariances) between these observations. Suitable mixed models lead to more appropriate estimates of fixed effects and can investigate the nature of these covariances.

Random effects models, random coefficients models and covariance pattern models form three categories of mixed models. Mixed models can also be defined with combinations of random effects, random coefficient effects and covariance patterns. The choice will depend on the application and the objectives of the analysis.

1.5.2   Why use mixed models?

To stimulate further interest we now mention some potential advantages that can be gained by using a mixed model. In some situations a mixed model may simply be the most plausible model for a given data structure. For example, it is clearly desirable to take account of correlations between measurements in repeated measures data. In other circumstances the choice is less obvious between a fixed effects model and a mixed model. Factors influencing the decision will depend partly on the structure of the data. For example, in a multi-centre trial (as in Section 1.3) the decision depends mainly on the interpretation to be put on the results. When centre and centre-treatment effects are fitted as fixed, inference can only formally be applied to the centres observed, but if they are fitted as random, inference can be applied with more confidence to a wider population of centres.

Some potential advantages that can be gained by using a mixed model are as follows:

- *Fitting covariance pattern models leads to more appropriate fixed effect estimates and standard errors.* This type of model is of particular use for analysing repeated measures data. An important advantage is that the presence of missing data does not pose the major problems for analysis that can occur with a traditional analysis. The covariance parameter estimates may also uncover additional information about the data.

- *Results from a mixed model may be more appropriate to the required inference when the data structure is hierarchical.* For example, by fitting centre-treatment effects as random in a multi-centre trial analysis (as in Section 1.3), treatment effects are allowed to vary randomly across centres and the treatment standard error increases to allow for this. Inference can then be applied to the full population of centres. However, if centre and centre-treatment effects were fitted as fixed, treatment effects would be specific to the centres observed and inference should only be applied to these centres.

- *In a cross-over trial estimates of treatment effects can become more accurate in datasets where there are missing data* (as in Section 1.2). The degree of benefit from using a mixed model in this situation will depend on the amount of missing data. If the original trial design were balanced and only occasional values were missing, there would be little to be gained. However, if several values were missing, treatment estimates could become notably more accurate.

- *In a random effects model estimates of random effects are 'shrunken' compared with their fixed effects counterparts;* that is, their mean values are closer to the overall mean than if they were fitted as fixed. This helps to avoid the potential problem of extreme parameter estimates occurring due to chance when the estimates are based on small numbers. For example, in Section 1.1

we introduced an example on surgical audit. If failure rates from a particular type of operation were measured at several hospitals, a model fitting hospitals as fixed would produce unreliable failure rates for hospitals performing a small number of operations. Sometimes these would appear as outliers compared with other hospitals, purely due to chance variation. A model fitting hospitals as random would estimate failure rates that were shrunken towards the overall failure rate. The shrinkage is greatest for hospitals performing fewer operations because less is known about them, and so, misleading outliers are avoided.

- *Different variances can be fitted in a mixed model for each treatment group.* Such different variances for the treatment groups often arise in clinical trials comparing active treatments with a placebo, but they are rarely accounted for in fixed effects analyses.

- *Problems caused by missing data when fitting fixed effects models do not arise in mixed models*, provided that missing data can be assumed missing at random. This applies particularly in repeated measures trials, as noted above, and in cross-over trials.

Although we have listed several advantages to mixed models, there is a potential disadvantage. This is that more distributional assumptions are made and approximations are used to estimate certain model parameters. Consequently, the conclusions are dependent on more assumptions being valid and there will be some circumstances where parameter estimates are biased. These difficulties are addressed in Section 2.4.

1.5.3   Communicating results

Statistical methods have been defined as those which elucidate data affected by a multiplicity of causes. A problem with methods of increasing complexity can be difficulty in communicating the results of the analysis to the practitioner. There is the danger of obfuscating rather than elucidating. Estimation methods for mixed models are more complex than those used for fixed effects models and results can therefore be more difficult to justify to non-statistical colleagues. It is not usually realistic to describe the exact methodology. However, a satisfactory explanation can often be given by emphasising the key point that mixed models take account of the covariance structure or interdependence of the data, whereas more conventional fixed effects methods assume that all observations are independent. Mixed models may therefore provide results that are more appropriate to the study design. A (hypothetical) statistical methods section in a medical journal might read.

> *The trial was analysed using a mixed model (see Brown and Prescott, 1999) with centres and the centre-treatment interaction fitted as random, so that possible differences in the size of the treatment effect across centres could be assessed.*

1.5.4   Mixed models in medicine

Frequently, there are advantages to be gained from using mixed models in medical applications. Data in medical studies are often clustered; for example, data may be recorded at several centres, hospitals or general practices. This design can be described as hierarchical and wider inferences can be made by fitting the clustering effect as random. Repeated measures designs are also often used in medicine and it is not uncommon for some of the observations to be missing. There are then advantages to be gained from using a mixed model analysis which make allowance for the missing data. Another consideration is that it is ethically desirable to use as few patients as possible and therefore any improvements in the accuracy of treatment estimates gained by using a mixed model are particularly important. Although several examples of using mixed models in medicine have appeared in the literature (for example, Brown and Kempton, 1994), their use has not yet become routine.

1.5.5   Mixed models in perspective

It is interesting to see the application of mixed models in its historical context. In doing so, we will have to use occasional technical terms which have not yet been introduced in this book. They will, however, be met later on, and readers for whom some of the terms are unfamiliar may wish to return to this section after reading subsequent chapters.

The idea of attributing random variation to different sources by fitting random effects is not new. Fisher (1925), in his book *Statistical Methods for Research Workers*, outlined the basic method for estimating variance components by equating the mean squares from an analysis of variance table to their expected values (as described in Section 1.2). However, this method was only appropriate for balanced data. Yates (1940) and Henderson (1953) showed how Fisher's technique could be extended to unbalanced data, but their method did not always lead to unique variance components estimates. Hartley and Rao (1967) showed that unique estimates could be obtained using the method of maximum likelihood (see Section 2.2.1 for details on maximum likelihood). However, the estimates of the variance components are generally biased downwards because the method assumes that the fixed effects are known, rather than being estimated from the data. This problem of bias was overcome by Patterson and Thompson (1971) who proposed a method known as residual maximum likelihood (REML) (see Section 2.2.1), which automatically adjusted for the degrees of freedom corresponding to estimated fixed effects, as does analysis of variance (ANOVA) for balanced data. Many of the methods we describe in this book will be based on the REML method. Likelihood-based methods have only been adopted slowly because they are computationally intensive and this has limited their use until recently.

The problem of computational power has also been a factor in restricting the use of the Bayesian approach to analysis. While this approach is based on a different philosophy, it will often lead to superficially similar results to a conventional random effects model when used with uninformative priors. The increasing availability of good software to implement the Bayesian approach will undoubtedly lead to its wider use in future. The Bayesian approach to modelling is considered in Section 2.3.

In the past 20 years there have been developments in parallel, in the theory and practice of using the different types of mixed model which we described earlier. Random coefficients problems have sometimes in the past been handled in two stages: first, by estimating time slopes for each patient; and then by performing an analysis of the time slopes (e.g. Rowell and Walters, 1976). An early theoretical paper describing the fitting of a random coefficients model in a single stage, as we will do in this book, is by Laird and Ware (1982). We consider random coefficients models again in Section 6.5.

Covariance pattern models have developed largely from time series models. Jennrich and Schluchter (1986) described the use of different covariance pattern models for analysing repeated measures data and gave some indication of how to choose between them. These models are considered more fully in Section 6.2.

Random effects models have been frequently applied in agriculture. They have been used extensively in animal breeding to estimate heritabilities and predict genetic gain from breeding programmes (Meyer, 1986; Thompson, 1977). They have also been used for analysing crop variety trials. For example, Talbot (1984) used random effects models to estimate variance components for variety trialling systems carried out across several centres and years for different crops and was thus able to compare their general precision and effectiveness. The adoption of these models in medicine has been much slower, and a review of applications in clinical trials is given by Brown and Kempton (1994).

More recently, mixed models have become popular in the social sciences. However, they are usually described as multi-level or hierarchical models, and the terminology used for defining the models differs from that used in this book. This reflects parallel developments in different areas of application. However, the basic concept of allowing the data to have a covariance structure is the same. Two books published in this area are *Multilevel Statistical Models* by Harvey Goldstein (1995) and *Random Coefficients Models* by Nick Longford (1993).

1.6   SOME USEFUL DEFINITIONS

We conclude this introductory chapter with some definitions. The terms we are introducing here will recur frequently within subsequent chapters, and the understanding of these definitions and their relevance should increase as their applications are seen in greater detail. The terms we will introduce are containment, balance and error strata. In the analyses we will be presenting, we usually wish to concentrate on estimates of treatment effects. With the help of the definitions we are introducing we will be able to distinguish between situations where the treatment estimates are identical whether fixed effects models or mixed models are fitted. We will also be able to identify the situations where the treatment estimates will coincide with the simple average calculated from all observations involving that treatment. The first term we need to define is containment.

1.6.1   Containment

Containment occurs in two situations. First, consider the repeated measures data encountered in Section 1.4. In that hypertension trial, DBP was recorded at four visits after treatment had been started. In the analysis of that study, the residual variance will reflect variation *within* patients at individual visits. However, in this trial the patients receive the same treatment throughout, and so all the observations on a patient will reflect the effect of that one treatment on the patient. It can therefore perhaps be appreciated intuitively that it is the variation in response *between* patients which is appropriate for assessing the accuracy of the estimates of treatment effects rather than the residual or *within*-patient variation. We can see this more dramatically with a totally artificial set of data which might have arisen from this trial.

Patient	Treatment	Post-treatment visits			
		1	**2**	**3**	**4**
1	A	80	80	80	80
2	B	85	85	85	85
3	B	85	85	85	85
4	A	91	91	91	91

In this situation there is no within-patient variation and the residual variance is zero. Thus, if the residual variance were used in the determination of the precision of treatment estimates, we would conclude that this data showed convincingly that treatment B produced lower DBPs than treatment A. Common sense tells us that this conclusion is ridiculous with this data, and that between-patient variation must form the basis for any comparison.

In the above situation we say that treatment effects are contained within patient effects.

The second situation where we can meet containment can also be illustrated with data from the hypertension trial, this time concentrating on the multicentre aspect of the design. In Section 1.3 we actually met containment for the first time when dealing with Model E, and both centre effects and the centre·treatment effects were fitted as random. We say in this context that the treatment effects are contained within centre·treatment effects. In fact, there is no requirement for the centre·treatment effects to be random for the definition of containment to hold.

Thus, similarly in Model D, where the centre-treatment effects were regarded as fixed, we can still refer to the treatment effects as being contained within centre-treatment effects. It applies in general to any data with a hierarchical structure in which the fixed effects (treatment) appears in interaction terms with other effects.

1.6.2   Balance

In many statistical textbooks which discuss the concept of balance, it is never defined but, rather, left to the intuitive feel of the reader to determine whether an experimental design is balanced. Some authors (e.g. Searle, 1992) have defined balance as occurring when there are equal numbers of observations per *cell*. Cells are formed by all possible combinations of the levels of all the effects in the model, otherwise known as the crossing between all effects fitted in the model. For example, if we fit centre effects and treatment effects in the analysis of a multi-centre trial, and we suppose that there are four centres and two treatments, then each of the eight combinations of centre and treatment require the same number of patients to achieve balance.

When there is balance according to this definition, the estimate of a fixed effect mean will equal the mean of all the observations at that fixed effect level. To make this clearer, if we call one of the treatments in the above example A, then the estimate of the mean response to treatment A will simply be the average of all of the observations for all patients who received treatment A. In general, this will not happen when there is imbalance. Consider the dataset illustrated below. If all of the observations are present, then the estimated means for treatments A and B are 85.0 and 95.0, respectively, corresponding to their means.

Centre	Treatment A	Treatment B
1	90	100
	80	90
2	90	100
	80	(90)

If the figure in parentheses is missing, however, so that there is no longer balance, then the mean treatment estimates will be 85.0 and 97.0, compared with their means of 85.0 and 96.7.

Although the condition of equal numbers in all cells is a sufficient condition for the fixed effects mean estimates to equal their 'raw' means, it is not a necessary condition. In the multi-centre trial, for example, as long as we do not fit centre-treatment effects it does not matter if the numbers differ across centres, provided the treatments are allocated evenly within the centres. The following dataset produces treatment mean estimates which equal their raw means.

Centre	Treatment A	Treatment B
1	90	100
	80	
2	85	95
	85	90
	80	
	80	

Another anomaly is the cross-over trial, which is always unbalanced by the Searle definition if period effects are fitted as well as patient and treatment effects. This leads to empty cells because we cannot have both treatments given in the same period to any patient. Nevertheless, in a simple two-period, cross-over trial, if every patient receives every treatment, equal numbers of patients receive each sequence of treatments, and no covariates are fitted, the treatment mean estimates will equal their raw means.

We suggest, therefore, an alternative definition of balance whereby the fixed effects means will equal their raw means whenever data are balanced but not (in general) when they are unbalanced. Balance occurs for a fixed effect when both of the following conditions are met:

- Within each category of the fixed effect (e.g. treatment), observations occur in equal proportions among categories of every other effect which is fitted at the same containment level (see the previous section).
- If the fixed effect (e.g. treatment) is contained within a random effect (e.g. centre·treatment), then an equal number of observations are required in each category of the containing effect.

Balance across random effects

It is of importance in this book to identify the situations in which the fixed effects means (usually treatments) will differ depending on whether a fixed effects model or a mixed model is used. When balance, as defined above, is achieved, then the fixed effects mean estimates will equal the raw means, whether a fixed effects model or a mixed model has been applied. There are other situations when the fixed effects mean estimates will not equal their raw means, but the same estimates will be obtained whether the fixed effects approach or mixed model approach is followed. This occurs when both of the following conditions apply and we have a situation which we define as balance across random effects:

- Within each category of the specific effect (e.g. treatment), observations are allocated in equal proportions among categories of every random effect (e.g. patient) which is fitted at the same containment level.

- If the effect (e.g. treatment) is contained within a random effect (e.g. centre·treatment), then an equal number of observations are required in each category of the containing effect.

An example of the subtle distinction between these two definitions is provided by the cross-over trial example. If there were an equal number of patients on the AB and BA sequence of treatments, with no missing values, then our definition of balance would be satisfied, as described earlier. If there were no missing values, but the numbers differed between the AB and BA sequences, then there would be balance over random effects. This is true because the only random effect is patients and within each category of the containing effect (i.e. within individual patients), each treatment occurs once and hence the definition is satisfied. Thus, the treatment estimates will be identical whether the patient effect is fitted as fixed or random, but these estimates will (in general) differ from the raw means.

The above definition has been applied in the context of one particular type of mixed model; namely, the random effects model. In random coefficients models, the random coefficient blocking effect (usually patients) can be substituted for 'random effect' in the definition. In covariance pattern models the blocking effect within which the covariance pattern is defined (again usually patients) can be substituted for 'random effect'.

Assessing balance

It can sometimes be difficult to gain an immediate feel for when balance is achieved from the above definitions. The three following common situations are easily classified.

- If any observations are missing, then imbalance across random effects occurs (except for simple parallel group situations).
- If a continuous effect is fitted, then imbalance will occur (unless identical means for the effect happen to occur within each fixed effect category). However, balance across the random effects may still be achieved.
- If an equal number of observations occur in every cell and no continuous covariate is fitted, then all fixed effects will be balanced.

1.6.3   Error strata

In the random effects model, an error stratum or error level is defined by each random effect and by the residual. For example, if patients are fitted as random in a cross-over trial, there are error strata corresponding to the patients and to the residual. The *containment stratum* for a particular fixed effect is defined by the residual stratum, unless the effect is contained within a random effect in a random effects model or a blocking effect (see Section 6.2) in a random coefficients or covariance pattern model, in which case it is that of the containing effect. For example, in a repeated measures study, treatments are contained within patients and thus the patient error stratum forms the containment stratum for treatments. Usually an effect has only one containment stratum and examples in this book will be restricted to this more usual situation. However, situations could be conceived where this is not the case. For example, if clinics and GPs were recorded in a trial and GP·treatment and clinic·treatment effects were fitted as random, then both of these effects would form containing strata for the treatment effect.

Higher level strata are defined by any random effects that are contained within the containment stratum. For example, in a cross-over trial the containment stratum for treatment effects is the residual stratum and the patient stratum is a higher level stratum (see Figure 1.3(a)). In a multi-centre trial in which centre and centre·treatment effects are fitted as random, the centre·treatment stratum forms the containment stratum for treatment effects and the centre stratum forms a higher level stratum (see Figure 1.3(b)). Whenever higher level strata are present and data are not balanced across random effects, a fixed effect will be estimated using information from these strata as well as from the containment stratum (i.e. information is *recovered* from the higher level strata).

Thus, in a cross-over trial with missing values, information is recovered from the patient level, as we saw in Section 1.2. The same occurs with missing values in a repeated measures trial where a covariance pattern is fitted. In random coefficients models, information is recovered from the patient level except in

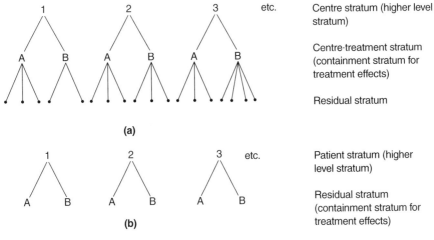

Figure 1.3   (a) Error strata for a multi-centre trial analysis fitting centre and centre·treatment effects as random; (b) error strata for a cross-over trial analysis fitting patient effects as random. A = Treatment A; B = Treatment B.

highly unusual circumstances of equal numbers of observations at the same set of time points for all patients.

In random coefficients and covariance pattern models, error strata are not defined quite as easily because correlations occur between the random coefficients or residuals. However, random coefficients and blocking effects have a similar role to error strata although their properties are not quite the same.

2

Normal Mixed Models

In this chapter we discuss in more detail the mixed model with normally distributed errors. We will refer to this as the 'normal mixed model'. Of course, this does not imply that values of the response variables follow normal distributions, because they are, in fact, mixtures of effects with different means. In practice, though, if a variable appears to have a normal distribution, the assumption of normal residuals and random effects is often reasonable.

In the examples introduced in Sections 1.1–1.4 we defined several mixed models using a notation chosen to suit to each situation. In Section 2.1 we define the mixed model using a general matrix notation which can be used for all types of mixed model. Matrix notation may at first be unfamiliar to some readers and it is outwith the scope of this book to teach matrix algebra. A good introductory guide is *Matrices for Statistics* by Healy (1986). Once grasped though, matrix notation can make the overall theory underlying mixed models easier to comprehend. Mixed models methods based on classical statistical techniques are described in Section 2.2, and in Section 2.3 the Bayesian approach to fitting mixed models will be introduced. These two sections can be omitted by readers who do not desire a detailed understanding of the more theoretical aspects of mixed models. In Section 2.4 some practical issues related to the use and interpretation of mixed models are considered, and a worked example illustrating several of the points made in Section 2.4 is described in Section 2.5.

2.1 MODEL DEFINITION

Here, the mixed model is defined using a general matrix notation which provides a compact means to specify all types of mixed model. We start by defining the fixed effects model and then extend this notation to encompass the mixed model.

2.1.1 The fixed effects model

All fixed effects models can be specified in the general form

$$y_i = \mu + \alpha_1 x_{i1} + \alpha_2 x_{i2} + \cdots + \alpha_p x_{ip} + e_i,$$

$$\text{var}(e_i) = \sigma^2.$$

For example, in Section 1.2, Model B was presented as

$$y_{ij} = \mu + p_i + t_j + e_{ij}.$$

This model used a subscript i to denote results from the ith patient, and a subscript j to denote results on the jth treatment, in the context of a cross-over trial. In the general model notation, every observation is denoted separately with a single subscript. Thus, y_1 and y_2 could represent the observations from patient 1; y_3 and y_4 the observations from patient 2, etc. The α terms in the general model will correspond to p_1, p_2, p_3, p_4, p_5 and p_6 and to t_1 and t_2 and are constants giving the size of the patient and treatment effects. The terms x_{i1}, x_{i2}, ..., x_{i8} are used in this example to indicate the patient and treatment to which the observation y_i belongs and will take the values one or zero. If y_1 is the observation from patient 1 receiving treatment 1, x_{11} then will equal 1 (corresponding to α_1, which represents the first patient effect), x_{12} to x_{16} will equal 0 (as this observation is not from patients 2 to 6), x_{17} will equal 1 (corresponding to α_7, representing the first treatment effect) and x_{18} will equal 0. A further example to follow shortly should clarify this notation further.

The model described above fits $p + 1$ fixed effects parameters, α_1 to α_p, and an intercept term, μ. If there are n observations, then these may be written as

$$y_1 = \mu + \alpha_1 x_{11} + \alpha_2 x_{12} + \cdots + \alpha_p x_{1p} + e_1,$$
$$y_2 = \mu + \alpha_1 x_{21} + \alpha_2 x_{22} + \cdots + \alpha_p x_{2p} + e_2,$$
$$\vdots$$
$$y_n = \mu + \alpha_1 x_{n1} + \alpha_2 x_{n2} + \cdots + \alpha_p x_{np} + e_n;$$
$$\text{var}(e_1) = \sigma^2,$$
$$\vdots$$
$$\text{var}(e_n) = \sigma^2.$$

This can be expressed more concisely in matrix notation as

$$\mathbf{y} = \mathbf{X}\boldsymbol{\alpha} + \mathbf{e},$$
$$\mathbf{V} = \text{var}(\mathbf{y}) = \sigma^2 \mathbf{I},$$

where

$$\mathbf{y} = (y_1, y_2, y_3, \ldots, y_n)' = \text{observed values},$$
$$\boldsymbol{\alpha} = (\mu, \alpha_1, \alpha_2, \ldots, \alpha_p)' = \text{fixed effects parameters},$$
$$\mathbf{e} = (e_1, e_2, e_3, \ldots, e_n)' = \text{residuals},$$
$$\sigma^2 \qquad\qquad\qquad = \text{residual variance},$$
$$\mathbf{I} \qquad\qquad\qquad = n \times n \text{ identity matrix}.$$

The parameters in $\boldsymbol{\alpha}$ may encompass several variables. In the example above they covered patient effects and treatment effects. Both of these are qualitative or categorical variables and we will refer to such effects as categorical effects. They are also sometimes referred to as factor effects. More generally, categorical effects are those where observations will belong to one of several classes. There may also be several covariate effects (such as age or baseline measurement) contained in α. These relate to variables which are measured on a quantitative scale. Several parameters may be required to model categorical effects, but just one parameter is needed to model a covariate effect.

$\mathbf{X}$ is known as the *design matrix* and has the dimension $n \times p$ (i.e. n rows and p columns). It specifies values of fixed effects corresponding to each parameter for each observation. For categorical effects the values of zero and one are used to denote the absence and presence of effect categories, and for covariate effects the variable values themselves are used in $\mathbf{X}$.

We will exemplify the notation with the following data, which are the first nine observations in a multi-centre trial of two treatments to lower blood pressure.

Centre	Treatment	Pre-treatment systolic BP	Post-treatment systolic BP
1	A	178	176
1	A	168	194
1	B	196	156
1	B	170	150
2	A	165	150
2	B	190	160
3	A	175	150
3	A	180	160
3	B	175	160

The observation vector $\mathbf{y}$ is formed from the values of the post-treatment systolic blood pressure:

$$\mathbf{y} = (176, 194, 156, 150, 150, 160, 150, 160, 160)'.$$

If pre-treatment blood pressure and treatment were fitted in the analysis model as fixed effects (ignoring centres for the moment), then the design matrix would be

$$\mathbf{X} = \begin{pmatrix} \mu & \alpha_1 & \alpha_2 & \alpha_3 \\ 1 & 178 & 1 & 0 \\ 1 & 168 & 1 & 0 \\ 1 & 196 & 0 & 1 \\ 1 & 170 & 0 & 1 \\ 1 & 165 & 1 & 0 \\ 1 & 190 & 0 & 1 \\ 1 & 175 & 1 & 0 \\ 1 & 180 & 1 & 0 \\ 1 & 175 & 0 & 1 \end{pmatrix},$$

where the columns of the design matrix correspond to the parameters,

$\mu = $ intercept,

$\alpha_1 = $ pre-treatment blood pressure parameter,

$\alpha_2, \alpha_3 = $ parameters for treatments A and B.

We note here that the design matrix, **X**, is overparameterised. This means that there are linear dependencies between the columns; e.g. we know that α_3 will be 0 if $\alpha_2 = 1$, and 1 if $\alpha_2 = 0$. **X** could alternatively be specified omitting the α_3 column to correspond with the number of parameters actually modelled. However, the overparameterised form is used here since it is used for specifying contrasts by SAS procedures such as PROC MIXED (this procedure will be used to analyse most of the examples in this book).

V is a matrix containing the variances and covariances of the observations. In the usual fixed effects model, variances for all observations are equal and no observations are correlated. Thus, **V** is simply $\sigma^2 \mathbf{I}$.

2.1.2   The mixed model

Extending our fixed effects model to incorporate random effect, the mixed model may be specified as

$$y_i = \mu + \alpha_1 x_{i1} + \alpha_2 x_{i2} + \cdots + \alpha_p x_{ip}$$
$$+ \beta_1 z_{i1} + \beta_2 z_{i2} + \cdots + \beta_q z_{iq} + e_i$$

for a model fitting p fixed effect parameters and q random effect (or coefficient) parameters. It will be recalled from Chapter 1 that random effects are assumed to follow a distribution, whereas fixed effects are regarded as fixed constants. The model can be expressed in matrix notation as

$$\mathbf{y} = \mathbf{X}\boldsymbol{\alpha} + \mathbf{Z}\boldsymbol{\beta} + \mathbf{e},$$

where **y**, **X**, $\boldsymbol{\alpha}$ and **e** are as defined in the fixed effects model, and

$$\boldsymbol{\beta} = (\beta_1, \beta_2, \ldots, \beta_q)' = \text{random effect/coefficient parameters.}$$

Z is a second design matrix with dimension $n \times q$ giving the values of random effects corresponding to each observation. It is specified in exactly the same way as **X** was for the fixed effects, except that an intercept term is not included. If centres were fitted as random in the multi-centre example given above, the $\boldsymbol{\beta}$ vector would then consist of three parameters, β_1, β_2 and β_3, corresponding to the three centres and the **Z** matrix would be

$$
\begin{array}{ccc}
\beta_1 & \beta_2 & \beta_3 \\
\end{array}
$$

$$
\mathbf{Z} = \begin{pmatrix}
1 & 0 & 0 \\
1 & 0 & 0 \\
1 & 0 & 0 \\
1 & 0 & 0 \\
0 & 1 & 0 \\
0 & 1 & 0 \\
0 & 0 & 1 \\
0 & 0 & 1 \\
0 & 0 & 1 \\
\end{pmatrix}.
$$

Alternatively, if both the centre and the centre·treatment effects were fitted as random, then the vector of random effect parameters, $\boldsymbol{\beta}$, would consist of the three centre parameters, plus six centre·treatment interaction parameters β_4, β_5, β_6, β_7, β_8 and β_9. The $\mathbf{Z}$ matrix would then be

$$
\begin{array}{ccccccccc}
\beta_1 & \beta_2 & \beta_3 & \beta_4 & \beta_5 & \beta_6 & \beta_7 & \beta_8 & \beta_9 \\
\end{array}
$$

$$
\mathbf{Z} = \begin{pmatrix}
1 & 0 & 0 & 1 & 0 & 0 & 0 & 0 & 0 \\
1 & 0 & 0 & 1 & 0 & 0 & 0 & 0 & 0 \\
1 & 0 & 0 & 0 & 1 & 0 & 0 & 0 & 0 \\
1 & 0 & 0 & 0 & 1 & 0 & 0 & 0 & 0 \\
0 & 1 & 0 & 0 & 0 & 1 & 0 & 0 & 0 \\
0 & 1 & 0 & 0 & 0 & 0 & 1 & 0 & 0 \\
0 & 0 & 1 & 0 & 0 & 0 & 0 & 1 & 0 \\
0 & 0 & 1 & 0 & 0 & 0 & 0 & 1 & 0 \\
0 & 0 & 1 & 0 & 0 & 0 & 0 & 0 & 1 \\
\end{pmatrix}.
$$

Again, note that this matrix is overparameterised due to linear dependencies between the columns. It could alternatively have been written using four columns: $3 - 1 = 2$ for the centre effects, and $(3 - 1) \times (2 - 1) = 2$ for the centre·treatment effects.

Covariance matrix, V

We saw in the fixed effects model that all observations have equal variances, and the observations are uncorrelated. This leads to the $\mathbf{V}$ matrix being diagonal. When random effects are fitted, we saw in Section 1.2 that this results in correlated observations. In the context of the cross-over trial, we saw that observations on the same patient were correlated (with covariance equal to the patient variance component), while observations on different patients were uncorrelated. We now generalise this result, using the matrix notation.

The covariance of $\mathbf{y}$, $\mathrm{var}(\mathbf{y}) = \mathbf{V}$ can be written as

$$
\mathbf{V} = \mathrm{var}(\mathbf{X}\boldsymbol{\alpha} + \mathbf{Z}\boldsymbol{\beta} + \mathbf{e}).
$$

Since we assume that the random effects and the residuals are uncorrelated,

$$\mathbf{V} = \mathrm{var}(\mathbf{X}\boldsymbol{\alpha}) + \mathrm{var}(\mathbf{Z}\boldsymbol{\beta}) + \mathrm{var}(\mathbf{e}).$$

Since $\boldsymbol{\alpha}$ describes the fixed effects parameters, $\mathrm{var}(\mathbf{X}\boldsymbol{\alpha}) = \mathbf{0}$. Also, $\mathbf{Z}$ is a matrix of constants. Therefore,

$$\mathbf{V} = \mathbf{Z}\,\mathrm{var}(\boldsymbol{\beta})\mathbf{Z}' + \mathrm{var}(\mathbf{e}).$$

We will let $\mathbf{G}$ denote $\mathrm{var}(\boldsymbol{\beta})$, and since the random effects are assumed to follow normal distributions we may write $\boldsymbol{\beta} \sim \mathrm{N}(\mathbf{0}, \mathbf{G})$. Similarly, we write $\mathrm{var}(\mathbf{e}) = \mathbf{R}$, the residual covariance matrix and $\mathbf{e} \sim \mathrm{N}(\mathbf{0}, \mathbf{R})$. Hence,

$$\mathbf{V} = \mathbf{Z}\mathbf{G}\mathbf{Z}' + \mathbf{R}.$$

In the following three sections we will define the structure of the $\mathbf{G}$ and $\mathbf{R}$ matrices in random effects models, random coefficients models and in covariance pattern models.

2.1.3   The random effects model covariance structure

The G matrix

The dimension of $\mathbf{G}$ is $q \times q$, where q is equal to the total number of random effects parameters.

In random effects models, $\mathbf{G}$ is always diagonal (i.e. random effects are assumed uncorrelated). If just centre effects were fitted as random in the simple multi-centre example with three centres, then $\mathbf{G}$ would have the form

$$\mathbf{G} = \begin{pmatrix} \sigma_c^2 & 0 & 0 \\ 0 & \sigma_c^2 & 0 \\ 0 & 0 & \sigma_c^2 \end{pmatrix},$$

where σ_c^2 is the centre variance component. If both centre and centre·treatment effects were fitted as random, then $\mathbf{G}$ would have the form

$$\mathbf{G} = \begin{pmatrix} \sigma_c^2 & 0 & 0 & 0 & 0 & 0 & 0 & 0 & 0 \\ 0 & \sigma_c^2 & 0 & 0 & 0 & 0 & 0 & 0 & 0 \\ 0 & 0 & \sigma_c^2 & 0 & 0 & 0 & 0 & 0 & 0 \\ 0 & 0 & 0 & \sigma_{ct}^2 & 0 & 0 & 0 & 0 & 0 \\ 0 & 0 & 0 & 0 & \sigma_{ct}^2 & 0 & 0 & 0 & 0 \\ 0 & 0 & 0 & 0 & 0 & \sigma_{ct}^2 & 0 & 0 & 0 \\ 0 & 0 & 0 & 0 & 0 & 0 & \sigma_{ct}^2 & 0 & 0 \\ 0 & 0 & 0 & 0 & 0 & 0 & 0 & \sigma_{ct}^2 & 0 \\ 0 & 0 & 0 & 0 & 0 & 0 & 0 & 0 & \sigma_{ct}^2 \end{pmatrix},$$

where σ_{ct}^2 is the centre·treatment variance component.

The R matrix

The residuals are uncorrelated in random effects models and $\mathbf{R} = \sigma^2\mathbf{I}$:

$$\mathbf{R} = \begin{pmatrix} \sigma^2 & 0 & 0 & 0 & 0 & 0 & 0 & 0 & 0 \\ 0 & \sigma^2 & 0 & 0 & 0 & 0 & 0 & 0 & 0 \\ 0 & 0 & \sigma^2 & 0 & 0 & 0 & 0 & 0 & 0 \\ 0 & 0 & 0 & \sigma^2 & 0 & 0 & 0 & 0 & 0 \\ 0 & 0 & 0 & 0 & \sigma^2 & 0 & 0 & 0 & 0 \\ 0 & 0 & 0 & 0 & 0 & \sigma^2 & 0 & 0 & 0 \\ 0 & 0 & 0 & 0 & 0 & 0 & \sigma^2 & 0 & 0 \\ 0 & 0 & 0 & 0 & 0 & 0 & 0 & \sigma^2 & 0 \\ 0 & 0 & 0 & 0 & 0 & 0 & 0 & 0 & \sigma^2 \end{pmatrix}.$$

The V matrix

We showed earlier that the variance matrix, $\mathbf{V}$, has the form $\mathbf{V} = \mathbf{ZGZ'} + \mathbf{R}$. $\mathbf{ZGZ'}$ specifies the covariance due to the random effects. If just centre effects are fitted as random, then we obtain

$$\mathbf{ZGZ'} = \begin{pmatrix} \sigma_c^2 & \sigma_c^2 & \sigma_c^2 & \sigma_c^2 & 0 & 0 & 0 & 0 & 0 \\ \sigma_c^2 & \sigma_c^2 & \sigma_c^2 & \sigma_c^2 & 0 & 0 & 0 & 0 & 0 \\ \sigma_c^2 & \sigma_c^2 & \sigma_c^2 & \sigma_c^2 & 0 & 0 & 0 & 0 & 0 \\ \sigma_c^2 & \sigma_c^2 & \sigma_c^2 & \sigma_c^2 & 0 & 0 & 0 & 0 & 0 \\ 0 & 0 & 0 & 0 & \sigma_c^2 & \sigma_c^2 & 0 & 0 & 0 \\ 0 & 0 & 0 & 0 & \sigma_c^2 & \sigma_c^2 & 0 & 0 & 0 \\ 0 & 0 & 0 & 0 & 0 & 0 & \sigma_c^2 & \sigma_c^2 & \sigma_c^2 \\ 0 & 0 & 0 & 0 & 0 & 0 & \sigma_c^2 & \sigma_c^2 & \sigma_c^2 \\ 0 & 0 & 0 & 0 & 0 & 0 & \sigma_c^2 & \sigma_c^2 & \sigma_c^2 \end{pmatrix}.$$

This matrix could be obtained by the laborious process of matrix multiplication but it always has the same form. The resulting matrix has a *block diagonal* form with the size of blocks corresponding to the number of observations at each random effect category. The total variance matrix, $\mathbf{V} = \mathbf{ZGZ'} + \mathbf{R}$, is then

$$\mathbf{V} = \begin{pmatrix} \sigma_c^2 + \sigma^2 & \sigma_c^2 & \sigma_c^2 & \sigma_c^2 & 0 & 0 & 0 & 0 & 0 \\ \sigma_c^2 & \sigma_c^2 + \sigma^2 & \sigma_c^2 & \sigma_c^2 & 0 & 0 & 0 & 0 & 0 \\ \sigma_c^2 & \sigma_c^2 & \sigma_c^2 + \sigma^2 & \sigma_c^2 & 0 & 0 & 0 & 0 & 0 \\ \sigma_c^2 & \sigma_c^2 & \sigma_c^2 & \sigma_c^2 + \sigma^2 & 0 & 0 & 0 & 0 & 0 \\ 0 & 0 & 0 & 0 & \sigma_c^2 + \sigma^2 & \sigma_c^2 & 0 & 0 & 0 \\ 0 & 0 & 0 & 0 & \sigma_c^2 & \sigma_c^2 + \sigma^2 & 0 & 0 & 0 \\ 0 & 0 & 0 & 0 & 0 & 0 & \sigma_c^2 + \sigma^2 & \sigma_c^2 & \sigma_c^2 \\ 0 & 0 & 0 & 0 & 0 & 0 & \sigma_c^2 & \sigma_c^2 + \sigma^2 & \sigma_c^2 \\ 0 & 0 & 0 & 0 & 0 & 0 & \sigma_c^2 & \sigma_c^2 & \sigma_c^2 + \sigma^2 \end{pmatrix}.$$

This also has a block diagonal form with the covariances for observations at the same centre equal to the random effect variance component, σ_c^2, and variance terms on the diagonal equal to the sum of the centre and residual variance components, $\sigma_c^2 + \sigma^2$. (We note that this corresponds to the results from the cross-over trial example introduced in Section 1.2, where the random effect was patient rather than centre.) If both centre and centre·treatment effects had been fitted as random, then

ZGZ′

$$
= \begin{pmatrix}
\sigma_c^2+\sigma_{ct}^2 & \sigma_c^2+\sigma_{ct}^2 & \sigma_c^2 & \sigma_c^2 & 0 & 0 & 0 & 0 & 0 \\
\sigma_c^2+\sigma_{ct}^2 & \sigma_c^2+\sigma_{ct}^2 & \sigma_c^2 & \sigma_c^2 & 0 & 0 & 0 & 0 & 0 \\
\sigma_c^2 & \sigma_c^2 & \sigma_c^2+\sigma_{ct}^2 & \sigma_c^2+\sigma_{ct}^2 & 0 & 0 & 0 & 0 & 0 \\
\sigma_c^2 & \sigma_c^2 & \sigma_c^2+\sigma_{ct}^2 & \sigma_c^2+\sigma_{ct}^2 & 0 & 0 & 0 & 0 & 0 \\
0 & 0 & 0 & 0 & \sigma_c^2+\sigma_{ct}^2 & \sigma_c^2 & 0 & 0 & 0 \\
0 & 0 & 0 & 0 & \sigma_c^2 & \sigma_c^2+\sigma_{ct}^2 & 0 & 0 & 0 \\
0 & 0 & 0 & 0 & 0 & 0 & \sigma_c^2+\sigma_{ct}^2 & \sigma_c^2+\sigma_{ct}^2 & \sigma_c^2 \\
0 & 0 & 0 & 0 & 0 & 0 & \sigma_c^2+\sigma_{ct}^2 & \sigma_c^2+\sigma_{ct}^2 & \sigma_c^2 \\
0 & 0 & 0 & 0 & 0 & 0 & \sigma_c^2 & \sigma_c^2 & \sigma_c^2+\sigma_{ct}^2
\end{pmatrix}
$$

and

$$
\mathbf{V} = \begin{pmatrix}
\theta & \sigma_c^2+\sigma_{ct}^2 & \sigma_c^2 & \sigma_c^2 & 0 & 0 & 0 & 0 & 0 \\
\sigma_c^2+\sigma_{ct}^2 & \theta & \sigma_c^2 & \sigma_c^2 & 0 & 0 & 0 & 0 & 0 \\
\sigma_c^2 & \sigma_c^2 & \theta & \sigma_c^2+\sigma_{ct}^2 & 0 & 0 & 0 & 0 & 0 \\
\sigma_c^2 & \sigma_c^2 & \sigma_c^2+\sigma_{ct}^2 & \theta & 0 & 0 & 0 & 0 & 0 \\
0 & 0 & 0 & 0 & \theta & \sigma_c^2 & 0 & 0 & 0 \\
0 & 0 & 0 & 0 & \sigma_c^2 & \theta & 0 & 0 & 0 \\
0 & 0 & 0 & 0 & 0 & 0 & \theta & \sigma_c^2+\sigma_{ct}^2 & \sigma_c^2 \\
0 & 0 & 0 & 0 & 0 & 0 & \sigma_c^2+\sigma_{ct}^2 & \theta & \sigma_c^2 \\
0 & 0 & 0 & 0 & 0 & 0 & \sigma_c^2 & \sigma_c^2 & \theta
\end{pmatrix},
$$

where $\theta = \sigma_c^2 + \sigma_{ct}^2 + \sigma^2$. Thus, **V** again has a block diagonal form with a slightly more complicated structure. The centre·treatment variance component is added to the covariance terms for observations at the same centre and with the same treatment.

2.1.4 The random coefficients model covariance structure

The statistical properties of random coefficients models were described in the repeated measures example introduced in Section 1.4. Here, we define their covariance structure in terms of the general matrix notation we have just introduced for mixed models. Random coefficients models will be discussed in more detail in Section 6.5.

The following data will be used to illustrate the covariance structure. They represent measurement times for the first three patients in a repeated measures trial of two treatments.

Patient	Treatment	Time (days)
1	A	t_{11}
1	A	t_{12}
1	A	t_{13}
1	A	t_{14}
2	B	t_{21}
2	B	t_{22}
3	A	t_{31}
3	A	t_{32}
3	A	t_{33}

If patient and patient·time effects were fitted as random coefficients, then there would be six random coefficients: $\beta_{p,1}$, $\beta_{pt,1}$, $\beta_{p,2}$, $\beta_{pt,2}$, $\beta_{p,3}$ and $\beta_{pt,3}$ (allowing an intercept and slope to be calculated for each of the three patients). The $\mathbf{Z}$ matrix would then be

$$\mathbf{Z} = \begin{array}{cccccc} \beta_{p,1} & \beta_{pt,1} & \beta_{p,2} & \beta_{pt,2} & \beta_{p,3} & \beta_{pt,3} \\ \left(\begin{array}{cccccc} 1 & t_{11} & 0 & 0 & 0 & 0 \\ 1 & t_{12} & 0 & 0 & 0 & 0 \\ 1 & t_{13} & 0 & 0 & 0 & 0 \\ 1 & t_{14} & 0 & 0 & 0 & 0 \\ 0 & 0 & 1 & t_{21} & 0 & 0 \\ 0 & 0 & 1 & t_{22} & 0 & 0 \\ 0 & 0 & 0 & 0 & 1 & t_{31} \\ 0 & 0 & 0 & 0 & 1 & t_{32} \\ 0 & 0 & 0 & 0 & 1 & t_{33}. \end{array} \right) \end{array}$$

The R matrix

As in random effects models the residuals are uncorrelated and the residual covariance matrix is

$$\mathbf{R} = \begin{pmatrix} \sigma^2 & 0 & 0 & 0 & 0 & 0 & 0 & 0 & 0 \\ 0 & \sigma^2 & 0 & 0 & 0 & 0 & 0 & 0 & 0 \\ 0 & 0 & \sigma^2 & 0 & 0 & 0 & 0 & 0 & 0 \\ 0 & 0 & 0 & \sigma^2 & 0 & 0 & 0 & 0 & 0 \\ 0 & 0 & 0 & 0 & \sigma^2 & 0 & 0 & 0 & 0 \\ 0 & 0 & 0 & 0 & 0 & \sigma^2 & 0 & 0 & 0 \\ 0 & 0 & 0 & 0 & 0 & 0 & \sigma^2 & 0 & 0 \\ 0 & 0 & 0 & 0 & 0 & 0 & 0 & \sigma^2 & 0 \\ 0 & 0 & 0 & 0 & 0 & 0 & 0 & 0 & \sigma^2 \end{pmatrix}.$$

The G matrix

In a random coefficients model the patient effects (intercepts) are correlated with the random patient·time effects (slopes). Correlation occurs only for coefficients on the same patient (i.e. between $\beta_{p,i}$ and $\beta_{pt,i}$) and coefficients on different patients are uncorrelated. Thus, the G matrix would be

$$
\mathbf{G} =
\begin{pmatrix}
\sigma_p^2 & \sigma_{p,pt} & 0 & 0 & 0 & 0 \\
\sigma_{p,pt} & \sigma_{pt}^2 & 0 & 0 & 0 & 0 \\
0 & 0 & \sigma_p^2 & \sigma_{p,pt} & 0 & 0 \\
0 & 0 & \sigma_{p,pt} & \sigma_{pt}^2 & 0 & 0 \\
0 & 0 & 0 & 0 & \sigma_p^2 & \sigma_{p,pt} \\
0 & 0 & 0 & 0 & \sigma_{p,pt} & \sigma_{pt}^2
\end{pmatrix},
$$

where σ_p^2 and σ_{pt}^2 are the patient and patient·time variance components and $\sigma_{p,pt}$ is the covariance between the random coefficients. Note that $\mathbf{G}$ has dimension 6×6 because the model includes six random coefficients.

The V matrix

Again, $\mathbf{V}$ is obtained as $\mathbf{V} = \mathbf{ZGZ'} + \mathbf{R}$. $\mathbf{ZGZ'}$ specifies the covariance due to the random coefficients and for our data $\mathbf{V}$ has the form

$$
\mathbf{ZGZ'} =
\begin{pmatrix}
v_{1,11} & v_{1,12} & v_{1,13} & v_{1,14} & 0 & 0 & 0 & 0 & 0 \\
v_{1,12} & v_{1,22} & v_{1,23} & v_{1,24} & 0 & 0 & 0 & 0 & 0 \\
v_{1,13} & v_{1,23} & v_{1,33} & v_{1,34} & 0 & 0 & 0 & 0 & 0 \\
v_{1,14} & v_{1,24} & v_{1,34} & v_{1,44} & 0 & 0 & 0 & 0 & 0 \\
0 & 0 & 0 & 0 & v_{2,11} & v_{2,12} & 0 & 0 & 0 \\
0 & 0 & 0 & 0 & v_{2,12} & v_{2,22} & 0 & 0 & 0 \\
0 & 0 & 0 & 0 & 0 & 0 & v_{3,11} & v_{3,12} & v_{3,13} \\
0 & 0 & 0 & 0 & 0 & 0 & v_{3,12} & v_{3,22} & v_{3,23} \\
0 & 0 & 0 & 0 & 0 & 0 & v_{3,13} & v_{3,23} & v_{3,33}
\end{pmatrix},
$$

where $v_{i,jk} = \sigma_p^2 + (t_{ij} + t_{ik})\sigma_{p,pt} + t_{ij}t_{ik}\sigma_{pt}^2$.

Thus, $\mathbf{ZGZ'}$ has a block diagonal form with the size of blocks corresponding to the number of observations on each patient. It is added to the diagonal $\mathbf{R} = \sigma^2\mathbf{I}$ to form the total covariance matrix, $\mathbf{V} = \mathbf{ZGZ'} + \mathbf{R}$, which will also have a block diagonal form. It may appear that covariances will increase with time and that a different origin for time would lead to different results. However, $\mathbf{V}$ is invariant to time origin and although the covariance parameters alter, we still obtain the same overall results (see further discussion in Section 6.5 and examples in Section 6.6).

Note that the covariance structure in random coefficients models is induced by the random coefficients. This differs from covariance pattern models (below) where covariance parameters in the $\mathbf{R}$ (or occasionally $\mathbf{G}$) matrix are chosen to reflect a particular pattern in the data.

2.1.5   The covariance pattern model covariance structure

In the repeated measures example in Section 1.4 the idea of modelling the covariances between observations was introduced. Here, we show how covariance patterns fit into the general mixed model definition using matrix notation. In covariance pattern models the covariance structure of the data is not defined by specifying random effects or coefficients, but by specifying a pattern for the covariance terms directly in the **R** (or, occasionally, **G**) matrix. Observations within a chosen *blocking variable* (e.g. patients) are allowed to be correlated and a pattern for their covariances is specified. This pattern is usually chosen to depend on a variable such as time or the visit number. **R** will have a block diagonal form and can be written

$$
\mathbf{R} = \begin{pmatrix}
\mathbf{R}_1 & \mathbf{0} & \mathbf{0} & \mathbf{0} & \mathbf{0} & \mathbf{0} & \mathbf{0} & \cdot & \cdot & \cdot \\
\mathbf{0} & \mathbf{R}_2 & \mathbf{0} & \mathbf{0} & \mathbf{0} & \mathbf{0} & \mathbf{0} & \cdot & \cdot & \cdot \\
\mathbf{0} & \mathbf{0} & \mathbf{R}_3 & \mathbf{0} & \mathbf{0} & \mathbf{0} & \mathbf{0} & \cdot & \cdot & \cdot \\
\mathbf{0} & \mathbf{0} & \mathbf{0} & \mathbf{R}_4 & \mathbf{0} & \mathbf{0} & \mathbf{0} & \cdot & \cdot & \cdot \\
\mathbf{0} & \mathbf{0} & \mathbf{0} & \mathbf{0} & \mathbf{R}_5 & \mathbf{0} & \mathbf{0} & \cdot & \cdot & \cdot \\
\mathbf{0} & \mathbf{0} & \mathbf{0} & \mathbf{0} & \mathbf{0} & \mathbf{R}_6 & \mathbf{0} & \cdot & \cdot & \cdot \\
\mathbf{0} & \mathbf{0} & \mathbf{0} & \mathbf{0} & \mathbf{0} & \mathbf{0} & \mathbf{R}_7 & \cdot & \cdot & \cdot \\
\cdot & \cdot & \cdot & \cdot & \cdot & \cdot & \cdot & \cdot & \cdot & \cdot \\
\cdot & \cdot & \cdot & \cdot & \cdot & \cdot & \cdot & \cdot & \cdot & \cdot \\
\cdot & \cdot & \cdot & \cdot & \cdot & \cdot & \cdot & \cdot & \cdot & \cdot
\end{pmatrix} .
$$

The submatrices, $\mathbf{R}_i$, are covariance blocks corresponding to the ith blocking effect (the ith patient, say). They have dimension equal to the number of repeated measurements on each patient. The **0**'s represent matrix blocks of 0's giving zero covariances for observations on different patients. Different ways to define covariance patterns in the $\mathbf{R}_i$ matrix blocks will be considered in Section 6.2. We now give two examples of **R** matrices using a small hypothetical dataset. We assume that the first three patients in a repeated measures trial attended at the following visits.

Patient	Visit
1	1
1	2
1	3
2	1
2	2
2	3
2	4
3	1
3	2

Then, using patients as the blocking effect, an **R** matrix where a separate correlation is allowed for each pair of visits (this can be described as a 'general' covariance pattern) is given by

$$\mathbf{R} = \begin{pmatrix} \sigma_1^2 & \theta_{12} & \theta_{13} & 0 & 0 & 0 & 0 & 0 & 0 \\ \theta_{12} & \sigma_2^2 & \theta_{23} & 0 & 0 & 0 & 0 & 0 & 0 \\ \theta_{13} & \theta_{23} & \sigma_3^2 & 0 & 0 & 0 & 0 & 0 & 0 \\ 0 & 0 & 0 & \sigma_1^2 & \theta_{12} & \theta_{13} & \theta_{14} & 0 & 0 \\ 0 & 0 & 0 & \theta_{12} & \sigma_2^2 & \theta_{23} & \theta_{24} & 0 & 0 \\ 0 & 0 & 0 & \theta_{13} & \theta_{23} & \sigma_3^2 & \theta_{34} & 0 & 0 \\ 0 & 0 & 0 & \theta_{14} & \theta_{24} & \theta_{34} & \sigma_4^2 & 0 & 0 \\ 0 & 0 & 0 & 0 & 0 & 0 & 0 & \sigma_1^2 & \theta_{12} \\ 0 & 0 & 0 & 0 & 0 & 0 & 0 & \theta_{12} & \sigma_2^2 \end{pmatrix}.$$

Alternatively, a simpler pattern assuming a constant correlation between each visit pair (known as the 'compound symmetry' pattern) is given by

$$\mathbf{R} = \begin{pmatrix} \sigma^2 & \rho\sigma^2 & \rho\sigma^2 & 0 & 0 & 0 & 0 & 0 & 0 \\ \rho\sigma^2 & \sigma^2 & \rho\sigma^2 & 0 & 0 & 0 & 0 & 0 & 0 \\ \rho\sigma^2 & \rho\sigma^2 & \sigma^2 & 0 & 0 & 0 & 0 & 0 & 0 \\ 0 & 0 & 0 & \sigma^2 & \rho\sigma^2 & \rho\sigma^2 & \rho\sigma^2 & 0 & 0 \\ 0 & 0 & 0 & \rho\sigma^2 & \sigma^2 & \rho\sigma^{a2} & \rho\sigma^2 & 0 & 0 \\ 0 & 0 & 0 & \rho\sigma^2 & \rho\sigma^2 & \sigma^2 & \rho\sigma^2 & 0 & 0 \\ 0 & 0 & 0 & \rho\sigma^2 & \rho\sigma^2 & \rho\sigma^2 & \sigma^2 & 0 & 0 \\ 0 & 0 & 0 & 0 & 0 & 0 & 0 & \sigma^2 & \rho\sigma^2 \\ 0 & 0 & 0 & 0 & 0 & 0 & 0 & \rho\sigma^2 & \sigma^2 \end{pmatrix},$$

where ρ = the correlation between observations on the same patient.

Commonly, in the analysis of repeated measures data no random effects are fitted, in which case the variance matrix $\mathbf{V} = \mathbf{R}$. Otherwise $\mathbf{R}$ is added to $\mathbf{ZGZ}'$ to form the full variance matrix for the data, $\mathbf{V}$.

Covariance patterns in the G matrix

It is also possible, although less usual, to fit a covariance pattern in the $\mathbf{G}$ matrix so that the random effects are correlated within a blocking effect. For example, consider a repeated measures trial in which each patient is assessed at a number of visits and where several measurements are made at each visit. One may wish to model the correlation between visits, within patients, as well as modelling the correlation between observations at the same visit. To achieve this, it is necessary to specify covariance patterns in the $\mathbf{G}$ matrix as well as for the $\mathbf{R}$ matrix. We will return to this type of covariance structure in the example given in Section 8.1 (Model 3).

2.2   MODEL FITTING METHODS

In this section the numerical methods for fitting mixed models will be described. This material is not essential to those who only wish to apply mixed models without gaining a theoretical understanding and we will assume some knowledge of likelihood in our presentation. We showed in Chapter 1 that in some simple circumstances the random effects model could be fitted by using an ANOVA table. However, in general, a more sophisticated method is required to fit the mixed model. In Section 2.2.1 the likelihood function for the mixed model is specified and different methods for maximising it are introduced. The different approaches to maximising the likelihood lead to different estimates for the model parameters and their standard errors. The model fitting process has three distinctive components: estimating fixed effects (i.e. α), estimating random effects (i.e. β), and estimating variance parameters (i.e. variance components or covariance terms). In Sections 2.2.2, 2.2.3 and 2.2.4 we will see how the various fitting methods apply to each of these components, respectively.

2.2.1 The likelihood function and approaches to its maximisation

The mixed model can be fitted by maximising the *likelihood function* for values of the data. The likelihood function, L, measures the likelihood of the model parameters given the data and is defined using the density function of the observations. In models where the observations are assumed independent (e.g. fixed effects models), the likelihood function is simply the product of the density functions for each observation. However, observations in a mixed model are not independent and the likelihood function therefore needs to be based on a multivariate density function for the observations. The likelihood for the variance parameters and the fixed effects can be defined using the multivariate normal distribution for **y** (the term 'variance parameters' encompasses all parameters in the **G** and **R** matrix, i.e. variance components and the covariance parameters). As random effects have expected values of zero and therefore do not affect the mean, this distribution has a mean vector $\mathbf{X}\alpha$ and a covariance matrix **V**. The likelihood function based on the multivariate normal density function is then

$$L = \frac{\exp[-\frac{1}{2}(\mathbf{Y} - \mathbf{X}\alpha)'\mathbf{V}^{-1}(\mathbf{Y} - \mathbf{X}\alpha)]}{(2\pi)^{(1/2)n}|\mathbf{V}|^{(1/2)}}$$

In practice, the log likelihood function is usually used in place of the likelihood function since it is simpler to work with and its maximum value coincides with that of the likelihood. The log likelihood is given by

$$\log(L) = k - \frac{1}{2}[\log|\mathbf{V}| + (\mathbf{Y} - \mathbf{X}\alpha)'\mathbf{V}^{-1}(\mathbf{Y} - \mathbf{X}\alpha)],$$

where

$K = -\frac{1}{2}n\log(2\pi)$ (a constant that can be ignored in the maximisation process),

n = number of observations.

The values of the model parameters which maximise the log likelihood can then be determined. We now introduce briefly several approaches to fitting the mixed model which are all based (directly or indirectly) on maximising the likelihood function. As we shall see, the methods are not all equivalent, and can lead to different estimates of the model parameters. Following this introduction to the methods, we then look in more detail at separate aspects of the fitting process: estimation of fixed effects; estimation of random effects; and estimation of variance parameters.

Maximum likelihood (ML)

This method is based on the concept of maximising the log likelihood with respect to the variance parameters while treating the fixed effects, $\boldsymbol{\alpha}$, as constants. Having obtained the variance parameter estimates, the fixed effects estimates are then obtained by treating the variance parameters as fixed and finding the values of $\boldsymbol{\alpha}$ which maximise the log likelihood. This method has the effect of producing variance parameter estimates that are biased downwards to some degree. This can be illustrated with a very simple example. Suppose we have a simple random sample, $x_1, x_2, \ldots, x_n$ and wish to estimate the mean and variance. If $\hat{\mu}$ is the sample mean, then the ML variance estimator would be $\Sigma_i(x_i - \hat{\mu})^2/n$ rather than the unbiased estimator $\Sigma_i(x_i - \hat{\mu})^2/(n-1)$. The bias is greatest when small numbers of degrees of freedom are used for estimating the variance parameters.

Residual maximum likelihood (REML)

Residual maximum likelihood (sometimes referred to as restricted maximum likelihood) was first suggested by Patterson and Thompson (1971). In this approach, the parameter $\boldsymbol{\alpha}$ is eliminated from the log likelihood so that it is defined only in terms of the variance parameters. We outline the method below.

First, we obtain a likelihood function based on the residual terms, $\mathbf{y} - \mathbf{X}\hat{\boldsymbol{\alpha}}$. This contrasts with the likelihood initially defined which is based directly on the observations, $\mathbf{y}$. You will notice that these residuals differ from the ordinary residuals, $\mathbf{e} = \mathbf{y} - \mathbf{X}\hat{\boldsymbol{\alpha}} - \mathbf{Z}\hat{\boldsymbol{\beta}}$, in that $\mathbf{Z}\hat{\boldsymbol{\beta}}$ is not deducted. Some authors (including those of REML) also refer to the $\mathbf{y} - \mathbf{X}\hat{\boldsymbol{\alpha}}$ as residuals. This is not unreasonable since they can be considered as error terms that include all sources of random variation. Here, we will refer to $\mathbf{y} - \mathbf{X}\hat{\boldsymbol{\alpha}}$ as the *full residuals* in order to differentiate them from the ordinary residuals. The full residuals, $\mathbf{y} - \mathbf{X}\hat{\boldsymbol{\alpha}}$, are, in fact, a linear combination of $\mathbf{y}$ as we will see in Section 2.2.2 where we show how to produce the estimates, $\hat{\boldsymbol{\alpha}}$. It can also be shown that $\mathbf{y} - \mathbf{X}\hat{\boldsymbol{\alpha}}$ and $\hat{\boldsymbol{\alpha}}$ are independent (see Diggle *et al.*, 1994, Section 4.5) and therefore the joint likelihood for $\boldsymbol{\alpha}$ and the variance parameters, $\boldsymbol{\gamma}$, can be expressed as a product of the likelihoods based on $\mathbf{y} - \mathbf{X}\hat{\boldsymbol{\alpha}}$ and $\hat{\boldsymbol{\alpha}}$:

$$L(\boldsymbol{\gamma}, \boldsymbol{\alpha}; \mathbf{y}) = L(\boldsymbol{\gamma}; \mathbf{y} - \mathbf{X}\hat{\boldsymbol{\alpha}})L(\boldsymbol{\alpha}; \hat{\boldsymbol{\alpha}}, \boldsymbol{\gamma}).$$

Thus, the likelihood for $\boldsymbol{\gamma}$ based on $\mathbf{y} - \mathbf{X}\hat{\boldsymbol{\alpha}}$ is given by

$$L(\boldsymbol{\gamma}; \mathbf{y} - \mathbf{X}\hat{\boldsymbol{\alpha}}) = L(\boldsymbol{\gamma}, \boldsymbol{\alpha}; \mathbf{y}) / L(\boldsymbol{\alpha}; \hat{\boldsymbol{\alpha}}, \boldsymbol{\gamma}).$$

Now, from the above,

$$L(\boldsymbol{\gamma}, \boldsymbol{\alpha}; \mathbf{y}) \propto |\mathbf{V}|^{-1/2} \exp(-\tfrac{1}{2}(\mathbf{y} - \mathbf{X}\boldsymbol{\alpha})'\mathbf{V}^{-1}(\mathbf{y} - \mathbf{X}\boldsymbol{\alpha}))$$

and $\hat{\boldsymbol{\alpha}}$ has a multivariate normal distribution with mean and variance given by the maximum likelihood estimates which will be obtained in Section 2.2.2. Hence,

$$L(\boldsymbol{\alpha}; \hat{\boldsymbol{\alpha}}, \boldsymbol{\gamma}) \propto |\mathbf{X}'\mathbf{V}^{-1}\mathbf{X}|^{1/2} \exp(\tfrac{1}{2}(\hat{\boldsymbol{\alpha}} - \boldsymbol{\alpha})'\mathbf{X}\mathbf{V}^{-1}\mathbf{X}(\hat{\boldsymbol{\alpha}} - \boldsymbol{\alpha})).$$

Taking the ratio of these two likelihoods we obtain the REML likelihood as

$$L(\boldsymbol{\gamma}; \mathbf{y} - \mathbf{X}\hat{\boldsymbol{\alpha}}) \propto |\mathbf{X}'\mathbf{V}^{-1}\mathbf{X}|^{-1/2}|\mathbf{V}|^{-1/2} \exp(-\tfrac{1}{2}(\mathbf{y} - \mathbf{X}\hat{\boldsymbol{\alpha}})'\mathbf{V}^{-1}(\mathbf{y} - \mathbf{X}\hat{\boldsymbol{\alpha}})),$$

and the REML log likelihood as

$$\log(L(\boldsymbol{\gamma}; \mathbf{y} - \mathbf{X}\hat{\boldsymbol{\alpha}})) = K - \tfrac{1}{2}\{\log|\mathbf{V}| - \log|\mathbf{X}'\mathbf{V}^{-1}\mathbf{X}|^{-1} + (\mathbf{y} - \mathbf{X}\hat{\boldsymbol{\alpha}})'\mathbf{V}^{-1}(\mathbf{y} - \mathbf{X}\hat{\boldsymbol{\alpha}})\}.$$

Although $\hat{\boldsymbol{\alpha}}$ still appears it does so as a function of the variance parameters ($\hat{\boldsymbol{\alpha}}$ is derived in Section 2.2.2 as $\hat{\boldsymbol{\alpha}} = (\mathbf{X}'\mathbf{V}^{-1}\mathbf{X})^{-1}\mathbf{X}'\mathbf{V}^{-1}\mathbf{y}$). The parameter $\boldsymbol{\alpha}$ does not appear. Note that the difference between the REML log likelihood and the ordinary log likelihood is caused by the extra term $\log|\mathbf{X}'\mathbf{V}^{-1}\mathbf{X}|^{-1}$ which is the log of the determinant of var($\hat{\boldsymbol{\alpha}}$). The REML likelihood is equivalent to having integrated $\boldsymbol{\alpha}$ out of the likelihood for $\boldsymbol{\alpha}$ and $\boldsymbol{\gamma}$, and for this reason REML is sometimes referred to as a 'marginal' method. Because the REML likelihood takes account of the fact that $\boldsymbol{\alpha}$ is a parameter and not a constant, the resulting variance parameter estimates are unbiased. As with ML, $\boldsymbol{\alpha}$ is then estimated by treating the variance parameters as fixed, and finding the values of $\boldsymbol{\alpha}$ which maximise the REML log likelihood.

Iterative generalised least squares (IGLS)

This method can be used iteratively to fit a mixed model and the results will be the same as those obtained using maximum likelihood. This approach obtains estimates of the fixed effects parameters, $\boldsymbol{\alpha}$, by minimising the product of the full residuals weighted by the inverse of the variance matrix, $\mathbf{V}^{-1}$. The residual product is given by $(\mathbf{y} - \mathbf{X}\boldsymbol{\alpha})'\mathbf{V}^{-1}(\mathbf{y} - \mathbf{X}\boldsymbol{\alpha})$.

The variance parameters are obtained by setting the matrix of products of the full residuals $(\mathbf{y} - \mathbf{X}\boldsymbol{\alpha})$ equal to the variance matrix, $\mathbf{V}$, specified in terms of the variance parameters. This gives

$$(\mathbf{y} - \mathbf{X}\hat{\boldsymbol{\alpha}})(\mathbf{y} - \mathbf{X}\hat{\boldsymbol{\alpha}})' = \mathbf{V}.$$

This leads to a set of $n \times n$ simultaneous equations (one for each element in the $n \times n$ matrices) that can be solved iteratively for the variance parameters

(n = number of observations). The equations do not take account of the fact that $\boldsymbol{\alpha}$ will be estimated and is not known. Therefore, resulting variance parameter estimates are biased downwards and are the same as the ML estimates.

An adaption to IGLS which leads to the unbiased REML variance parameter estimates is *restricted iterative generalised least squares (RIGLS)*. It is described by Goldstein (1989) who notes that because $\hat{\boldsymbol{\alpha}}$ is estimated and not known, then

$$E(\mathbf{y} - \mathbf{X}\hat{\boldsymbol{\alpha}})'(\mathbf{y} - \mathbf{X}\hat{\boldsymbol{\alpha}}) = \mathbf{V} - \mathrm{var}(\mathbf{X}\hat{\boldsymbol{\alpha}}) = \mathbf{V} - \mathbf{X}(\mathbf{X}'\mathbf{V}^{-1}\mathbf{X})^{-1}\mathbf{X}',$$

which leads to an alternative set of $n \times n$ equations to solve for the variance parameters

$$(\mathbf{y} - \mathbf{X}\hat{\boldsymbol{\alpha}})'(\mathbf{y} - \mathbf{X}\hat{\boldsymbol{\alpha}}) = \mathbf{V} - \mathbf{X}(\mathbf{X}'\mathbf{V}^{-1}\mathbf{X})^{-1}\mathbf{X}'.$$

Since the observed full residuals, $\mathbf{y} - \mathbf{X}\hat{\boldsymbol{\alpha}}$, depend on the fixed effects parameter estimates, iteration is required between the fixed effects and the variance parameter estimates to obtain the IGLS solution. Further detail on this method can be found in Goldstein (1995, Section 2.5).

Variance parameter bias

We have stated that estimates of variance parameters are biased in ML and unbiased in REML. We believe that lack of bias is an important property and therefore REML will be used to analyse most of the examples in this book. Fixed effects estimates are unlikely to differ greatly between ML and REML analyses, but their standard errors will always be biased downwards if the variance parameters are biased (because they are calculated as weighted sums of the variance parameters). This will be most noticeable when the degrees of freedom used to estimate the variance parameters are small.

2.2.2   Estimation of fixed effects

ML and REML

The fixed effects solution can be obtained by maximising the likelihood (or REML likelihood) by differentiating the log likelihood with respect to $\boldsymbol{\alpha}$ and setting the resulting expression to zero. This leads to a solution which is expressed in terms of the variance parameters:

$$\mathbf{X}'\mathbf{V}^{-1}(\mathbf{y} - \mathbf{X}\boldsymbol{\alpha}) = \mathbf{0}.$$

Rearrangement then gives

$$\hat{\boldsymbol{\alpha}} = (\mathbf{X}'\mathbf{V}^{-1}\mathbf{X})^{-1}\mathbf{X}'\mathbf{V}^{-1}\mathbf{y}$$

and the variance of $\hat{\boldsymbol{\alpha}}$ is obtained as

$$\mathrm{var}(\hat{\boldsymbol{\alpha}}) = (\mathbf{X}'\mathbf{V}^{-1}\mathbf{X})^{-1}\mathbf{X}'\mathbf{V}^{-1}\,\mathrm{var}(\mathbf{y})\mathbf{V}^{-1}\mathbf{X}(\mathbf{X}'\mathbf{V}^{-1}\mathbf{X})^{-1}$$

$$= (\mathbf{X}'\mathbf{V}^{-1}\mathbf{X})^{-1}\mathbf{X}'\mathbf{V}^{-1}\mathbf{V}\mathbf{V}^{-1}\mathbf{X}(\mathbf{X}'\mathbf{V}^{-1}\mathbf{X})^{-1}$$
$$= (\mathbf{X}'\mathbf{V}^{-1}\mathbf{X})^{-1}.$$

This formula is based on the assumption that $\mathbf{V}$ is known. Because $\mathbf{V}$ is, in fact, estimated it can be shown that there will be some downward bias in $\mathrm{var}(\hat{\boldsymbol{\alpha}})$, although this is usually very small (see Section 2.4.3). In Section 2.3 we will see that the Bayesian approach avoids having to make this assumption so that the problem of bias does not arise.

Iterative generalised least squares (IGLS)

The same maximum likelihood solution for $\boldsymbol{\alpha}$ can alternatively be obtained using *generalised least squares*. With this approach the product of the full residuals, weighted by the inverse of the variances, $(\mathbf{y} - \mathbf{X}\boldsymbol{\alpha})'\mathbf{V}^{-1}(\mathbf{y} - \mathbf{X}\boldsymbol{\alpha})$, is minimised by differentiation with respect to $\boldsymbol{\alpha}$:

$$\frac{\delta(\mathbf{y} - \mathbf{X}\boldsymbol{\alpha})'\mathbf{V}^{-1}(\mathbf{y} - \mathbf{X}\boldsymbol{\alpha})}{\delta\boldsymbol{\alpha}} = -2\mathbf{X}'\mathbf{V}^{-1}(\mathbf{Y} - \mathbf{X}\boldsymbol{\alpha}).$$

By setting this expression to zero we find that the residual product is minimised when $\mathbf{X}'\mathbf{V}^{-1}\mathbf{y} = \mathbf{X}'\mathbf{V}^{-1}\mathbf{X}\boldsymbol{\alpha}$ giving the solution for $\boldsymbol{\alpha}$ as

$$\hat{\boldsymbol{\alpha}} = (\mathbf{X}'\mathbf{V}^{-1}\mathbf{X})^{-1}\mathbf{X}'\mathbf{V}^{-1}\mathbf{y},$$

again with variance

$$\mathrm{var}(\hat{\boldsymbol{\alpha}}) = (\mathbf{X}'\mathbf{V}^{-1}\mathbf{X})^{-1}.$$

This solution is sometimes referred to as the *generalised least squares* solution.

In unweighted least squares, where $\mathbf{V} = \sigma^2\mathbf{I}$, the solution will be $\hat{\boldsymbol{\alpha}} = (\mathbf{X}'\mathbf{X})^{-1}\mathbf{X}'\mathbf{y}$ with variance $(\mathbf{X}'\mathbf{X})^{-1}\sigma^2$. This is the solution obtained from fitting fixed effects models using *ordinary least squares (OLS)*; for example, by using PROC GLM in SAS. The difference in the mixed model estimate is due to the use of the inverse variance matrix, $\mathbf{V}^{-1}$, in the formula for $\hat{\boldsymbol{\alpha}}$. When the data are unbalanced it is this difference that causes information on the fixed effects to be combined from different error strata.

2.2.3 Estimation (or prediction) of random effects and coefficients

In general, random effects and coefficients are defined to have normal distributions with zero means and the specific values they take must be thought of as realisations of a sample from a distribution. Thus, their expected values are, by definition, zero. Nonetheless, as we saw earlier, in Section 1.2 in the context of patient effects in a cross-over trial, it is possible to obtain predictions of them. We will now outline how these predictions are obtained.

ML and REML

To predict the random effects, $\boldsymbol{\beta}$ (for simplicity we use the term random effects to refer to either random effects or coefficients) we define a likelihood function in terms of $\boldsymbol{\alpha}$, $\boldsymbol{\beta}$ and $\boldsymbol{\gamma}$ ($\boldsymbol{\gamma}$ is the vector of variance parameters). This can be written as the product of the likelihoods for $\boldsymbol{\beta}$ and $\mathbf{y}$ conditional on the value of $\boldsymbol{\beta}(\mathbf{y}|\boldsymbol{\beta})$.

$$L(\boldsymbol{\alpha}, \boldsymbol{\beta}, \boldsymbol{\gamma}; \mathbf{y}) = L(\boldsymbol{\alpha}, \boldsymbol{\gamma}_{\mathbf{R}}; \mathbf{y}|\boldsymbol{\beta})L(\boldsymbol{\gamma}_{\mathbf{G}}; \boldsymbol{\beta}),$$

where

$\boldsymbol{\gamma}_{\mathbf{R}}$ = variance parameters in the $\mathbf{R}$ matrix,
$\boldsymbol{\gamma}_{\mathbf{G}}$ = variance parameters in the $\mathbf{G}$ matrix.

Using multivariate normal distributions for $\mathbf{y}|\boldsymbol{\beta}$ and $\boldsymbol{\beta}$ we have

$$L(\boldsymbol{\alpha}, \boldsymbol{\beta}, \boldsymbol{\gamma}; \mathbf{y}) \propto |\mathbf{R}|^{-1/2} \exp\{-1/2(\mathbf{y} - \mathbf{X}\boldsymbol{\alpha} - \mathbf{Z}\boldsymbol{\beta})'\mathbf{R}^{-1}(\mathbf{y} - \mathbf{X}\boldsymbol{\alpha} - \mathbf{Z}\boldsymbol{\beta})\}$$
$$\times |\mathbf{G}|^{-1/2} \exp\{-1/2\boldsymbol{\beta}'\mathbf{G}^{-1}\boldsymbol{\beta}\},$$

giving the corresponding log likelihood as

$$\log(L) = -1/2\{\log |\mathbf{R}| + (\mathbf{y} - \mathbf{X}\boldsymbol{\alpha} - \mathbf{Z}\boldsymbol{\beta})'\mathbf{R}^{-1}(\mathbf{y} - \mathbf{X}\boldsymbol{\alpha} - \mathbf{Z}\boldsymbol{\beta})$$
$$+ \log |\mathbf{G}| + \boldsymbol{\beta}'\mathbf{G}^{-1}\boldsymbol{\beta}\} + K.$$

The maximum likelihood solution for $\boldsymbol{\beta}$ can be obtained by differentiating this log likelihood with respect to $\boldsymbol{\beta}$ and setting the resulting expression to zero.

$$\delta\log(L)/\delta\boldsymbol{\beta} = \mathbf{Z}'\mathbf{R}^{-1}(\mathbf{y} - \mathbf{X}\boldsymbol{\alpha} - \mathbf{Z}\boldsymbol{\beta}) - \mathbf{G}^{-1}\boldsymbol{\beta}$$
$$= -(\mathbf{Z}'\mathbf{R}^{-1}\mathbf{Z} + \mathbf{G}^{-1})\boldsymbol{\beta} + \mathbf{Z}'\mathbf{R}^{-1}(\mathbf{y} - \mathbf{X}\boldsymbol{\alpha}).$$

Setting to zero gives

$$\hat{\boldsymbol{\beta}}(\mathbf{Z}'\mathbf{R}^{-1}\mathbf{Z} + \mathbf{G}^{-1}) = \mathbf{Z}'\mathbf{R}^{-1}(\mathbf{y} - \mathbf{X}\boldsymbol{\alpha}),$$
$$\hat{\boldsymbol{\beta}} = (\mathbf{Z}'\mathbf{R}^{-1}\mathbf{Z} + \mathbf{G}^{-1})^{-1}\mathbf{Z}'\mathbf{R}^{-1}(\mathbf{y} - \mathbf{X}\boldsymbol{\alpha}). \qquad (A)$$

As discussed in Chapter 1, estimates are 'shrunken' compared with what they would have been if fitted as fixed. Note that since the estimates are centred about zero, the intercept estimate would need to be added in order to obtain mean random effects estimates. In random effects models the $\mathbf{R}$ matrix is diagonal, $\mathbf{R} = \sigma^2\mathbf{I}$, and we can alternatively write

$$\hat{\boldsymbol{\beta}} = (\mathbf{Z}'\mathbf{Z} + \mathbf{G}^{-1}/\sigma^2)^{-1}\mathbf{Z}'(\mathbf{y} - \mathbf{X}\boldsymbol{\alpha})$$

Compared with the OLS solution for a fixed effects model, $\hat{\alpha} = (\mathbf{X'X})^{-1}\mathbf{X'y}$, we notice the additional term, $\mathbf{G}^{-1}/\sigma^2$, in the denominator. It is this term that causes the estimates to be shrunken towards zero.

An alternative, more compact, form for $\hat{\beta}$ can be obtained from the solution (A) above using matrix manipulation and recalling that $\mathbf{V} = \mathbf{ZGZ'} + \mathbf{R}$:

$$\hat{\beta} = \mathbf{GZ'V}^{-1}(\mathbf{y} - \mathbf{X\alpha}).$$

The variance of $\hat{\beta}$ can be obtained as

$$\text{var}(\hat{\beta}) = \mathbf{GZ'V}^{-1}\mathbf{ZG} - \mathbf{GZ'V}^{-1}\mathbf{X}(\mathbf{X'V}^{-1}\mathbf{X})^{-1}\mathbf{X'V}^{-1}\mathbf{ZG}.$$

As with $\text{var}(\hat{\alpha})$, this formula is based on the assumption that $\mathbf{V}$ is known. Because $\mathbf{V}$ is, in fact, estimated there will be some downward bias in $\text{var}(\hat{\beta})$, although this is usually small (see Section 2.4.3). Again, the Bayesian approach (Section 2.3) avoids having to make this assumption and the problem of bias does not arise.

Iterative generalised least squares (IGLS)

We note that $\hat{\beta}$ cannot be obtained directly from the usual least squares equations used by *iterative generalised least squares (IGLS)*. However, once estimates for the variance parameters are obtained using IGLS, the above formulae can then be applied to obtain $\hat{\beta}$.

2.2.4   Estimating variance parameters

In this Section we consider the numerical procedures used to apply maximum likelihood and least-squares-based methods for estimating variance parameters. These are usually embedded in the statistical packages used to perform the analysis, so knowledge of their details is not necessary for application. We present them here for completeness.

ML and REML

Both of these methods work by obtaining variance parameter estimates that maximise a likelihood function. A solution cannot be specified by a single equation as it was for the fixed and random effects because the derivatives of the log likelihood with respect to the variance parameters are non-linear. An iterative process such as the widely applied *Newton–Raphson algorithm* is therefore required. This works by repeatedly solving a quadratic approximation to the log likelihood function. We will now illustrate this algorithm by firstly showing how a quadratic function is solved in terms of its first and second derivatives, and then showing how this solution is used to define the iterative Newton–Raphson method for maximising a likelihood function.

Solving a quadratic   To express the solution of a quadratic function of θ in terms of its first and second derivatives, we first write the function in matrix notation in the general quadratic form as

$$f(\theta) = a + \mathbf{b}'\theta + \tfrac{1}{2}\theta'\mathbf{C}\theta.$$

Note that the first derivative of $f(\theta)$ will be a vector and the second derivative will be a square matrix. For example, if $\theta = (\theta_1, \theta_2, \theta_3)$, then the first derivative of $f(\theta)$ is a vector of length three and the second derivative of the log likelihood is a 3×3 matrix. The first derivative is given by

$$f'(\theta) = \mathbf{b} + \mathbf{C}\theta,$$

and the second derivative by

$$f''(\theta) = \mathbf{C}.$$

The solution for θ , which maximises $f(\theta)$, $\hat{\theta}$, is obtained by setting the first derivative to zero:

$$f'(\theta) = \mathbf{b} + \mathbf{C}\theta = \mathbf{0},$$

to give,

$$\hat{\theta} = -\mathbf{C}^{-1}\mathbf{b}.$$

By adding and deducting an arbitrary value, θ_i say, this solution to the quadratic function can then be expressed as

$$\hat{\theta} = \theta_i - \mathbf{C}^{-1}(\mathbf{b} + \mathbf{C}\theta_i),$$

and can be rewritten in terms of θ_i and the first and second derivatives of $f(\theta)$ evaluated at θ_i:

$$\hat{\theta} = \theta_i - f''^{-1}(\theta_i)f'(\theta_i)$$

We see below that this mathematical trick is the key to the use of the Newton–Raphson algorithm.

The Newton–Raphson iteration   We are seeking the value θ to maximise the likelihood function $f(\theta)$. If we start with an initial approximate solution θ_1, we make the assumption that the function will be approximately quadratic to obtain an improved approximation θ_2, using the formula obtained above. The process is then repeated using θ_2 as the approximate solution, to obtain an improved approximation θ_3. Although the function will not, in general, be quadratic, in the region of the maximum likelihood solution the quadratic approximation is usually quite good, and the Newton–Raphson iterative procedure will usually converge appropriately. Convergence is obtained when parameter values change very little between successive iterations. The iterative process can be defined by

$$\theta_{i+1} = \theta_i - f''^{-1}(\theta_i) \times f'(\theta_i),$$

where $f''(\theta_i)$ and $f'(\theta_i)$ are the actual (unapproximated) derivatives of $f(\theta)$ evaluated at θ_i. The matrix of second derivatives, $f''(\theta)$, is often referred to as the *Hessian matrix*. In mixed models, $f(\theta)$ is taken to be the log likelihood expressed in terms of the variance parameters, θ.

The need to evaluate the derivatives at each iteration can make the Newton–Raphson algorithm computationally intensive. Computation can be made easier by using a matrix known as the information matrix in place of $f''(\theta_i)$ in the iterative process. The information matrix is the expected value of the Hessian matrix and it is easier to compute than the Hessian matrix because some of the correlation terms are zero. When it is used the process can be referred to as the *method of scoring* or *Fisher scoring*. This method has been shown to be more robust to poor starting values than the Newton–Raphson algorithm. Within SAS, PROC MIXED uses Fisher scoring for the first iteration and then Newton–Raphson for the remaining iterations, as the default fitting method.

Covariances of variance parameters　　An indication of the precision of the variance parameters can be obtained from an estimate of their variance and their degree of correlation from the covariances. However, this estimate is based on standard asymptotic (large sample) theory. The asymptotic covariances of the variance parameters are given by the negative of the expected values of second partial derivatives of the log likelihood (see, for example, Searle *et al.*, 1992, Section 3.7):

$$\hat{\text{var}}(\theta_i) = -\text{E}\{\delta^2 \log(L)/\delta\theta_i\delta\theta_i\},$$

$$\hat{\text{cov}}(\theta_i, \theta_j) = -\text{E}\{\delta^2 \log(L)/\delta\theta_i\delta\theta_j\}.$$

Since the resulting covariances are based on asymptotic theory and are related to the estimated variance parameter values themselves, they should be interpreted cautiously. Also, remember that the distribution of variance parameters is not usually symmetrical.

Iterative generalised least squares (IGLS)

This method estimates the variance parameters by setting the full residual products equal to the variance matrix, $\mathbf{V}$, specified in terms of the variance parameters and solving the resulting equations. This leads to a set of $n \times n$ simultaneous equations that can be solved iteratively for the variance parameters ($n =$ number of observations). In ordinary IGLS which gives variance parameters that are biased downwards the equations are

$$\mathbf{V} = (\mathbf{y} - \mathbf{X}\alpha)(\mathbf{y} - \mathbf{X}\alpha)'.$$

To illustrate the structure of the equations more clearly we will show their form for a small hypothetical dataset. We assume that the first two patients in a repeated measures trial attended at the following visits.

Patient	Visit
1	1
1	2
1	3
2	1
2	2

The equations in a model using a separate covariance term for each pair of visits (i.e. with a 'general' covariance structure) are given by

$$
\begin{pmatrix}
\sigma_1^2 & \theta_{12} & \theta_{13} & 0 & 0 \\
\theta_{12} & \sigma_2^2 & \theta_{23} & 0 & 0 \\
\theta_{13} & \theta_{23} & \sigma_3^2 & 0 & 0 \\
0 & 0 & 0 & \sigma_1^2 & \theta_{12} \\
0 & 0 & 0 & \theta_{12} & \sigma_2^2
\end{pmatrix}
$$

$$
=
\begin{pmatrix}
(y_1 - \mu_1) & (y_1 - \mu_1) & (y_1 - \mu_1) & (y_1 - \mu_1) & (y_1 - \mu_1) \\
\times (y_1 - \mu_1) & \times (y_2 - \mu_2) & \times (y_3 - \mu_3) & \times (y_4 - \mu_4) & \times (y_5 - \mu_5) \\
(y_2 - \mu_2) & (y_2 - \mu_2) & (y_2 - \mu_2) & (y_2 - \mu_2) & (y_2 - \mu_2) \\
\times (y_1 - \mu_1) & \times (y_2 - \mu_2) & \times (y_3 - \mu_3) & \times (y_4 - \mu_4) & \times (y_5 - \mu_5) \\
(y_3 - \mu_3) & (y_3 - \mu_3) & (y_3 - \mu_3) & (y_3 - \mu_3) & (y_3 - \mu_3) \\
\times (y_1 - \mu_1) & \times (y_2 - \mu_2) & \times (y_3 - \mu_3) & \times (y_4 - \mu_4) & \times (y_5 - \mu_5) \\
(y_4 - \mu_4) & (y_4 - \mu_4) & (y_4 - \mu_4) & (y_4 - \mu_4) & (y_4 - \mu_4) \\
\times (y_1 - \mu_1) & \times (y_2 - \mu_2) & \times (y_3 - \mu_3) & \times (y_4 - \mu_4) & \times (y_5 - \mu_5) \\
(y_5 - \mu_5) & (y_5 - \mu_5) & (y_5 - \mu_5) & (y_5 - \mu_5) & (y_5 - \mu_5) \\
\times (y_1 - \mu_1) & \times (y_2 - \mu_2) & \times (y_3 - \mu_3) & \times (y_4 - \mu_4) & \times (y_5 - \mu_5)
\end{pmatrix}
$$

In this simple example, equating corresponding terms from the left-hand side and right-hand side of this equation gives

$$\theta_{13} = (y_1 - \mu_1)(y_3 - \mu_3),$$

$$\theta_{23} = (y_2 - \mu_2)(y_3 - \mu_3).$$

There are two equations for θ_{12}, and we may obtain an estimate from their average:

$$\theta_{12} = \{(y_1 - \mu_1)(y_2 - \mu_2) + (y_5 - \mu_5)(y_4 - \mu_4)\}/2.$$

Thus, in this artificially simple dataset the covariance terms are calculated over the average of the observed covariances for just one or two subjects. In a genuine dataset, the covariances will be estimated from the average of the observed covariances from many more subjects. The variance terms can be estimated in a similar way.

The approach extends to other covariance patterns. Suppose that with the same artificial dataset we wish to fit a simpler correlation pattern, with a constant covariance between each visit pair (i.e. compound symmetry pattern). Then

$$\begin{pmatrix} \sigma^2 & \theta & \theta & 0 & 0 \\ \theta & \sigma^2 & \theta & 0 & 0 \\ \theta & \theta & \sigma^2 & 0 & 0 \\ 0 & 0 & 0 & \sigma^2 & \theta \\ 0 & 0 & 0 & \theta & \sigma^2 \end{pmatrix}$$

$$= \begin{pmatrix} (y_1 - \mu_1) & (y_1 - \mu_1) & (y_1 - \mu_1) & (y_1 - \mu_1) & (y_1 - \mu_1) \\ \times(y_1 - \mu_1) & \times(y_2 - \mu_2) & \times(y_3 - \mu_3) & \times(y_4 - \mu_4) & \times(y_5 - \mu_5) \\ (y_2 - \mu_2) & (y_2 - \mu_2) & (y_2 - \mu_2) & (y_2 - \mu_2) & (y_2 - \mu_2) \\ \times(y_1 - \mu_1) & \times(y_2 - \mu_2) & \times(y_3 - \mu_3) & \times(y_4 - \mu_4) & \times(y_5 - \mu_5) \\ (y_3 - \mu_3) & (y_3 - \mu_3) & (y_3 - \mu_3) & (y_3 - \mu_3) & (y_3 - \mu_3) \\ \times(y_1 - \mu_1) & \times(y_2 - \mu_2) & \times(y_3 - \mu_3) & \times(y_4 - \mu_4) & \times(y_5 - \mu_5) \\ (y_4 - \mu_4) & (y_4 - \mu_4) & (y_4 - \mu_4) & (y_4 - \mu_4) & (y_4 - \mu_4) \\ \times(y_1 - \mu_1) & \times(y_2 - \mu_2) & \times(y_3 - \mu_3) & \times(y_4 - \mu_4) & \times(y_5 - \mu_5) \\ (y_5 - \mu_5) & (y_5 - \mu_5) & (y_5 - \mu_5) & (y_5 - \mu_5) & (y_5 - \mu_5) \\ \times(y_1 - \mu_1) & \times(y_2 - \mu_2) & \times(y_3 - \mu_3) & \times(y_4 - \mu_4) & \times(y_5 - \mu_5) \end{pmatrix}$$

The solution is then given by averaging the observed covariances over all pairs of time points so that

$$\theta = \{(y_1 - \mu_1)(y_2 - \mu_2) + (y_4 - \mu_4)(\mu_5 - \mu_5)$$
$$+ (y_1 - \mu_1)(y_3 - \mu_3) + (y_2 - \mu_2)(y_3 - \mu_3)\}/4,$$

and $\quad \sigma^2 = \sum_{i=1}^{5} (y_i - \mu_i)^2 / 5.$

In random effects and coefficients models, each linear equation may involve more than one parameter. Simple averaging will then not be sufficient to obtain the parameter estimates and standard methods for solving sets of linear equations can be applied.

For restricted iterative generalised least squares (RIGLS) which gives unbiased variance parameters (as in REML) the equations are

$$(\mathbf{y} - \mathbf{X}\hat{\boldsymbol{\alpha}})'(\mathbf{y} - \mathbf{X}\hat{\boldsymbol{\alpha}}) = \mathbf{V} - \mathbf{X}(\mathbf{X}'\mathbf{V}^{-1}\mathbf{X})^{-1}\mathbf{X}'.$$

Further details on generalised iterative least squares can be found in Goldstein (1995, Section 2.5).

2.3   THE BAYESIAN APPROACH

The Bayesian approach to fitting a mixed model provides an interesting alternative to the classical methods we have already described which are based on maximising the likelihood function. Some statisticians have difficulties in accepting the philosophy of the Bayesian approach and will not be willing to use such an analysis. We feel that such an attitude is misguided and the Bayesian approach to mixed models is introduced here because it has several potential advantages over maximum likelihood methods. It must be recognised though that the application of Bayesian methods in medicine is relatively uncommon and may be unfamiliar to many of the potential 'consumers' of the analysis. There may therefore be a greater communication problem in reporting such analyses.

Bayesian methods have also been hampered in the past by the fact that they can require very large amounts of computer power and time. However, recently this has become less of a restriction and their use has become much more widespread. In the context of mixed models, Bayesian methods have been developed most fully for use with random effects models and we therefore concentrate primarily on these models. As some of the ideas underlying Bayesian modelling are quite different from those underlying classical statistical approaches, we will first spend some time giving a brief introduction to Bayesian concepts. For a more indepth introduction, the book *Bayesian Inference in Statistical Analysis* by Box and Tiao (1973) is recommended. An example of a Bayesian analysis will be given in Section 2.5.

2.3.1   Introduction

In a Bayesian analysis the distribution of the model parameters is used to obtain parameter estimates (e.g. of the mean treatment effects). This contrasts with classical statistical methods which use the distribution of the data, not of the parameters, to estimate parameters. The distribution of the model parameters is obtained by combining the likelihood function with a *prior* distribution for the parameters to obtain what is called the *posterior* distribution. The prior distribution can either be informative (based on prior knowledge of the parameter) or non-informative (containing no prior information on the parameter). While there can be good reason to use informative priors, particularly when they only use basic knowledge of the context of the problem (e.g. range for human temperature is $30-45°C$, or a parameter is positive), they can often be difficult to elicit accurately and model correctly, particularly when there are several parameters. For this reason, here we will only consider the use of non-informative priors.

Bayesian methods are often considered to be quite different from maximum likelihood methods (e.g. ML and REML). However, when a non-informative prior is used in a Bayesian analysis the posterior density has a similar shape to the likelihood function. Thus, a Bayesian analysis is usually almost equivalent to evaluating the likelihood function over its full parameter space. In fact, when a flat prior is used it is exactly equivalent and the maximum likelihood estimators occur at the mode of the posterior distribution. Thus, the main difference between a Bayesian analysis (with non-informative priors) and a maximum likelihood analysis is that the posterior density (which is similar to the likelihood function) is fully evaluated, whereas in maximum likelihood only the parameter values that maximise the likelihood and their standard errors are obtained.

The advantage of fully evaluating the posterior density is that exact posterior standard deviations and probability intervals can be obtained from the posterior distributions for each model parameter. These statistics are analagous to the standard errors and confidence intervals derived using maximum likelihood. We will also show how posterior distributions can be used to yield exact 'Bayesian' p-values which are analogous to p-values resulting from classical significance tests (see Section 2.3.3). We believe that the potential to obtain such exact statistics is a major advantage of using a Bayesian approach over maximum likelihood where parameter standard errors, confidence intervals and p-values are computed using formulae which assume that the variance parameters are known (see Sections 2.2.2 and 2.4.3).

A potential disadvantage of using a Bayesian approach is that the techniques used to obtain the posterior density usually rely on simulation and it can be difficult to define exactly when a simulated density has converged to the true density. In contrast, it is usually much easier to conclude that a maximum likelihood method has converged. The problems associated with defining convergence when simulation techniques are used are currently the subject of much research interest.

2.3.2   Determining the posterior density

In a Bayesian model it is assumed that parameters have a prior distribution which reflects knowledge (or lack of knowledge) about the parameters before the analysis commences, that this distribution can be modified to take account of observed data to form a posterior distribution, and that the posterior distribution can be used to make inferences about the model parameters. The posterior density function for the model parameters is proportional to the product of the likelihood function and the prior density of all the model parameters, $p(\theta)$:

$$p(\theta; \mathbf{y}) = p(\theta)L(\theta; \mathbf{y})/K,$$

where

$$K = \int p(\theta)L(\theta; \mathbf{y}) \, d\theta.$$

The denominator integral is necessary to ensure that the posterior integrates to one. The likelihood function can be based on any distribution and hence Bayesian methods can be applied to data with either normal or non-normal distributions. We note that by using a flat prior, $p(\theta) \propto c(c = \text{constant})$, the posterior density becomes

$$p(\theta; \mathbf{y}) = L(\theta; \mathbf{y})/K,$$

which is the *standardised likelihood function*. A Bayesian analysis is then equivalent to evaluating the likelihood function over its full parameter space. However, as we will see later, somewhat surprisingly a flat prior cannot always be regarded as non-informative.

A mixed model will usually contain several parameters. For example, consider a two-way, cross-over trial model fitting treatments and periods as fixed and patients as random. We let

$\mu = \text{intercept},$
$t = \text{treatment difference},$

p = period difference,

$s_1, s_2, \ldots, s_{n-1}$ = subject effects (n subjects),

σ^2 = residual variance component,

σ_s^2 = subject variance component.

(Note that the redundant parameters for the second treatment and period, and the final subject, are omitted in this parameterisation.)

We assume that estimation of the subject effects is not of interest. Taking the Bayesian approach, a prior distribution, $p(\mu, t, p, \sigma^2, \sigma_s^2)$ say, is first specified for the model parameters. The posterior distribution of μ, t, p, σ^2 and σ_p^2 is then obtained using the product of the prior distribution and the likelihood function:

$$p(\mu, t, p, \sigma^2, \sigma_s^2; \mathbf{y}) = p(\mu, t, p, \sigma^2, \sigma_s^2) \times L(\mu, t, p, \sigma^2, \sigma_s^2; \mathbf{y})/K,$$

where K is the standardising constant used to ensure that the distribution integrates to one.

$$K = \int\int\int\int\int p(\mu, t, p, \sigma^2, \sigma_s^2) \times L(\mu, t, p, \sigma^2, \sigma_s^2; \mathbf{y})\delta\mu \, \delta t \, \delta p \, \delta\sigma^2 \, \delta\sigma_s^2.$$

The posterior distribution can then be used to estimate each of the model parameters. More detail on how this is done is given in the next section.

2.3.3   Parameter estimation, probability intervals and *p*-values

The full posterior distribution $p(\mu, t, p, \sigma^2, \sigma_s^2; \mathbf{y})$ is not often useful in itself for making inferences about individual parameters. The human brain does not readily handle such complicated information. To make inferences about one parameter, t say, the posterior is usually integrated over all the other parameters (μ, p, σ^2 and σ_s^2) to form a *marginal posterior distribution* for t:

$$p_t(t; \mathbf{y}) = \int\int\int\int p(\mu, t, p, \sigma^2, \sigma_s^2; \mathbf{y}) \, \mathrm{d}\mu \, \mathrm{d}p \, \mathrm{d}\sigma^2 \, \mathrm{d}\sigma_s^2.$$

This distribution can then be used to calculate estimators for the treatment parameter. With this approach we note that the problem of biased fixed and random effects standard errors will not occur as it does in maximum likelihood (see Section 2.4.3). This is because variances can be obtained from exact parameter distributions, rather than by using formulae which assume that the variance components are known.

Parameter estimation

There are no strict rules about which estimator should be used for a given parameter. For location parameters (i.e. fixed and random effects) the posterior distribution can often be conveniently summarised by the posterior mean and its variance. The square root of var(t) gives the standard deviation of t. Although this is analogous to the standard error of t given by maximum likelihood, Bayesian statisticians usually prefer to quote standard deviations of parameters rather than standard errors of means.

The mean value is not usually the most appropriate estimate for a variance component because the marginal distribution is not symmetrical. The median or mode are better choices when judged by the 'average closeness to the true value' measured, say, by the mean squared error. Another estimator sometimes used is the posterior mean of the square root of the variance component. However, our own preference is to use the median. Box and Tiao (1973, Appendix A5.6) discuss the choice of variance component estimators in more detail. However, in many applications the variance component estimate is not of particular interest and may not even need to be obtained.

Probability intervals

Exact probability intervals for model parameters can be calculated directly from their marginal posterior distributions. These are obtained such that

- The probability density of points within the interval is higher than all points outside it.
- The total probability for the region is equal to a specified $1 - \alpha$.

Probability intervals are analogous to the confidence intervals calculated by maximum likelihood methods. However, unlike confidence intervals, they are calculated directly from the posterior distribution and do not rely on estimates of parameter standard errors.

Bayesian *p*-values

The concept of the significance test as it appears in 'classical' statistics does not fall within the Bayesian philosophy. In classical statistics a hypothesis is tested by constructing an appropriate test statistic and obtaining the distribution of this statistic under a null hypothesis (e.g. the treatment difference is zero). Acceptance of the null hypothesis is then obtained from the position of the test statistic within this 'null' distribution. Specifically, we calculate the probability under the null hypothesis of obtaining values of the test statistic which are as, or more, extreme than the observed value. Probabilities below 0.05 are often seen as sufficient evidence to reject a null hypothesis. The closest equivalent using Bayesian methods is achieved through the use of probability intervals. The value of α for the probability interval which has zero on the boundary can be used to provide a 'Bayesian' *p*-value to examine the plausibility that a parameter is zero. This is equivalent to a two-sided 'classical' *p*-value. However, it has the advantage of being exact and there are no potential inaccuracies in obtaining a test statistic (based on standard error estimates) or the degrees of freedom for its distribution.

2.3.4 Specifying non-informative prior distributions

We have given little indication so far of the distributional form that non-informative priors should take. The requirement is that a non-informative prior for a parameter should have minimal influence on the results obtained for that parameter. The theoretical background of how we set non-informative priors is not easily accessible to those without a background in Bayesian statistics, so we will outline the methods which can be used in the following section. Prior to that though (excusing the pun) we will take the pragmatic approach and simply describe some distributions which have been suggested to provide non-informative priors for mixed models.

For the fixed effects (μ, t and p in the cross-over model), there are (at least) two suitable non-informative priors:

- Uniform distribution $(-\infty, \infty)$, $p(\theta) \propto c$, i.e. a flat prior,
- Normal $(0, K)$, where K is a very large number.

We note that as K tends to ∞, so the normal distribution tends to the Uniform $(-\infty, \infty)$ distribution. For the practitioner there is the question of how big a number is very large? This depends on the scale on which observations are being recorded. Recording distances in millimetres gives larger numbers than recording in kilometres. The choice of K should be so that its square root is at least an order of magnitude larger than any of the observations.

For the variance components (σ^2 and σ_s^2 in the cross-over model), any of the following distributions may provide suitable non-informative priors:

- Uniform $(-\infty, \infty)$, $p(\theta) \propto c$,
- Reciprocal distribution, $p(\theta) \propto c/\theta$ ($c = $ constant),
- Inverse gamma distribution (K, K), where K is a very small number.

In this book we will not describe the inverse gamma distribution, other than to note that it is a two-parameter distribution, and that as the parameters tend to zero, the distribution tends to the reciprocal distribution. The practical guidance to the choice of K is again that it should be at least an order of magnitude smaller than the observations.

In practice, it often makes little difference to the results obtained from the posterior distribution, whichever of these priors is chosen. An exception is when the true value of a variance component is close to zero. Under these circumstances the posterior distribution arising from the uniform prior will differ with the alternative choices of the prior. We note, though, that many statisticians would be unlikely to choose the uniform prior for variance components, because it is known that variance components cannot be negative. However, it is this prior which leads to a posterior density that is exactly proportional to the likelihood!

We now introduce in more detail a general approach to the setting of non-informative priors. This section will be of greatest interest to those readers who wish to extend their knowledge of the Bayesian approach.

Setting a non-informative prior

At first sight it might appear that simply using a flat distribution, $p(\theta) \propto c$, would provide a non-informative prior for a parameter, θ. However, this is not always the case. To obtain a non-informative prior for θ, it is first necessary to find a transformation of θ, $h(\theta)$ say, for which the likelihood has the same shape when plotted against it regardless of any changes to the data. Since the shape of the likelihood distribution is then unaffected by the value of $h(\theta)$, a flat density for $h(\theta)$, $p(h(\theta)) \propto c$, will give a non-informative prior for $h(\theta)$. From this, a non-informative prior distribution of θ can be obtained as (see Box and Tiao, 1973, Section 1.3)

$$p(\theta) = p(h(\theta))|dh(\theta)/d\theta|$$

$$\propto c|dh(\theta)/d\theta|.$$

Consider a sample from the normal distribution, $N(\mu, \sigma^2)$. If $L(\mu; \sigma, \mathbf{y})$ is plotted against μ the same shape always arises regardless of the data, only the location is dependent on the data (see Figure 2.1). Thus, in this case h is the identity function, $h(\mu) = \mu$, and we obtain $p(\mu) \propto c|dh(\mu)/d\mu| = c$, a flat prior.

However, when $L(\sigma; \mu, \mathbf{y})$ is plotted against σ we find that its shape varies with the data (see Figure 2.2). Thus, h is not now the identity function and we need to consider alternative functions. It turns out that plotting $L(\sigma; \mu, \mathbf{y})$ against $\log(\sigma)$

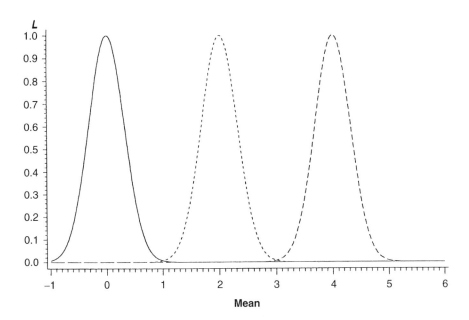

Figure 2.1   Likelihood vs. mean. Mean: ——— 0; - - - - - 2; – – – 4. (L is proportional to likelihood).

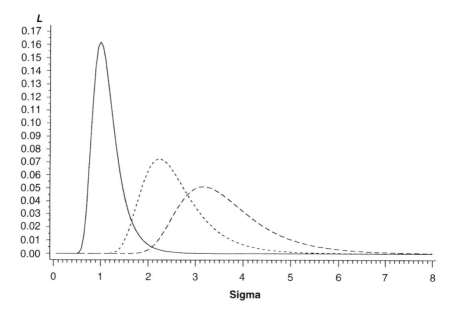

Figure 2.2   Likelihood vs. sigma. Variance: ———— 1; - - - - - 5; — — — 10. (*L* is proportional to likelihood).

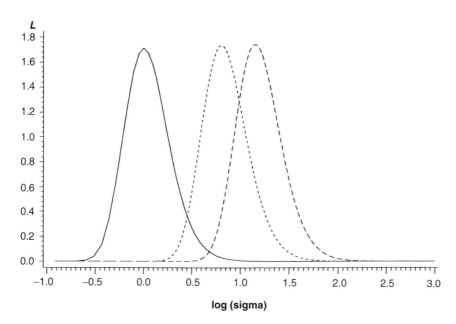

Figure 2.3   Likelihood vs. log (sigma). Variance: ———— 1; - - - - - 5; — — — 10. (*L* is proportional to likelihood).

gives likelihoods which have the same shape regardless of the data values (see Figure 2.3). Thus, $h(\sigma) = \log(\sigma)$ and we obtain $p(\sigma) \propto c|\mathrm{d}\log(\sigma)/\mathrm{d}\sigma| = c/\sigma$.

In general, it is not necessary to go to the trouble of plotting the likelihood against various transformations of a parameter to find the function h. h can instead be obtained by writing the likelihood for θ in the form $g(h(\theta) - t(\mathbf{y}))$. For example, the likelihood for μ for the normal sample can be written

$$L(\mu; \sigma, \mathbf{y}) \propto (1/\sigma^n) \exp\{-n(\mu - \bar{y})^2/2\sigma^2\}.$$

This gives $g(z) = \exp(-z^2/2\sigma^2)$, $h(\mu) = \mu$ and $t(\mathbf{y}) = -\bar{y}$.

The likelihood for σ is obtained by rearranging the likelihood function as

$$L(\sigma; \mu, \mathbf{y}) \propto \exp\{-n(\log(\sigma) - \log(s)) - n/2 \exp(-2(\log(\sigma) - \log(s)))\}$$

to give $g(z) = \exp(-nz - \frac{1}{2}\exp(-2z))$, $h(\sigma) = \log(\sigma)$ and $t(\mathbf{y}) = \log(s)$.

Above we have assumed that μ and σ are uncorrelated, although this is not in fact the case for the normal distribution. When there are several model parameters, prior specification is often simplified by assuming independence. Thus, in the cross-over model introduced in Section 2.3.2 we could write

$$p(\mu, t, p, \sigma^2, \sigma_s^2) = p_\mu(\mu) \times p_t(t) \times p_p(p) \times p_{\sigma^2}(\sigma^2) \times p_{\sigma_s^2}(\sigma_s^2).$$

Alternatively, a joint prior that takes account of the correlations between parameters can be obtained using Jeffreys' method (Jeffreys, 1961). Jeffreys' method is also helpful in situations where the likelihood cannot be arranged in the required form, $g(h(\sigma) - t(\mathbf{y}))$. It is the default prior used for the variance components when the PRIOR statement in PROC MIXED is used. Details of Jeffreys' method and further discussion on setting non-informative priors can be found in Box and Tiao (1973, Section 1.3), Tanner (1996, Section 2.2.1) or Carlin and Louis (1996, p. 33).

Properties for prior distributions

We now define some of the properties that are considered when setting priors and discuss their relevance in mixed models.

Proper priors   Ideally the prior distribution should integrate to one and it is then described as a proper prior. A flat prior, $p(\theta) \propto c$ is not proper because the integral $\int_{-\infty}^{\infty} p(\theta)\,\mathrm{d}\theta$ does not exist no matter how small k is. Likewise the non-informative prior suggested for variance components, $p(\sigma) \propto c/\sigma$, is also improper. However, the normal distributions with zero mean and very large variances suggested for fixed and random effects, and the inverse gamma distributions with very small parameters for variance components, are integrable and hence can be described as 'proper'. It is for this reason that these priors are sometimes preferred. However, as we noted earlier, when the spreads of these distributions are taken to their extremes, we obtain $N(0, \infty) = \mathrm{Uniform}(-\infty, \infty)$ and $\mathrm{IG}(0,0) = $ reciprocal distribution $(p(\theta) \propto c/\theta)$ which are improper priors.

In practice, it is not always important for a prior to be proper. Provided the integral of the likelihood over all the model parameters is finite, there is not a problem. In mixed models applications the integral will often be finite even when improper priors are used. However, one situation where this may not be the case, which we will meet later, is when there are uniform fixed or random effect categories in generalised linear mixed models (these will be defined in Section 3.3.2).

Conjugacy   When a prior distribution is conjugate it leads to a posterior distribution that is of the same type as the prior. For example, a particular distribution called the beta distribution could be chosen as the prior for a binomial parameter, and this would lead to a posterior distribution which is also a beta distribution. The beta distribution is then described as conjugate to the binomial distribution. Likewise, conjugate distributions can be obtained for all other distributions from the exponential family.

When a model uses more than one parameter a joint conjugate distribution should ideally be used. However, in practice, independence is often assumed between the parameters and the joint prior is taken as the product of the conjugate priors for each parameter (obtained assuming the other parameters are fixed). For example, in setting a joint prior density for the normal distribution parameterised by μ and σ^2, the product of a normal prior density (conjugate for μ) and an inverse gamma prior density (conjugate for σ^2) could be taken if independence between μ and σ^2 were assumed.

We have already noted above that distributions parameterised to have very large (but not infinite) spreads are often used to create proper non-informative priors. These distributions in fact are chosen as the conjugate distributions for each parameter. Hence, using normal prior distributions for fixed and random effects parameters is expected to lead to normal posterior distributions for the parameters. Likewise, inverse gamma prior distributions for variance components are expected to lead to inverse gamma posterior distributions. However, this is based on assuming independence of the parameters in setting the priors. In practice, the model parameters are not independent, so these posterior distributions will not be obtained exactly.

2.3.5   Evaluating the posterior distribution

Evaluation of the posterior distribution and using it to obtain marginal posterior distributions for individual parameters relies heavily on integration. However, in most situations algebraic integration is not possible and thus a numerical method of integration is required.

An alternative approach that does not evaluate the full posterior distribution is known as *empirical Bayes*. We mention it here because it has been used a lot in the past due to its ease of implementation. It works by first estimating the variance components from their marginal posterior distributions. The fixed effects are then obtained by conditioning the joint posterior distribution on the variance component estimates (i.e. evaluating the posterior distribution with the variance component parameters fixed at the estimated values). If a flat prior were assumed for all the model parameters, then the marginal posterior distribution for the variance components would be proportional to the REML likelihood and its mode would coincide with the REML variance component estimates. Thus, in this situation, a REML analysis is equivalent to an analysis using empirical Bayes. However, the benefits associated with using Bayesian models (e.g.exact standard deviations, probability intervals and significance test p-values) do not occur with empirical Bayes because the full conditional posterior distribution is not evaluated.

The most popular methods of evaluating the posterior now rely on simulation as a means of performing the integration. Such methods can be described as *Monte Carlo* methods and with the recent increased availability of computer power they have become much more feasible. Instead of seeking to obtain the marginal posterior distribution directly, we take random samples from the joint posterior density of all the model parameters. Each sample provides us with a set of values of our model parameters, (e.g. μ, t, p, σ^2 and σ_s^2 in our cross-over example). If we are interested in the marginal distribution of t, say, then we simply ignore the other parameter values and consider only the randomly sampled values for t. If we take a sufficiently large number of samples from the joint posterior density, then we will be able to characterise the marginal distribution of t, to whatever level of detail we wish. Of course, as well as estimating the marginal distributions (usually our main purpose) we can use the full set of values of our model parameters to evaluate the full posterior density.

A Monte Carlo method can either be non-iterative and simply sample the posterior distribution using a random process, or, alternatively, it can be iterative so that samples are taken from the posterior distribution conditioned on the previous set of sampled values. Iterative approaches are often referred to as *Markov chain Monte Carlo (MCMC)* methods, the description 'Markov chain' being used because parameter values are sampled from a distribution depending only on parameter values sampled at the preceding iteration.

We will first describe a non-iterative Monte Carlo method known as rejection sampling, and then a Markov chain Monte Carlo method known as Gibbs sampling. More detail can be found on Monte Carlo methods in the following books: *Markov Chain Monte Carlo in Practice* by Gilks *et al.* (1995); *Bayes and Empirical Bayes Methods for Data Analysis* by Carlin and Louis (1996); and *Methods for Exploration of Posterior Distributions and Likelihood Functions* by Tanner (1996).

Rejection sampling

We will first describe this method in general terms before showing in the next section how it can be applied to mixed models. Monte Carlo methods are based on being able to randomly sample the joint posterior density of the model parameters.

However, often it is not easy to define a process for sampling directly from the posterior density, $p(\theta; \mathbf{y})$ (θ = the vector of model parameters). One way to get around this difficulty is to define an alternative density for the model parameters, $g(\theta; \mathbf{y})$ (the *base density*), that is easier to sample. Parameters are sampled from the base density but are only accepted with probability proportional to $p(\theta; \mathbf{y})/g(\theta; \mathbf{y})$. In practice, the method usually works by

- Obtaining the upper bound, K, for the ratio of the joint density to the base density $p(\theta; \mathbf{y})/g(\theta; \mathbf{y})$;
- Sampling values of θ from $g(\theta; \mathbf{y})$ and accepting them only if a uniform variate, u, sampled from Uniform(0,1) is less than $p(\theta; \mathbf{y})/Kg(\theta; \mathbf{y})$ (K having been set so that $p(\theta; \mathbf{y})/Kg(\theta; \mathbf{y})$ will never be more than one).

A large number of samples are taken and their frequencies form a simulated joint density for the model parameters. The process is called 'rejection sampling' because it depends on rejecting some of the sampled values. Justification for the method is given by Ripley (1987).

Using rejection sampling to fit a mixed model

To apply rejection sampling in the mixed model we use the fact that the posterior density can be written as a product of the posterior density for the variance parameters, γ, and the posterior density for the fixed effects

$$p(\gamma, \alpha; \mathbf{y}) = p(\gamma; \mathbf{y} - \mathbf{X}\hat{\alpha})p(\alpha; \mathbf{y}, \gamma).$$

This is similar to REML where the likelihood function was written as a product of likelihoods for the variance parameters and the fixed effects (see Section 2.2.1). Again, note that $\mathbf{y} - \mathbf{X}\hat{\alpha}$ is a linear function of $\mathbf{y}$ and does not depend on α.

The posterior distribution, $p(\gamma, \alpha; \mathbf{y})$, can be obtained by sampling both $p(\gamma; \mathbf{y} - \mathbf{X}\hat{\alpha})$ and $p(\alpha; \mathbf{y}, \gamma)$. The posterior for the variance components, $p(\gamma; \mathbf{y} - \mathbf{X}\hat{\alpha})$, does not follow a known distribution and is not straightforward to sample directly. However, it can be sampled using rejection sampling using an appropriate base density, $g(\gamma; \mathbf{y} - \mathbf{X}\hat{\alpha})$. In PROC MIXED the base density for the variance components is obtained by deriving approximate inverse gamma distributions for each of the variance components, and taking the product of these distributions as the base density. The posterior for the fixed effects, $p(\alpha; \mathbf{y}, \gamma)$, has a multivariate normal distribution defined conditionally on the variance components (assuming a non-informative flat prior):

$$p(\alpha; \mathbf{y}, \gamma) = p(\alpha)L(\alpha; \mathbf{y}, \gamma)/K_\alpha \quad (K_\alpha = \int p(\alpha)L(\alpha; \mathbf{y}, \gamma)\, d\alpha)$$

$$= \mathrm{N}((\mathbf{X}'\mathbf{V}^{-1}\mathbf{X})^{-1}\mathbf{X}'\mathbf{V}^{-1}\mathbf{y}, |\mathbf{X}'\mathbf{V}^{-1}\mathbf{X}|^{-1}) \quad \text{assuming } p(\gamma) \propto c.$$

Note that the mean and variance of this distribution are given by the maximum likelihood fixed effects solution (see Section 2.2.2). This distribution is conditioned on the variance parameters, γ, and a random process can be defined to sample from this distribution directly. The two posteriors are sampled in turn. Each time the variance parameters, γ, are sampled they are used to define the multivariate normal distribution for the fixed effects, $p(\alpha; \mathbf{y}, \gamma)$, from which the fixed effects are sampled.

If, additionally, random effects estimates are required then we note that these also have a multivariate normal distribution defined conditionally on the variance components and fixed effects:

$$p(\beta; \mathbf{y}, \gamma, \alpha) = N(\mathbf{GZ'V}^{-1}(\mathbf{y} - \mathbf{X}\alpha), \mathbf{GZ'V}^{-1}\mathbf{ZG} - \mathbf{GZ'V}^{-1}\mathbf{X}(\mathbf{X'V}^{-1}\mathbf{X})^{-1}\mathbf{X'V}^{-1}\mathbf{ZG})$$

and we take a sample of β following each sample of γ and α.

The PRIOR statement in PROC MIXED uses rejection sampling in this way to fit random effects models.

Gibbs sampling

This is a popular Markov chain Monte Carlo method. Parameters are sampled one by one from the joint density function which is conditioned on the previous set of parameters sampled. Sampling is continued until the simulated posterior is considered to have converged to the true posterior. We now define the basic procedure for using the Gibbs sampler to fit a random effects model in a two-way, cross-over trial. As before we let

$\mu = $ intercept,
$t = $ treatment difference,
$p = $ period difference,
$\sigma^2 = $ residual variance component,
$\sigma_s^2 = $ subject variance component.

The Gibbs sampler then proceeds using the following steps:

1. Specify initial values for the parameters — $\mu_0, t_0, p_0, \sigma_0^2$ and $\sigma_{s,0}^2$, say. (These are entirely arbitrary, and their choice will not affect the ultimate convergence of the method. However, their convergence is likely to be quicker if they are 'sensible' values.)

2. Sample each parameter from its posterior distribution, conditioned on the previous values sampled for the other parameters.

μ_1 from $p_\mu(u; t = t_0, p = p_0, \sigma^2 = \sigma_0^2, \sigma_s^2 = \sigma_{s,0}^2, \mathbf{y})$,
t_1 from $p_t(t; \mu = \mu_1, p = p_0, \sigma^2 = \sigma_0^2, \sigma_s^2 = \sigma_{s,0}^2, \mathbf{y})$,
p_1 from $p_p(p; \mu = \mu_1, t = t_1, \sigma^2 = \sigma_0^2, \sigma_s^2 = \sigma_{s,0}^2, \mathbf{y})$,
σ_1^2 from $p_{\sigma^2}(\sigma_1^2; \mu = \mu_1, t = t_1, p = p_1, \sigma_s^2 = \sigma_{s,0}^2, \mathbf{y})$,
$\sigma_{s,1}^2$ from $p_{\sigma_s^2}(\sigma_{p,1}^2; \mu = \mu_1, t = t_1, p = p_1, \sigma^2 = \sigma_1^2, \mathbf{y})$.

These values form the first iteration.

3. Sample a second set of parameters from their posterior distributions conditioned on the last set of parameters, to form the second iteration:

$$\mu_2 \text{ from } p_\mu(\mu; t = t_1, p = p_1, \sigma^2 = \sigma_1^2, \sigma_s^2 = \sigma_{s,1}^2, \mathbf{y}),$$
$$t_2 \text{ from } p_t(t; \mu = \mu_2, p = p_1, \sigma^2 = \sigma_1^2, \sigma_s^2 = \sigma_{s,1}^2, \mathbf{y}),$$
$$p_2 \text{ from } p_p(p; \mu = \mu_2, t = t_2, \sigma^2 = \sigma_1^2, \sigma_s^2 = \sigma_{s,1}^2, \mathbf{y}),$$
$$\sigma_2^2 \text{ from } p_{\sigma^2}(\sigma^2; \mu = \mu_2, t = t_2, p = p_2, \sigma_s^2 = \sigma_{s,1}^2, \mathbf{y}),$$
$$\sigma_{s,2}^2 \text{ from } p_{\sigma_s^2}(\sigma_p^2; \mu = \mu_2, t = t_2, p = p_2, \sigma^2 = \sigma_2^2, \mathbf{y}).$$

4. Continue until a large number of samples are generated (e.g. 10 000).

The samples obtained will then provide the full conditional distribution of the parameters. However, in practice, the procedure is not usually quite this simple and checks are necessary to ensure that convergence has been achieved. Also, it is likely that the first batch of iterations will be unstable and therefore these are often discounted. Typically, 1000 might be discounted.

Note that here we have assumed that the sampling of subject effects is not of interest; however, they can, if required, be sampled at each iteration.

Gibbs sampling can be carried out using a package known as BUGS developed by the Biostatistics Unit of the Medical Research Council. It is currently available free of charge and information on how to obtain it can be found on the Web page: www.mrc-bsu.cam.ac.uk/bugs.

Using the simulated posterior to estimate parameters

The simulated samples provide the joint distribution for the model parameters. As we described earlier, the marginal distribution for any parameter is obtained by simply using the sampled values for that parameter. The sampled values can be used directly to estimate the mean, standard deviation, probability intervals and Bayesian p-values for each parameter .

2.4 PRACTICAL APPLICATION AND INTERPRETATION

So far in this chapter we have considered how to specify and fit mixed models. In this section we look in some depth at points relating to the practical use of mixed models and their interpretation: negative variance components estimates (2.4.1); variance parameter accuracy (2.4.2); bias in fixed effect standard errors (2.4.3); significance testing (2.4.4); confidence intervals (2.4.5); model checking (2.4.6); and handling missing data (2.4.7). Readers may find that the material presented becomes most helpful once they have gained some experience of applying mixed models and have a need for considering particular aspects more closely. The worked example in Section 2.5 will illustrate many of the points made and readers may find it helpful to read this section in conjunction with the example.

Additional practical points relating specifically to covariance pattern and random coefficients models will be made in Sections 6.2 and 6.5.

2.4.1   Negative variance components

Variance components, by their definition, are non-negative. Nevertheless, the methods for estimating variance components will sometimes produce negative values which are not permissible. Such an estimate will usually be an underestimate of a variance component whose true value is small or zero. The chances of obtaining a negative variance component estimate when the true variance component is positive are increased when

- The ratio of the true variance component to the residual is small.
- The number of random effect categories is small (e.g. a small number of centres in a multi-centre trial where centre and centre-treatment effects are random).
- The number of observations per random effect category is small (e.g. a two-period, cross-over trial with patients fitted as random).

It is not usually straightforward to calculate the probability with which a negative variance component will occur. However, it is possible when balance is achieved across the random effects (see Section 1.6) and there are an equal number of observations per random effect category. The probability of obtaining a negative variance component estimate is then obtained by reference to the F distribution:

$$P(\text{negative variance component}) = P(F_{\text{DF1,DF2}} > 1 + n\gamma),$$

where

$\quad n =$ observations per category,
$\quad \gamma =$ true variance component/residual,
$\quad \text{DF1} =$ residual DF,
$\quad \text{DF2} =$ effect DF.

The graphs in Figure 2.4 show how the probability of obtaining a negative variance component is affected by γ, the number of random effect categories, and the number of observations per category. When there are only a few observations per category (e.g. there are only two observations per patient in two-way, cross-over trials), there is a reasonable chance that a negative variance component may occur as an underestimate of a true positive variance component. However, the variance components can be constrained to be non-negative, which would lead to the patient variance component estimate becoming zero. This, in turn, will modify the residual variance estimate, relative to the unconstrained situation, in that the negative patient variance component will be absorbed, resulting in a lower residual variance. The residual variance from the unconstrained random effects model will be the same as fitting a fixed effects model.

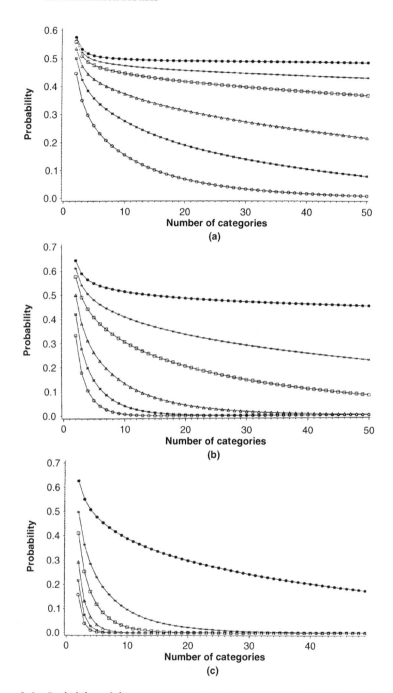

Figure 2.4 Probability of obtaining a negative variance component. (a) Observations per category=2; (b) observations per category=5; (c) observations per category=25. Gamma: ●–●–● 0.01; ✕✕✕✕ 0.05; ⊟–⊟–⊟ 0.10; △–△–△ 0.25; ✶✶✶✶ 0.50; ○–○–○ 1.00 .

How to handle a negative variance component

If a variance component is negative, the usual action would be either to remove the corresponding random effect from the model or to fix the variance component at zero (PROC MIXED sets negative variance components to zero by default). Either of these models will lead to the same parameter estimates for the fixed and random effects; however, the degrees of freedom (DF) for significance tests will differ between the models. In some situations the option of setting a negative variance component to zero but retaining its DF may be preferable. This would seem appropriate when an aspect of the study design is modelled by the random effect. For example, cross-over trials are designed to allow for patient effects and the DF corresponding to patients might be retained even if the patient variance component is fixed at zero. Similarly, if a multi-centre trial analysis produced a negative centre variance component, but a positive centre-treatment variance component, the centre effect DF might be retained. However, if the random effect does not form part of the study design, then there is more reason to justify removing it from the model, hence excluding its DF.

We note in the previous section that setting a negative variance component estimate to zero will lead to a different residual variance estimate to an equivalent fixed effects model. This is because a fixed effects model effectively allows negative variance components to occur (they are indicated whenever F is less than one). We will consider now the effect that this has on a cross-over trial with complete data. When the patient variance component estimate is positive, identical treatment effect estimates and standard errors will be obtained regardless of whether patients are fitted as random or fixed. However, when the patient variance component is negative and set to zero in the random effects model, the residual variance will be smaller than that obtained in the fixed effects model. This, in turn, will lead to smaller treatment standard errors. We believe that the residual estimate from the random effects model is therefore preferable since the intention of fitting patients is only to improve the precision of the treatment estimate.

Modelling negative correlation

Usually, a negative variance component is an underestimate of a small or zero variance component. However, occasionally it can indicate negative correlation between observations within the same random effect category. This is an unlikely scenario in most clinical trials; for example, it would be hard to imagine a situation where observations taken on the same patient could be more variable than observations taken on different patients. However, in the following veterinary example, negative correlation is more feasible. Imagine an animal feeding experiment where animals are grouped in cages. Here, it is possible that the greediest animals in a cage eat more than other animals, causing animal weight to become more variable within cages than between cages. This would lead to a negative variance component for cage effects, indicating negative correlation between animal weights in the same cage.

To model this negative correlation the model can be redefined as a covariance pattern model in which negative covariance parameters are allowed. We illustrate this model redefinition using the multicentre data used to describe the mixed model in Section 2.1. Recall that a random effects model was specified with centre effects fitted as random and that the variance matrix, $\mathbf{V}$, was given by

$$\mathbf{V} = \mathbf{ZGZ'} + \mathbf{R}$$

$$= \begin{pmatrix} \sigma_c^2 + \sigma^2 & \sigma_c^2 & \sigma_c^2 & \sigma_c^2 & 0 & 0 & 0 & 0 & 0 \\ \sigma_c^2 & \sigma_c^2 + \sigma^2 & \sigma_c^2 & \sigma_c^2 & 0 & 0 & 0 & 0 & 0 \\ \sigma_c^2 & \sigma_c^2 & \sigma_c^2 + \sigma^2 & \sigma_c^2 & 0 & 0 & 0 & 0 & 0 \\ \sigma_c^2 & \sigma_c^2 & \sigma_c^2 & \sigma_c^2 + \sigma^2 & 0 & 0 & 0 & 0 & 0 \\ 0 & 0 & 0 & 0 & \sigma_c^2 + \sigma^2 & \sigma_c^2 & 0 & 0 & 0 \\ 0 & 0 & 0 & 0 & \sigma_c^2 & \sigma_c^2 + \sigma^2 & 0 & 0 & 0 \\ 0 & 0 & 0 & 0 & 0 & 0 & \sigma_c^2 + \sigma^2 & \sigma_c^2 & \sigma_c^2 \\ 0 & 0 & 0 & 0 & 0 & 0 & \sigma_c^2 & \sigma_c^2 + \sigma^2 & \sigma_c^2 \\ 0 & 0 & 0 & 0 & 0 & 0 & \sigma_c^2 & \sigma_c^2 & \sigma_c^2 + \sigma^2 \end{pmatrix}.$$

When the model is redefined as a covariance pattern model, centre effects are excluded from the model but covariance is allowed in the $\mathbf{R}$ matrix between observations at the same centre. A constant covariance can be obtained by using a compound symmetry covariance pattern which gives the $\mathbf{V}$ matrix as

$$\mathbf{V} = \mathbf{R} = \begin{pmatrix} \sigma^2 & \rho\sigma^2 & \rho\sigma^2 & \rho\sigma^2 & 0 & 0 & 0 & 0 & 0 \\ \rho\sigma^2 & \sigma^2 & \rho\sigma^2 & \rho\sigma^2 & 0 & 0 & 0 & 0 & 0 \\ \rho\sigma^2 & \rho\sigma^2 & \sigma^2 & \rho\sigma^2 & 0 & 0 & 0 & 0 & 0 \\ \rho\sigma^2 & \rho\sigma^2 & \rho\sigma^2 & \sigma^2 & 0 & 0 & 0 & 0 & 0 \\ 0 & 0 & 0 & 0 & \sigma^2 & \rho\sigma^2 & 0 & 0 & 0 \\ 0 & 0 & 0 & 0 & \rho\sigma^2 & \sigma^2 & 0 & 0 & 0 \\ 0 & 0 & 0 & 0 & 0 & 0 & \sigma^2 & \rho\sigma^2 & \rho\sigma^2 \\ 0 & 0 & 0 & 0 & 0 & 0 & \rho\sigma^2 & \sigma^2 & \rho\sigma^2 \\ 0 & 0 & 0 & 0 & 0 & 0 & \rho\sigma^2 & \rho\sigma^2 & \sigma^2 \end{pmatrix},$$

where ρ = the correlation between patients at same centre.

Thus, $\mathbf{V}$ has an identical form to the random effects model except that it is parameterised differently. A negative covariance of observations at the same centre, $\rho\sigma^2$, is now permissible.

The Bayesian approach

When the Bayesian approach is used, negative variance components estimates are usually avoided by choosing a prior distribution for the variance components that is restricted to have positive values only. However, we have found that this can sometimes cause peaks in the posterior distribution for the variance components close to zero. Therefore, use of an estimator such as the median or expected value for variance parameters will be preferable to using the mode.

2.4.2   Accuracy of variance parameters

It is important that variance parameters are estimated with a reasonable accuracy because of their effect on the calculation of fixed effects and their standard errors. The accuracy of the variance parameters is dependent on the number of DF used to estimate them. Although there are no hard and fast rules, it would seem inadvisable to fit an effect as random if less than about five DF were available for estimation (e.g. a multi-centre trial with five or less centres).

When an insufficient number of DF are available to estimate a variance parameter accurately, an alternative to resorting to a fixed effects model would be to utilise variance parameter estimates from a similar previous study. An approach that is specifically allowed for in PROC MIXED is to fix the variance parameters in the new analysis at their previous values. The fixed effects, $\hat{\boldsymbol{\alpha}} = (\mathbf{X}'\mathbf{V}^{-1}\mathbf{X})^{-1}\mathbf{X}'\mathbf{V}^{-1}\mathbf{y}$, are then calculated using a fixed $\mathbf{V}$ matrix. However, this has the weakness of not utilising information on the variance parameters contained in the current study. A more natural approach, using both the previous variance parameter estimates and information in the current study, would be to use an informative prior for the variance parameters in a Bayesian analysis. This can be achieved by using the previous posterior distribution of the variance parameters as the prior distribution in the current analysis.

2.4.3   Bias in fixed and random effects standard errors

Fixed and random effects standard errors are calculated using a formula that is based on a known $\mathbf{V}$ (e.g. $\mathrm{var}(\hat{\boldsymbol{\alpha}}) = (\mathbf{X}\mathbf{V}^{-1}\mathbf{X})^{-1}$ for fixed effects). When data are balanced the standard errors will not be not biased. However, because $\mathbf{V}$ is, in fact, estimated, it is known that in most situations we meet in clinical trials there will be some downward bias in the standard errors. Bias will occur when the data are not balanced across random effects and effects are estimated using information from several error strata. In most situations the bias will be small. It is most likely to be relevant when

- the variance parameters are imprecise;
- the ratio of the variance parameters to the residual variance is small; and
- there is a large degree of imbalance in the data.

However, there is not a simple way to determine how much bias there will be in a given analysis. Results from simulation studies for particular circumstances have been reported in the literature (e.g. Yates, 1940; Kempthorne, 1952; McLean and Sanders, 1988; Nabugoomu and Allen, 1994; and Kenward and Roger, 1997). Although information from these studies is not yet comprehensive enough to allow any firm rules to be defined, they indicate that the bias may be 5% or more if the number of random effect categories relating to a variance parameter is less than about 10 *and* the ratio of the variance parameter to the residual variance is less than one. In these situations a mixed model may not always be advisable.

Various adjustments for the bias have been suggested (e.g. Kacker and Harville, 1984; and Kenward and Roger, 1997). However, none of these are yet available in PROC MIXED. For models fitting covariance patterns in the **R** matrix (e.g. repeated measures models) an alternative, more robust, variance estimator using the observed correlations between residuals known as the *'empirical' variance estimator* (Liang and Zeger, 1986) has been suggested. It is calculated by

$$\text{var}(\hat{\boldsymbol{\alpha}}) = (\mathbf{X}'\mathbf{V}^{-1}\mathbf{X})^{-1}\mathbf{X}'\mathbf{V}^{-1}\,\text{cov}(\mathbf{y})\mathbf{V}^{-1}\mathbf{X}(\mathbf{X}'\mathbf{V}^{-1}\mathbf{X})^{-1},$$

where cov(**y**) can be taken as $(\mathbf{y} - \mathbf{X}\hat{\boldsymbol{\alpha}})(\mathbf{y} - \mathbf{X}\hat{\boldsymbol{\alpha}})'$. This estimator takes into account the observed covariance in the data and may help alleviate some of the small sample bias. It does, however, cause the fixed effects variances to reflect observed covariances in the data rather than those specified by the covariance pattern modelled, so it may not always be the best choice. We discuss this further in Section 6.2. The empirical variance is calculated in SAS by using the EMPIRICAL option in the PROC MIXED statement. It would appear that the empirical variance could also be used with random effects and random coefficients models; however, we have not explored this further.

Note that when the Bayesian approach is used (Section 2.3) exact standard deviations are obtained directly from the posterior distributions for each parameter and the problem of bias does not arise.

2.4.4   Significance testing

Testing fixed effects, random effects and random coefficients

Significance tests for fixed effects, random effects and random coefficients can be carried out using tests based on F or t distributions, as we show below. A test can be defined using a contrast; for example, $\mathbf{C} = \mathbf{L}'\hat{\boldsymbol{\alpha}} = \mathbf{0}$ for fixed effects or $\mathbf{C} = \mathbf{L}'\hat{\boldsymbol{\beta}} = \mathbf{0}$ for random effects/coefficients. For simple contrasts **L** will just have one column. For example, in a trial comparing three treatments A, B and C, a pairwise comparison of treatments A and B is given by

$$\mathbf{L}'\hat{\boldsymbol{\alpha}} = (\,0 \quad 1 \quad -1 \quad 0\,)\,\hat{\boldsymbol{\alpha}} = \hat{\alpha}_A - \hat{\alpha}_B.$$

The first term corresponds to the intercept effect and the other three to the treatment effects (we assume treatments are the only fixed effects fitted). Alternatively, in multiple contrasts (e.g. to test overall equality of treatments) **L** will have several columns. For example, equality of the three treatments might be tested using the multiple contrast

$$\mathbf{L}'\hat{\boldsymbol{\alpha}} = \begin{pmatrix} 0 & 1 & -1 & 0 \\ 0 & 1 & 0 & -1 \end{pmatrix}\hat{\boldsymbol{\alpha}} = \begin{pmatrix} \hat{\alpha}_A - \hat{\alpha}_B \\ \hat{\alpha}_A - \hat{\alpha}_C \end{pmatrix}.$$

The F test statistic for testing the null hypothesis that the contrast is zero is calculated from a statistic known as the Wald statistic, which is given by

$$W = (\mathbf{L}'\hat{\boldsymbol{\alpha}})'(\mathrm{var}(\mathbf{L}\hat{\boldsymbol{\alpha}}))^{-1}(\mathbf{L}'\hat{\boldsymbol{\alpha}})$$

$$= (\mathbf{L}'\hat{\boldsymbol{\alpha}})'(\mathbf{L}'\,\mathrm{var}(\hat{\boldsymbol{\alpha}})\mathbf{L})^{-1}(\mathbf{L}'\hat{\boldsymbol{\alpha}})$$

for fixed effects. Alternatively, for single comparisons a Wald z statistic is given by

$$z = (\mathbf{L}'\hat{\boldsymbol{\alpha}})/\mathrm{SE}(\mathbf{L}'\hat{\boldsymbol{\alpha}}).$$

For random effects and coefficients, $\hat{\boldsymbol{\beta}}$ and $\mathrm{var}(\hat{\boldsymbol{\beta}})$ are used in place of $\hat{\boldsymbol{\alpha}}$ and $\mathrm{var}(\hat{\boldsymbol{\alpha}})$ in these formulae.

The Wald statistic can be thought of as the square of the contrast divided by its variance. Since $\mathrm{var}(\hat{\boldsymbol{\alpha}}) = (\mathbf{X}'\mathbf{V}^{-1}\mathbf{X})^{-1}$, W can be written

$$W = (\mathbf{L}'\hat{\boldsymbol{\alpha}})'(\mathbf{L}'(\mathbf{X}'\mathbf{V}^{-1}\mathbf{X})^{-1}\mathbf{L})^{-1}(\mathbf{L}'\hat{\boldsymbol{\alpha}}).$$

Asymptotically, W follows a chi-squared distribution with DF equal to the DF of **L** (number of linearly independent rows of **L**), and z has a normal distribution. However, these distributions are derived on the assumption that there is no variation in the denominator term, $\mathrm{var}(\mathbf{L}'\hat{\boldsymbol{\alpha}})$. They are equivalent to F or t tests with an infinite denominator DF. Thus, they will only be accurate if the DF of all error strata from which the effect is estimated from are high (e.g. in an unbalanced cross-over trial a high patient and residual DF are required).

A better option is to use the Wald F statistic, which is calculated by

$$F_{\mathrm{DF1,DF2}} = W/\mathrm{DF1},$$

where DF1 is the contrast DF (number of linearly independent rows of L), and DF2 is the denominator DF corresponding to the DF of the contrast variance, $\mathbf{L}'\,\mathrm{var}(\hat{\boldsymbol{\alpha}})\mathbf{L}$. This statistic takes account of the fact that $\mathbf{L}'\,\mathrm{var}(\hat{\boldsymbol{\alpha}})\mathbf{L}$ is estimated and not known. The Wald t statistic for tests of single contrasts is given by

$$t_{\mathrm{DF2}} = (F_{1,\mathrm{DF2}})^{1/2} = W^{1/2}.$$

Wald F and t tests are produced by default in PROC MIXED.

The denominator DF for F tests   This corresponds to those of the variance of the contrast, $\mathbf{L}'\,\mathrm{var}(\hat{\boldsymbol{\alpha}})\mathbf{L}$, and should reflect all the error strata from which the fixed effects have been estimated. When a fixed effect is balanced across random effects (see Section 1.6) it is estimated from only one error stratum and the DF are simply those of this error stratum. However, when fixed effects are estimated from two or more error strata, it is less straightforward to calculate the appropriate DF and usually an approximation is used. A well-known approximation is given by Satterthwaite (1946):

$$\mathrm{DF} = 2(\mathbf{L}'\,\mathrm{var}(\hat{\boldsymbol{\alpha}})\mathbf{L})^2/\,\mathrm{var}(\mathbf{L}'\,\mathrm{var}(\hat{\boldsymbol{\alpha}})\mathbf{L}).$$

It is equal to twice the variance of the contrast divided by the variance of the variance of the contrast. A further approximation is usually required to calculate $\text{var}(\mathbf{L}' \, \text{var}(\hat{\boldsymbol{\alpha}})\mathbf{L})$. Giesbrecht and Burns (1985) show how Satterthwaite's approximation can be calculated to test single contrasts. SAS uses a generalisation of this technique to calculate Satterthwaite's approximation for multiple contrasts (obtained by using the option DDFM=SATTERTH in PROC MIXED). However, details of this generalisation are not available in the SAS manuals. If software is not available for calculating the correct denominator DF, a conservative strategy would be to use an F test taking the lowest DF of the error strata used for estimating the contrast as the denominator DF. This DF is usually less than Satterthwaite's approximation to the true DF. For example, if in an unbalanced cross-over trial the patient DF were five and the residual DF were 10, then treatment effects could be tested using F tests with a denominator DF of five.

Kenward and Roger (1997) and Elston (1998) have also both suggested suitable approximations for denominator DF for F and t tests. Elston (1998) is recommended reading for those who wish for a more detailed knowledge on this subject.

Testing variance parameters

The significance of a variance parameter can be tested by using a likelihood ratio test to compare the likelihoods of models including (L_1) and excluding (L_2), the parameter. It is a standard result that under the null hypothesis that the additional terms in the model have no effect, the difference in the log likelihoods are distributed as $1/2\chi_1^2$. Hence,

$$2(\log(L_1) - \log(L_2)) \sim \chi_1^2.$$

The notation $\sim\chi_1^2$ is used to show that the likelihood ratio test statistic has a chi-squared distribution with one DF. In general, χ_n^2 denotes a chi-squared distribution with n DF.

In covariance pattern models interest usually lies with testing whether a particular covariance pattern causes a significant improvement over another pattern, rather than with testing a single variance parameter. Likelihood ratio tests can again be applied when the models are 'nested', with the χ^2 distribution DF corresponding to the difference in number of model parameters. Further details will be given in Section 6.2.

The Bayesian approach

When the Bayesian approach is used, exact 'Bayesian' p-values can be obtained from the posterior distribution and there are no potential inaccuracies in obtaining a test statistic or the DF for its distribution. A p-value to test the null hypothesis that a parameter is zero is obtained as the probability of being outside the probability interval which has zero on the boundary (see Section 2.3.3). This corresponds to a two-sided, 'classical' *p*-value. However, note that such tests are not available for variance components if a prior distribution with a positive range has been used (e.g. the reciprocal or inverse gamma distribution).

2.4.5   Confidence intervals

The reporting of confidence intervals relating to the fixed effects of interest is, of course, of great importance. The only difficulty in applying the 'obvious' classical approach concerns specification of the DF as described above, e.g.

$$\text{lower } 95\% \text{ confidence limit} = \text{mean effect} - t_{DF, 0.975} \times SE,$$
$$\text{Upper } 95\% \text{ confidence limit} = \text{mean effect} + t_{DF, 0.975} \times SE.$$

2.4.6   Model checking

In this section model checking methods are considered for the random effects model where it is assumed that the residuals, $\mathbf{y} - \mathbf{X}\hat{\alpha} - \mathbf{Z}\hat{\beta}$, and also the random effect estimates, $\hat{\beta}$, are normally distributed about zero and are uncorrelated. In random coefficients models and covariance pattern models the residuals and random coefficients are correlated and model checking methods addressing this feature will be considered in Sections 6.2 and 6.5. Model checking methods for mixed models have not yet been developed in depth, and the consequences of violating model assumptions are not fully known. Here, we will consider some simple visual checks that are based on plots of the residuals.

Examining the residuals and random effects

A plot of the residuals and of each set of random effects (e.g. the centre effects in a multi-centre trial) against their corresponding predicted values can be used to

- provide a rough check of normality of residuals and random effects,
- check whether the residual variance is constant across observations,
- look for outliers.

The predicted values corresponding to the residuals are given by $\mathbf{X}\hat{\alpha} + \mathbf{Z}\hat{\beta}$. We choose the predicted values rather than the expected values, $\mathbf{X}\hat{\alpha}$, because one of the features we wish to check against in the plot is whether the size of the residual is associated with the magnitude of the underlying value, which should, of course, reflect both fixed and random effects.

Predicted values corresponding to the random effects can be obtained by calculating the means of the expected values, $\mathbf{X}\hat{\alpha}$, within each random effect category (e.g. calculate the means of $\mathbf{X}\hat{\alpha}$ within each centre).

The assumption of normality can be checked more carefully using normal probability plots (i.e. plotting the ordered residuals against their values expected from the standard normal distribution given their ranks, see Snedecor and Cochran, 1989, Section 4.13). If the data are normally distributed, then the residuals will roughly form a straight line on the normal plot. If the plotted data deviate markedly from a straight line, then it is likely that the data are not normally distributed. For example, a bow shape indicates a skewed distribution, and a sigmoid shape indicates a symmetrical but non-normal distribution.

Homogeneity of the residuals can be further assessed by comparing the variances of each set of residuals within fixed effect categories (e.g. within treatments).

An illustration of the use of residual plots is given in the worked example in Section 2.5.

Unsatisfactory residual plots

If a general lack of normality is indicated in the residuals, a transformation of the data can be considered. Alternatively, if a residual plot indicates outlying values, then checks should be made to determine any possible reasons for them being extreme. When a value is clearly wrong (e.g. recording error, punching error, machine error) it should be corrected if possible, or else removed from the analysis. In other situations the influence of the outliers can be assessed in an *ad hoc* manner by re-analysing the data with them removed to determine whether parameter estimates alter noticeably. If the estimates are similar, then the outliers can be retained. If they differ, then it would seem sensible to remove them provided there is a convincing external reason for doing so (e.g. measurement suspected to be inaccurate because a blood sample was clotted; centre did not follow protocol). If there is no apparent reason for the outlier, then two alternative analyses and sets of conclusions may need to be presented. We also note that another alternative would be to construct a robust method to reduce the influence of the outlier; however, we will not be considering these methods here. Huber (1981) gives an introduction to robust methods.

Standardised residuals

It is possible to standardise residuals to take account of the fact that the observed residuals have differing variances. This is because they are each calculated from a different combination of fixed effect parameters, and each fixed effect parameter has a different variance (given by the diagonals of $(\mathbf{X}\mathbf{V}^{-1}\mathbf{X})^{-1}$). Goldstein (1995, Appendix 2.2) shows how standardised residuals can be calculated in the mixed model. However, this can be tedious when suitable software is unavailable (as is the case with PROC MIXED). It is likely that using checks based on unstandardised residuals will be sufficient to detect any important deviations from model assumptions, though the use of standardised residuals is to be preferred.

Example   79

2.4.7   Missing data

Mixed models are much more flexible than fixed effects models in the treatment of missing values. For example, in a two-period, cross-over trial, information on subjects with one value missing is completely lost when a fixed effects analysis is used. In contrast, mixed models are capable of handling the imbalance caused by missing observations provided that they are missing at random. This is often a reasonable assumption to make. If a subject withdraws from a cross-over trial after receiving one treatment, then we may have no idea of how the subject would have responded to the other treatments, and to handle these non-observed periods as if they were missing at random seems eminently sensible. However, it is common for observations to be missing for non-random reasons. An obvious example would be if a patient discontinued in a trial because of lack of efficacy of a treatment or due to a side effect. An assessment should be made of whether missing observations are likely to be related to the study treatment or to other aspects of the study (e.g. dislike of clinical procedures; inconvenience of visits). The reasons for withdrawal can be examined and simple summaries of the frequency of missing values or of baseline characteristics by treatment can be helpful. Also, tests of randomness have been suggested by Diggle (1989) and Ridout (1991) and different mechanisms for missing data are described by Little and Rubin (1987).

When data cannot be considered to be missing at random, *ad hoc* approaches such as the 'last value carried forward' approach will sometimes be adequate. In this method, the last observed value of the response variable is substituted for every subsequent missing observation. As such, it may help to reduce the bias due to 'poor' responders having missing values. For some situations, methods are available that allow for non-random missing data, such as those described by Diggle and Kenward (1994) for fixed effects models. Handling missing data is, however, a substantial topic in its own right and we will not consider it further in this book.

It is important to be aware that mixed models do not overcome the problems caused by missing dependent variables; for example, when a fixed or random effect category is unknown or a baseline value missing. As with a fixed effects model, when this occurs the observation will be automatically deleted from the analysis unless a dummy value is used to denote the missing effect.

2.5   EXAMPLE

In Section 1.2 we introduced a multi-centre trial of treatments for hypertension. In this trial a new hypertensive drug (A) was compared with two standard drugs (B and C). Here, we will consider analyses of the trial in greater detail and cover some of the practical points made in the previous section. The SAS code used for each model will be given at the end of the section. We will follow this pattern of supplying SAS code following each example throughout the book.

2.5.1   Analysis models

The main response variable in the trial (DBP at the final visit) will be analysed. The last post-treatment visit attended is used for patients who do not complete the trial, forming an 'intention to treat' analysis. Analyses were carried out using the models listed below. Initial DBP (baseline) was included as a covariate in all models to reduce between-patient variation.

Model	Fixed effects	Random effects	Method
1	Baseline, treatment, centre	—	OLS*
2	Baseline, treatment	Centre	REML
3	Baseline, treatment	Centre, treatment·centre	REML
4	Baseline, treatment	Centre	Bayes
5	Baseline, treatment	Centre, treatment·centre	Bayes

*OLS = ordinary least squares. This is the fixed effects method used by PROC GLM (see Section 2.2.2).

Fixed effects models (Model 1)

The data were analysed first using the more conventional fixed effects approach. Initially, a full model including baseline, treatment, centre and centre·treatment effects was fitted to test whether centre·treatment effects were significant. They were found to be non-significant ($p = 0.19$) and were therefore omitted from the model to give Model 1. This is the usual action taken when a fixed effects approach is used, since inclusion of fixed centre·treatment effects will cause the overall treatment estimate to be an unweighted average of the individual centre estimates. However, note that in this instance it would not have been possible in any case to estimate overall treatment effects when centre·treatment effects were included, because all the treatments were not received at every centre.

When centre·treatment effects are included in a fixed effects analysis, this causes a separate treatment effect, specific to each centre included in the trial, to be fitted. The inferences strictly apply only to those centres which were included in the trial. If the model which omits the centre·treatment effect is used, the inferences again should still strictly apply only to the centres used. However, extrapolation of the inferences regarding the effect of treatment to other centres seems more reasonable when an assumption has been made that the treatment effect does not depend on the centre in which it has been applied.

Random effects models fitted using REML (Models 2 and 3)

In Model 2, centre effects are fitted as random and baseline and treatment effects as fixed, using REML. With this model, treatment effects are estimated using information from the residual error stratum and, additionally, from the centre error stratum since balance does not occur across random effects (centres). By

Example **81**

comparing the results with those from Model 1 it is possible to determine whether any additional information has been recovered on treatments from the centre error stratum. Note that only under the strong assumption that there is no centre-treatment interaction will inferences apply to the population of centres from which those in the trial can be considered a sample.

In Model 3, both centre and centre-treatment effects are fitted as random, and baseline and treatment effects as fixed, using REML. Unlike the fixed effects approach, centre-treatment effects are retained in the model provided their variance component is positive. The logic here is that use of the model implies a belief that treatment effects may differ between centres, and we wish this to be included in our estimates, even if the interaction is non-significant. Since centre-treatment effects are taken as random in this model, treatment effects are assumed to vary randomly between the centres and results can be related with some confidence to the wider population of centres.

Bayesian models (Models 4 and 5)

Models 4 and 5 are the same as Models 2 and 3, except that they are fitted using the Bayesian approach. They were fitted using rejection sampling (see Section 2.3) and 10 000 samples were taken using the PRIOR statement in PROC MIXED. A product of inverse gamma distributions was used by PROC MIXED for the base density and thus negative variance component samples could not be obtained. Frequencies of the sampled values were used directly to obtain the marginal distribution for each parameter. The first 15 sampled values obtained in Model 4 are listed below.

		Centre		Treatment	Treatment
Obs	Residual	Variance component	Baseline	A − C	B − C
1	4.78	77.99	0.38	2.97	2.56
2	4.25	59.31	0.46	1.43	2.72
3	18.22	63.06	0.42	1.94	3.42
4	9.65	65.01	0.47	3.08	2.50
5	12.79	58.17	0.31	2.86	0.67
6	4.23	68.03	0.47	1.50	1.68
7	9.42	71.98	0.18	1.19	1.81
8	7.88	81.53	0.14	1.94	0.93
9	11.35	72.41	0.36	2.20	2.67
10	10.89	68.71	0.39	1.78	1.29
11	4.56	70.05	0.30	5.21	2.45
12	6.92	84.02	0.31	2.93	3.07
13	5.45	71.44	0.49	2.12	2.11
14	6.05	71.02	0.26	3.84	2.88
15	13.85	74.04	0.33	2.78	3.04
etc.					

Fixed effects estimates were obtained by simply calculating the means and standard deviations of the sampled posterior distributions (i.e. of the relevant column of values in the above table). Samples for the treatment difference B − C are obtained by creating an extra column of differences between the A − C and B − C samples.

The variance components are of less direct interest in this example. However, they can be estimated by taking statistics such as the mode, median or mean of the sampled posterior (see Section 2.3.3). Here, we obtain an idea of their size by taking the medians of their posterior distributions. These estimates will not be exactly the same as the REML estimates, because the posterior density is not exactly proportional to the likelihood function (since a flat prior is not used), and also because REML estimates would correspond to modes rather than to medians. A measure of the spread of the variance components can be obtained by calculating probability intervals for the sampled values (see Section 2.3.3).

2.5.2   Results

The results obtained from each model are shown in Table 2.1. The variance components for centres and treatment·centres are much smaller than the residual in all models. This indicates that most of the variation in the data is due to differences between patients and not to differences between centres. The Bayesian probability intervals for the variance components illustrate the skewness of their distributions and the inadequacy of the asymptotic standard errors produced by the REML analyses.

The treatment standard errors are slightly smaller in Model 2 than in Model 1, indicating that only a small amount of information on treatments has been recovered from the centre error stratum. This is mainly because there was only a small degree of imbalance in the allocation of treatments within centres. We note, however, that the treatment comparisons involving B have been modified appreciably. When the treatment·centre interaction is included as a random effect (Models 3 and 5) treatment standard errors are increased. The amount of this increase is related to the size of the centre·treatment variance component and to the number of centres used. Although the centre·treatment variance component is small compared with the residual, its influence on the standard error is noticeable because it is related inversely to the number of centres, and not to the number of observations. (The relationship between the treatment effect standard errors and the variance components in a multi-centre analysis is defined more precisely in Section 5.2.)

The standard errors of the fixed effects are very similar between Models 2 and 4, but we see appreciable differences between Models 3 and 5. This arises because of inherent differences in the 'standard errors' obtained in the Bayesian analysis. They are based directly on the simulated distribution of $\boldsymbol{\alpha}$, which takes into account the distribution of the variance parameters. As such, they are related

Example   **83**

Table 2.1   Estimates of variance components and fixed effects.

Model	Fixed effects	Random effects	Method
1	Baseline, treatment, centre	—	OLS
2	Baseline, treatment	Centre	REML
3	Baseline, treatment	Centre, treatment·centre	REML
4	Baseline, treatment	Centre	Bayes
5	Baseline, treatment	Centre, treatment·centre	Bayes

Treatment effects (SEs)				
Model	**Baseline**	**A − B**	**A − C**	**B − C**
1	0.22 (0.11)	1.20 (1.24)	2.99 (1.23)	1.79 (1.27)
2	0.22 (0.11)	1.03 (1.22)	2.98 (1.21)	1.95 (1.24)
3	0.28 (0.11)	1.29 (1.39)	2.93 (1.37)	1.64 (1.40)
4^1	0.28 (0.11)	1.05 (1.23)	2.99 (1.23)	1.94 (1.25)
5^1	0.26 (0.11)	1.30 (1.51)	2.86 (1.49)	1.56 (1.54)

Variance components (SEs)			
Model	**Centre**	**Treatment·centre**	**Residual**
1	—	—	71.9 (−)
2	7.82 (3.99)	—	70.9 (6.2)
3	6.46 (4.35)	4.10 (5.33)	68.4 (6.5)
4^2	8.36 (3.30–18.18)	—	71.1 (61.9–82.2)
5^2	7.40 (1.49–19.25)	4.49 (0.78–18.07)	68.3 (58.7–78.9)

[1] Strictly speaking the SEs in these Bayesian models are parameter standard deviations.
[2] Medians are given with 5th and 95th centiles.

directly to the probability intervals. In contrast, the REML analysis calculates standard errors using a formula which assumes that the variance parameters are known $(\text{var}(\hat{\boldsymbol{\alpha}}) = (\mathbf{X'V}^{-1}\mathbf{X})^{-1})$. Confidence intervals are then obtained using multiples of the standard errors, based on the t distribution with appropriate DF (calculated using Satterthwaite's approximation). Thus, if confidence intervals and probability intervals are in agreement, we should only anticipate similar 'standard errors' from the two approaches when the DF with the REML approach are large (e.g. Model 2). When the DF are small, the standard error from the REML analysis will be less than the corresponding parameter standard deviation in the Bayesian analysis. This point is illustrated further in the following section.

2.5.3   Discussion of points from Section 2.4

Each of the practical points covered in Section 2.4 is discussed below in relation to the example.

Negative variance component estimates (2.4.1)

Negative variance component estimates were not obtained by any model in this example.

Variance parameter accuracy (2.4.2)

Since 29 centres were used in this study we would not be concerned that insufficient DF were available for estimating the variance components. However, it is perhaps surprising that their estimates have such wide probability intervals in the Bayesian analyses (see Table 2.1, Models 4 and 5).

Bias in fixed effect standard errors (2.4.3)

Again, the use of 29 centres leads us to expect the estimates of the fixed effects and their standard errors to be fairly accurate. A comparison of the results from Models 3 and 5 might give us cause to doubt this though. For example, the standard error of A − B is 1.39 from Model 3 and the corresponding parameter standard deviation in Model 5 is 1.51. As described above, they do have a different basis, however, and a more appropriate comparison is between the confidence intervals obtained with each method. For Model 3 this is obtained as the mean $\pm t \times$ SE, where t is based on the appropriate Satterthwaite DF, which, in this case, is 23.8. The 95% confidence limits are −1.59 to 4.16. For Model 5 we use the samples from the marginal posterior distribution to give us a probability interval of −1.55 to 4.14. Thus, there is good agreement between the alternative approaches in terms of the confidence and probability intervals.

Significance testing (2.4.4)

We demonstrate the use of significance tests for Models 3 and 5. In Model 3 the significance of treatment differences can be assessed with Wald F tests using Satterthwaite's approximation for the denominator DF. An $F_{2,25}$ value of 2.28 is obtained for the composite test of treatment equality which is non-significant ($p = 0.12$). Test statistics for pairwise treatment comparisons gave

$$A - B \quad t_{23.8} = 0.92, \ p = 0.36,$$

$$A - C \quad t_{25.6} = 2.13, \ p = 0.04,$$

$$B - C \quad t_{25.7} = 1.17, \ p = 0.25.$$

Thus, DBP appears to be significantly lower on treatment C than on treatment A on pairwise testing, although we advise very cautious interpretation because this is one of three pairwise tests and the overall F test was non-significant.

Exact p-values are obtained from Model 5 by calculating the proportion of sampled parameters with a value of less than (or greater than) zero, and then doubling the smaller of these values to obtain a two-sided p-value. The p-values obtained in this way are

$$A - B \quad p = 0.37,$$

$$A - C \quad p = 0.05,$$

$$B - C \quad p = 0.27.$$

Example **85**

These are slightly more conservative than the REML p-values above but lead to the same conclusions.

Confidence intervals (2.4.5)

95% confidence intervals for the differences between treatments using Model 3 are calculated by SAS using t distributions with Satterthwaite's DF:

$$A - B = (-1.59, \ 4.15),$$

$$A - C = (0.10, \ 5.75),$$

$$B - C = (-1.25, \ 4.53).$$

Model checking (2.4.6)

Checks of model assumptions are carried out below for Model 3 as an illustration of the techniques which can be used. We present plots of residuals and random effects against their predicted values, as well as normal plots. Additionally, the homogeneity of the variance components is checked by calculating the variances of the residuals and centre·treatment effects by treatment.

Residuals   First, we consider the residual plots. In Figure 2.5 the residuals are plotted against the predicted values (obtained from the PREDICTED option in SAS). The residual plots indicate one possible outlier with a residual of 40 (patient 314, visit 5). In practice, many statisticians would be happy to accept this degree of variation in the residuals. However, to illustrate the magnitude of the effect that an outlier can have we have examined the effect of removing this observation. When the data were reanalysed with this patient removed, the residual plots became visually acceptable. The parameter estimates, however, changed noticeably from their original estimates:

	Variance components		
Model	**Centre**	**Treatment·Centre**	**Residual**
With outlier	6.46	4.10	68.36
Without outlier	6.97	1.50	63.27

	Treatment effects (SEs)		
Model	**A − B**	**A − C**	**B − C**
With outlier	1.29 (1.39)	2.92 (1.37)	1.64 (1.40)
Without outlier	0.74 (1.24)	2.49 (1.22)	1.76 (1.25)

On closer examination it was likely, but not certain, that the DBP value (of 140) was due to a recording mistake. This is an extremely high reading, which is inherently unlikely, but the patient's baseline DBP was 113 mmHg, and he had dropped out of the trial at visit 5 due to an 'insufficient effect'. Under these

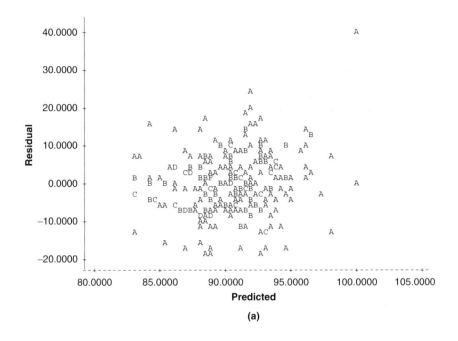

(a)

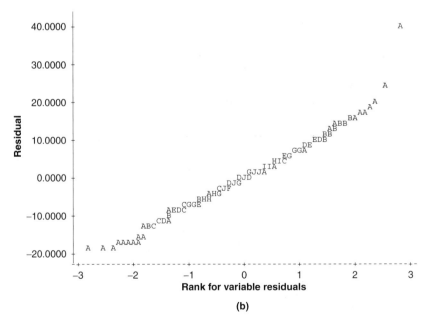

(b)

Figure 2.5 Plots of residuals: (a) Residuals against their predicted values. (b) Residuals–normal plot. A = 1 obs, B = 2 obs, etc.

Example **87**

circumstances we might wish to report results from analyses both including and excluding the outlier. The large changes in parameter estimates caused by the exclusion of one patient in a reasonably large trial illustrate the potential importance of at least basic checks of model assumptions.

Centre effects Predicted values are obtained for each centre by taking the means of the predicted observations, $\mathbf{X}\hat{\boldsymbol{\alpha}}$, for that centre. The plots (Figure 2.6) indicate one possible outlying centre (centre 31). When the analysis was repeated with centre 31 removed, plots of the centre effects became visually satisfactory and the following results were obtained:

Model	Variance components (SEs)		
	Centre	**Treatment·Centre**	**Residual**
With centre 31	6.46	4.10	68.4
Without centre 31	0.81	5.37	73.1

Model	Treatment effects (SEs)		
	A − B	**A − C**	**B − C**
With centre 31	1.29 (1.39)	2.92 (1.37)	1.64 (1.40)
Without centre 31	1.33 (1.55)	3.03 (1.52)	1.70 (1.56)

The variance component estimates have changed fairly noticeably from their previous values. However, the exclusion of this centre did not greatly change the treatment estimates, although their standard errors are increased. Thus, the centre could be retained in the analysis with reasonable confidence. However, possible reasons for the centre being outlying should be investigated, at least to check that the protocol was followed correctly.

Centre·treatment effects The plots (Figure 2.7) of centre·treatment effects do not indicate any noticeably outlying values.

Homogeneity of treatment variances The standard deviations of the residuals and the centre·treatment effects (below) were similar between the treatment groups indicating no strong evidence of non-homogeneity of variance.

Treatment	Residual	Centre·treatment effects
A	9.15	0.85
B	6.53	0.70
C	7.86	0.67

Missing data (2.4.7)

Nearly all missing data in the study were caused by dropouts. The numbers of patients dropping out following randomisation were A — 17, B — 12 and C — 3.

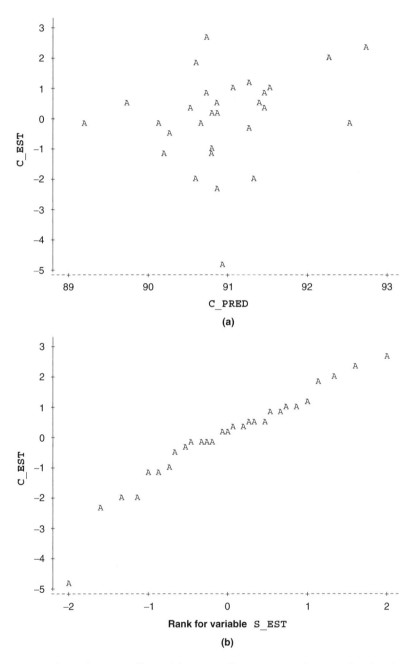

Figure 2.6 Plots of centre effects: (a) Centre effects against their predicted values. (b) Centre effects — normal plot.

Example 89

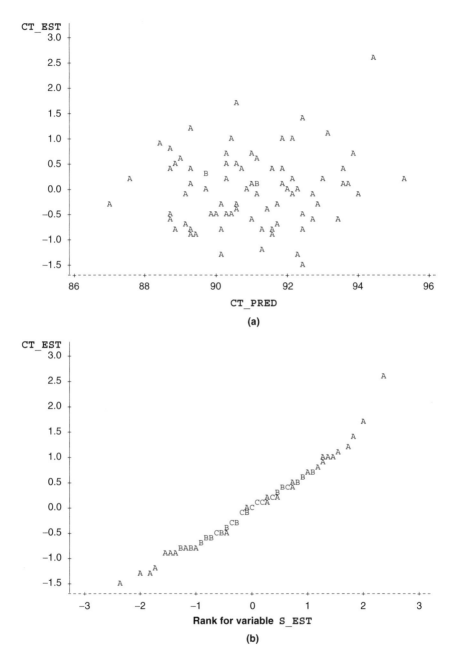

Figure 2.7 Plots of centre·treatment effects: (a) Centre·treatment effects against their predicted values. (b) Centre·treatment effects — normal plot. A = 1 obs, B = 2 obs, etc.

The larger numbers for the first two treatments indicate that the dropout rate has been influenced by treatment. Inclusion of values at the last visit helps to minimise any bias that this may cause.

SAS code and output

There now follows a description of the SAS code and the outputs which were produced in the analysis of the previous example. This is a pattern which will be followed throughout the rest of the book. Some readers may be relative novices at using SAS and may find some of the aspects of the code or output confusing. Anyone in this position is referred forward to Section 9.2 where the PROC MIXED procedure and the output it generates are introduced. The SAS code and datasets may be obtained electronically from Web page www.med.ed.ac.uk/phs/mixed.

The following example was analysed using SAS Release 6.12. The variable names used for the program are as follows:

centre   = centre number,
treat   = treatment (A,B,C),
patient = patient number,
dbp   = diastolic blood pressure at last attended visit,
dbp1   = baseline diastolic blood pressure.

SAS code only is given for Models 1 and 2 because their outputs have a similar form to that from Model 3. The full code and output are given for Model 3 to illustrate the use of PROC MIXED in fitting a mixed model using REML and also to show how model checking can be undertaken. The full code and output are given for Model 4 to demonstrate the use of SAS for performing a Bayesian analysis. The code for Model 5 is not given because it is very similar to that for Model 4.

Model 1

```
PROC MIXED; CLASS centre treat;
TITLE 'MODEL 1';
MODEL dbp = dbp1 treat centre;
LSMEANS treat/ DIFF PDIFF CL;
```

Model 2

```
PROC MIXED; CLASS centre treat;
TITLE 'MODEL 2';
MODEL dbp = dbp1 treat/ DDFM=SATTERTH;
RANDOM centre;
LSMEANS treat/ DIFF PDIFF CL;
```

Model 3

```
PROC MIXED; CLASS centre treat;
TITLE 'MODEL 3';
```

Example   **91**

```
MODEL dbp = dbp1 treat/ DDFM=SATTERTH;
RANDOM centre centre*treat;
LSMEANS treat/ DIFF PDIFF CL;
```

                        MODEL 3
                  The MIXED Procedure
                Class Level Information
        Class   Levels Values
        CENTRE   29   1 2 3 4 5 6 7 8 9 11 12 13 14
                          15 18 23 24 25 26 27 29 30 31
                          32 35 36 37 40 41
        TREAT   3   A B C

              REML Estimation Iteration History
        Iteration Evaluations   Objective   Criterion
            0   1   1550.3451721
            1   3   1533.6847949   0.00000322
            2   1   1533.6822800   0.00000000
                  Convergence criteria met.

          Covariance Parameter Estimates (REML)
          Cov Parm   Estimate
          CENTRE   6.46282024
          CENTRE*TREAT   4.09615054
          Residual   68.36773938

          Model Fitting Information for DBP
          Description   Value
          Observations   288.0000
          Res Log Likelihood   -1027.82
          Akaike's Information Criterion -1030.82
          Schwarz's Bayesian Criterion   -1036.29
          -2 Res Log Likelihood   2055.639

                 Tests of Fixed Effects
          Source   NDF   DDF Type III F   Pr>F
          DBP1   1   284   6.31   0.0125
          TREAT   2   25   2.28   0.1230

                 Least Squares Means

Effect	TREAT	LSMEAN	Std Error	DF	t	Pr>|t|	Alpha
TREAT	A	92.34913293	1.10348051	52	83.69	0.0001	0.05
TREAT	B	91.06321274	1.14545866	51.4	79.50	0.0001	0.05
TREAT	C	89.42174246	1.11231265	58.7	80.39	0.0001	0.05

                 Least Squares Means
                    Lower   Upper
                   90.1349   94.5634
                   88.7640   93.3624
                   87.1957   91.6477

Differences of Least Squares Means

Effect	TREAT	_TREAT	Difference	Std Error	DF	t	Pr>|t|	Alpha
TREAT	A	B	1.28592019	1.39135076	23.8	0.92	0.3646	0.05
TREAT	A	C	2.92739048	1.37270937	25.6	2.13	0.0427	0.05
TREAT	B	C	1.64147028	1.40487730	25.7	1.17	0.2534	0.05

Differences of Least Squares Means

Lower	Upper
-1.5867	4.1585
0.1036	5.7512
-1.2482	4.5312

Model 3 with model checking

```
PROC MIXED; CLASS centre treat;
TITLE 'MODEL 3';
MODEL dbp = dbp1 treat/ DDFM=SATTERTH PREDICTED PM;
RANDOM centre centre*treat/ SOLUTION;
LSMEANS treat/ DIFF PDIFF CL;
ID patient centre treat;
MAKE 'PREDICTED' OUT=resid noprint;
MAKE 'PREDMEANS' OUT=predm noprint;
MAKE 'SOLUTIONR' OUT=solut noprint;

DATA solut; SET solut;
centrex=centre*1; * obtain numeric centre variable;
drop centre;

DATA c_est(KEEP=centre c_est) ct_est(KEEP=centre treat
ct_est); SET solut;
centre=centrex;
IF _effect_='CENTRE' THEN DO;
 c_est=_est_;
 OUTPUT c_est;
END;
ELSE DO;
 ct_est=_est_;
 OUTPUT ct_est;
END;
PROC SORT DATA=ct_est; BY centre treat;

PROC SORT DATA=predm; BY centre treat;

DATA c_est; MERGE predm c_est; BY centre;
PROC MEANS NOPRINT; BY centre; ID c_est;
VAR_pred_; OUTPUT OUT=c_est MEAN=c_pred N=freq;

DATA ct_est; MERGE predm ct_est; BY centre treat;
PROC MEANS NOPRINT; BY centre treat; ID ct_est;
VAR_pred_; OUTPUT OUT=ct_est MEAN=ct_pred N=freq;

PROC PRINT NOOBS DATA=resid; VAR patient treat centre_resid_
_pred_;
```

Example **93**

```
TITLE 'RESIDUALS AND PREDICTED VALUES';
PROC PLOT DATA=resid; PLOT_resid_*_pred_;
TITLE 'RESIDUALS AGAINST THEIR PREDICTED VALUES';
PROC RANK DATA=resid OUT=norm NORMAL=TUKEY; VAR_resid_; RANKS
s_est;
PROC PLOT DATA=norm; PLOT_resid_*s_est;
TITLE 'RESIDUALS - NORMAL PLOT';

PROC PRINT NOOBS DATA=c_est; VAR centre c_est c_pred freq;
TITLE 'CENTRE EFFECTS AND PREDICTED VALUES';
PROC PLOT DATA=c_est; PLOT c_est*c_pred;
TITLE 'CENTRE EFFECTS AGAINST THEIR PREDICTED VALUES';
PROC RANK OUT=norm NORMAL=TUKEY DATA=c_est; VAR c_est; RANKS
s_est;
PROC PLOT DATA=norm; PLOT c_est*s_est;
TITLE 'CENTRE EFFECTS - NORMAL PLOT';

PROC PRINT NOOBS DATA=ct_est; VAR centre treat ct_est ct_pred
freq;
TITLE 'CENTRE.TREAT EFFECTS AND PREDICTED VALUES';
PROC PLOT DATA=ct_est; PLOT ct_est*ct_pred;
TITLE 'CENTRE.TREAT EFFECTS AGAINST THEIR PREDICTED VALUES';
PROC RANK OUT=norm NORMAL=TUKEY DATA=ct_est; VAR ct_est; RANKS
s_est;
PROC PLOT DATA=norm; PLOT ct_est*s_est;
TITLE 'CENTRE.TREAT EFFECTS - NORMAL PLOT';

* CHECK HETEROGENEITY OF RESIDUAL AND CENTRE*TREAT VARIANCE BY
TREATMENT;
PROC SORT DATA=resid; BY treat;
PROC MEANS DATA=resid; VAR_resid_; BY treat;
TITLE 'PROC MEANS TO CHECK RESIDUAL VARIANCE IS HOMOGENEOUS
ACROSS TREATMENTS';

PROC SORT DATA=ct_est; BY treat;
PROC MEANS DATA=ct_est; VAR ct_est; BY treat;
TITLE 'PROC MEANS TO CHECK CENTRE.TREAT EFFECT VARIANCE IS
HOMOGENEOUS ACROSS TREATMENTS';
```

This code may not at first sight be straightforward to understand. The steps used are summarised below.

- Fit Model 3.
- Use MAKE statements to: output the residuals and predicted values given by $X\hat{\alpha} + Z\hat{\beta}$ to dataset resid; output the predicted values given by $X\hat{\alpha}$ to dataset predm; output the random effects estimates to dataset solut. The ID statement causes the patient, centre and treat variables to be included in the datasets resid and predm.
- Create datasets c_est and ct_est containing random effects estimates for the centre and centre·treatment effects respectively.
- Merge datasets c_est and ct_est with the predicted values and calculate the mean predicted value within each random effect category.

- Print and plot the residuals and the centre and centre·treatment effect estimates.
- Calculate the standard deviations of the residuals and centre·treatment effects within each treatment group.

The output from PROC MIXED is identical to that given above. The residual plots produced by this output were given earlier in this section. The following output starts with the results of the use of PROC PRINT.

```
              RESIDUALS AND PREDICTED VALUES
    PATIENT    TREAT    CENTRE    _RESID_      _PRED_
       79        A         1       16.0823     91.9177
       82        A         1       24.0823     91.9177
      183        A         1       -3.9177     91.9177
      184        A         1      -11.5311     93.5311
      187        A         1      -16.6068     94.6068
      262        A         1        5.0067     92.9933
      263        A         1       -3.9177     91.9177
      ETC
```

```
        CENTRE EFFECTS AND PREDICTED VALUES
      CENTRE        C_EST        C_PRED        FREQ
         1         0.79661       90.7571        39
         2        -2.34406       90.8882        10
         3         1.97600       92.2499         8
         4        -0.14289       90.1324        12
         5         1.17788       91.2716        11
         6         0.40495       91.4442         5
         7        -0.97227       90.8046        18
         8         1.75095       90.6254         6
         9         0.51342       89.7299         1
        11        -0.33060       91.2752        12
        12         2.59830       90.7287        11
      ETC
```

```
      CENTRE.TREAT EFFECTS AND PREDICTED VALUES
    CENTRE     TREAT      CT_EST       CT_PRED      FREQ
       1         A       -0.46053      92.4090       13
       1         B        1.68999      90.6414       14
       1         C       -0.72457      89.1024       12
       2         A       -0.83297      92.4780        3
       2         B       -1.18755      91.2369        4
       2         C        0.53485      88.8335        3
       3         A        2.58960      94.3603        3
       3         B       -0.80488      91.6402        3
       3         C       -0.53232      89.9988        2
       4         A        0.08982      91.1783        4
       4         B       -0.77684      90.1613        4
       4         C        0.59646      89.0576        4
       5         A        1.00945      91.8505        4
       5         B        0.36116      91.8554        5
       5         C       -0.62407      88.6542        2
      ETC
```

Example **95**

PROC MEANS TO CHECK RESIDUAL VARIANCE IS HOMOGENEOUS ACROSS TREATMENTS

Analysis Variable : _RESID_ Residual

---------------------------- TREAT=A ----------------------------

N	Mean	Std Dev	Minimum	Maximum
100	6.536993E-15	9.1545453	-17.2589799	39.8192060

---------------------------- TREAT=B ----------------------------

N	Mean	Std Dev	Minimum	Maximum
93	1.466927E-14	6.5308257	-18.7823225	18.2781249

---------------------------- TREAT=C ----------------------------

N	Mean	Std Dev	Minimum	Maximum
95	-2.51308E-14	7.8623760	-18.7313617	17.3011936

PROC MEANS TO CHECK CENTRE.TREAT EFFECT VARIANCE IS HOMOGENEOUS ACROSS TREATMENT

Analysis Variable : CT_EST

---------------------------- TREAT=A ----------------------------

N	Mean	Std Dev	Minimum	Maximum
26	2.216176E-15	0.8457655	-1.4859312	2.5895978

---------------------------- TREAT=B ----------------------------

N	Mean	Std Dev	Minimum	Maximum
25	2.413625E-15	0.6983672	-1.3161783	1.6899927

---------------------------- TREAT=C ----------------------------

N	Mean	Std Dev	Minimum	Maximum
28	-5.18881E-15	0.6670323	-0.9090217	1.4170861

Model 4

The code below will work in SAS Release 6.11. However, it fails in SAS Release 6.12 for this example due to the percentage of rejected samples being too high. It is expected that this problem will not arise in later versions of SAS; however, for users of Release 6.12 we have provided alternative code (suggested by SAS) which will avoid the problem. This code appears after the SAS output.

```
PROC MIXED; CLASS centre treat;
TITLE 'MODEL 4';
MODEL dbp = dbp1 treat/ SOLUTION;
RANDOM centre;
PRIOR / NSAMPLE=11000;
MAKE 'SAMPLE' OUT=samples NOPRINT;
```

```
                            MODEL 4
                       The MIXED Procedure
                   Class Level Information
           Class      Levels  Values
           CENTRE         29  1 2 3 4 5 6 7 8 9 11 12 13 14
                              15 18 23 24 25 26 27 29 30 31
                              32 35 36 37 40 41
           TREAT           3  A B C
```

```
               REML Estimation Iteration History
        Iteration  Evaluations    Objective      Criterion
               0            1  1550.3451721
               1            3  1534.5449028     0.00003184
               2            1  1534.5190010     0.00000030
               3            1  1534.5187668     0.00000000
                   Convergence criteria met.
```

```
             Covariance Parameter Estimates (REML)
     Cov Parm      Ratio     Estimate  Std Error      Z   Pr>|Z|
     CENTRE    0.11032234   7.82475258 3.99186691   1.96   0.0500
     Residual  1.00000000  70.92627567 6.15939523  11.52   0.0001
```

```
               Model Fitting Information for DBP
              Description                    Value
              Observations                288.0000
              Variance Estimate            70.9263
              Standard Deviation Estimate   8.4218
              REML Log Likelihood        -1028.24
              Akaike's Information Criterion -1030.24
              Schwarz's Bayesian Criterion  -1033.89
              -2 REML Log Likelihood      2056.476
```

```
                   Solution for Fixed Effects
        Parameter     Estimate   Std Error  DDF      T   Pr>|T|
        INTERCEPT  60.51690576 11.11189813   28   5.45   0.0001
        DBP1        0.27997449  0.10708429  256   2.61   0.0095
```

```
TREAT A      2.98073456   1.21079135   256   2.46   0.0145
TREAT B      1.94737641   1.23865270   256   1.57   0.1171
TREAT C      0.00000000         .        .     .      .
```

```
                    Tests of Fixed Effects
         Source     NDF    DDF    Type III F      Pr>F
         DBP1         1     256         6.84    0.0095
         TREAT        2     256         3.10    0.0466
```

```
                 Posterior Sampling Information
       Description                 Value
       Prior                       Jeffreys
       Algorithm                   Rejection Sampling
       Base Density 1              IG(18.063199,2797.896725)
       Base Density 2              IG(134.632749,9506.512711)
       Ratio Bound                 131.981352
       Sample Size                 11000
       Seed                        45591
```

Thus, most of the output is identical to that obtained from the default REML analysis. Note that the output produced does *not* relate to the Bayesian parameter estimates. These will need to be obtained from the dataset created by the program, `samples`, which contains the sampled parameters (see below). Note also that the `NOPRINT` option was used to avoid 11 000 sets of parameter values being printed.

Code to avoid fitting problems in SAS Release 6.12

If a Bayesian analysis using the `PRIOR` statement fails in SAS `Release 6.12` because the proportion of rejected samples is too high, the code below can be substituted.

```
DATA t;
INPUT tcovp covp1 covp2;
DATALINES;
1    17.729        1
2        0         1
RUN;

PROC MIXED DATA=a; CLASS centre treat;
TITLE 'MODEL 4';
MODEL dbp = dbp1 treat/ SOLUTION;
RANDOM centre;
PRIOR / NSAMPLE=11000 PTRANS PSEARCH SEED=27148032 UPDATE=1000
TDATA=t;
MAKE 'SAMPLE' OUT=samples NOPRINT;
```

Estimation of model parameters

To obtain these it is necessary to summarise the simulated values in the posterior distribution which have been stored in dataset `bp2.bayes4` by the `MAKE` statement. This is done using the SAS code below which is in `Release 6.12`.

```
DATA a; SET samples;
* discard the first 1000 samples as they may not be reliable;
if _n_>=1000;
* calculate samples for treatment difference A-B;
a_b=beta3-beta4;
* define indicator variables for whether the sampled
differences are greater than or less than zero;
IF a_b<0 THEN a_b0=1; ELSE a_b0=0;
IF beta3<0 THEN a_c0=1; ELSE a_c0=0;
IF beta4<0 THEN b_c0=1; ELSE b_c0=0;

PROC MEANS; VAR beta2 a_b beta3 beta4 T3F2;
TITLE 'FIXED EFFECTS MEANS AND STANDARD ERRORS';
LABEL beta2='DBP1' a_b='A-B' beta3='A-C' beta4='B-C' t3f2='F';

PROC UNIVARIATE; VAR covp1 covp2;
TITLE 'SUMMARY OF MARGINAL POSTERIORS OF VARIANCE COMPONENTS';
LABEL covp1='CENTRE VC' covp2='RESID VC';

* count the numbers of samples greater than zero and their
denominators;
PROC MEANS NOPRINT DATA=a; VAR a_b0 a_c0 b_c0;
OUTPUT OUT=b SUM=r1 r2 r3 N=n1 n2 n3;

DATA b; SET b;
* calculate the p-values as twice the proportions of samples
greater than or less than 0 (whichever is  smaller);
IF r1/n1<0.5 THEN a_bp=2*r1/n1; ELSE a_bp=2*(n1-r1)/n1;
IF r2/n2<0.5 THEN a_cp=2*r2/n2; ELSE a_cp=2*(n2-r2)/n3;
IF r3/n3<0.5 THEN b_cp=2*r3/n3; ELSE b_cp=2*(n3-r3)/n3;
PROC PRINT NOOBS; VAR a_bp a_cp b_cp;
TITLE 'P-VALUES FOR TWO-SIDED TESTS OF TREATMENT PAIRS';

* to obtain 95% probability intervals - much output!;
PROC UNIVARIATE FREQ ROUND=0.01 DATA=a; VAR a_b beta3 beta4;
LABEL a_b='A-B' beta3='A-C' beta4='B-C';
```

```
                  FIXED EFFECTS MEANS AND STANDARD ERRORS
Variable Label      N       Mean     Std Dev     Minimum      Maximum
----------------------------------------------------------------------

BETA2     DBP1   10000   0.2758857   0.1075506   -0.1779219    0.6892840
A_B       A-B    10000   1.0535980   1.2266260   -3.5100084    5.6480213
BETA3     A-C    10000   2.9906960   1.2292824   -1.2925770    8.3232165
BETA4     B-C    10000   1.9370981   1.2466882   -2.7695870    6.6563721
T3F2      F      10000   3.0627002   2.0198709    0.0021348   18.4791587
----------------------------------------------------------------------
```

```
          SUMMARY OF MARGINAL POSTERIORS OF VARIANCE COMPONENTS
                          Univariate Procedure
Variable=COVP1   CENTRE VC
                               Moments
              N              10000   Sum Wgts          10000
              Mean        9.231821   Sum            92318.21
```

Example   99

```
Std Dev         4.73424   Variance     22.41303
Skewness       1.290669   Kurtosis     2.889594
USS             1076373   CSS          224107.9
CV             51.28176   Std Mean     0.047342
T:Mean=0       195.0011   Pr>|T|         0.0001
Num ^ = 0         10000   Num > 0         10000
M(Sign)            5000   Pr>=|M|        0.0001
Sgn Rank       25002500   Pr>=|S|        0.0001
```

                    Quantiles(Def=5)

```
100% Max       45.2241       99%   24.31401
 75% Q3          11.68       95%   18.17761
 50% Med       8.36429       90%   15.39085
 25% Q1       5.843674       10%   4.139059
  0% Min      0.362814        5%   3.298738
                              1%   2.105568
```

```
Range        44.86129
Q3-Q1        5.836322
Mode         0.362814
```

                          Extremes

Lowest	Obs	Highest	Obs
0.362814(	4378)	36.36254(	4596)
0.527271(	6144)	36.52679(	7861)
0.583303(	530)	39.91087(	3040)
0.623358(	9418)	41.90423(	5354)
0.67485 (	8037)	45.2241 (	8636)

SUMMARY OF MARGINAL POSTERIORS OF VARIANCE COMPONENTS
                    Univariate Procedure

Variable=COVP2 RESID VC

                          Moments

```
N                 10000   Sum Wgts        10000
Mean           71.47383   Sum           714738.3
Std Dev        6.200502   Variance     38.44622
Skewness        0.34657   Kurtosis     0.176429
USS            51469510   CSS          384423.8
CV             8.675205   Std Mean     0.062005
T:Mean=0        1152.71   Pr>|T|         0.0001
Num ^ = 0         10000   Num > 0         10000
M(Sign)            5000   Pr>=|M|        0.0001
Sgn Rank       25002500   Pr>=|S|        0.0001
```

                    Quantiles(Def=5)

```
100% Max      101.5628       99%   87.48869
 75% Q3       75.46201       95%   82.24455
 50% Med       71.1172       90%   79.55467
 25% Q1        67.1288       10%   63.7857
  0% Min      53.27083        5%   61.91752
                              1%   58.6116
```

```
Range        48.29197
Q3-Q1        8.333211
Mode         53.27083
```

```
                          Extremes
            Lowest    Obs       Highest    Obs
          53.27083(   1468)    94.85331(   5299)
          54.2183 (   9252)    95.0097 (   7589)
          54.22407(   3935)    98.27816(   5603)
          54.27935(   8716)    98.97346(   1822)
          54.62096(   3806)   101.5628 (   4520)
```

P-VALUES FOR TWO-SIDED TESTS OF TREATMENT PAIRS

```
              A_BP      A_CP      B_CP
             0.3894    0.0142    0.1252
```

Univariate Procedure

Variable=A_B   A-B

```
                          Moments
          N              10000    Sum Wgts        10000
          Mean         1.053598   Sum          10535.98
          Std Dev      1.226626   Variance     1.504611
          Skewness     -0.01245   Kurtosis      0.10083
          USS          26145.29   CSS          15044.61
          CV           116.4226   Std Mean     0.012266
          T:Mean=0     85.89399   Pr>|T|         0.0001
          Num ^ = 0      10000    Num > 0          8053
          M(Sign)         3053    Pr>=|M|        0.0001
          Sgn Rank     19432435   Pr>=|S|        0.0001
```

```
                     Quantiles(Def=5)
          100% Max    5.648021       99%    3.962346
           75% Q3     1.859476       95%    3.083847
           50% Med    1.061425       90%    2.614635
           25% Q1     0.245524       10%    -0.51003
            0% Min    -3.51001        5%    -0.96701
                                      1%    -1.89308
          Range       9.15803
          Q3-Q1       1.613952
          Mode        -3.51001
```

```
                          Extremes
            Lowest    Obs       Highest    Obs
          -3.51001(   5893)    5.034716(   4432)
          -3.42951(   5831)    5.154314(    911)
          -3.41027(   2114)    5.286033(   6855)
          -3.15554(   5858)    5.461446(   4409)
          -3.11196(   7901)    5.648021(   4009)
```

Variable=A_B   A-B

rounded to the nearest multiple of 0.01

Frequency Table

		Percents					Percents	
Value	Count	Cell	Cum		Value	Count	Cell	Cum
-3.51	1	0.0	0.0		-2.02	2	0.0	0.8
-3.43	1	0.0	0.0		-2.01	4	0.0	0.8

Example   101

-3.41	1	0.0	0.0	-2	1	0.0	0.8
-3.16	1	0.0	0.0	-1.98	3	0.0	0.9
-3.11	2	0.0	0.1	-1.97	3	0.0	0.9
.				.			
.				.			
.				.			
-1.34	3	0.0	2.5	-0.85	14	0.1	6.0
.				.			
.				.			
.				.			
3.5	6	0.1	97.5	4.03	3	0.0	99.2
.				.			
.				.			
.				.			

95% probability intervals can be derived from this table of cumulative frequencies produced by the UNIVARIATE procedure by searching for the values of a_b that are at the 2.5% and 97.5% centiles. A line is printed for each point to two decimal places so the full output is lengthy! Similar tables are produced for beta3 and beta4, but these are not printed here.

3

Generalised Linear Mixed Models

Up until now we have considered models with normally distributed errors. However, there are many situations where data are not of this type, e.g. where the presence/absence of an adverse event is recorded, and the normality assumption cannot be made. A class of models known as generalised linear models (GLMs) is available for fitting fixed effects models to such non-normal data. These models can be further extended to fit mixed models and are then referred to as generalised linear mixed models (GLMMs). Random effects, random coefficients or covariance patterns can be included in a GLMM in much the same way as in normal mixed models, and again either balanced or unbalanced data can be analysed. Although GLMMs can be used to analyse data from any distribution from the exponential family, binary data and Poisson data are most frequently encountered and for this reason this book will primarily concentrate on the use of GLMMs for these data types.

In introducing these topics, we will, of necessity, be less than comprehensive. In their excellent book *Generalised Linear Models*, McCullagh and Nelder (1989) take 500 pages to cover the subject and start by an assumption of 'a knowledge of matrix theory, including generalised inverses, together with basic ideas of probability theory, including orders of magnitude in probability'. At one end of the readership spectrum, therefore, those with no experience of generalised linear models may wish to skip all but the introductory paragraphs of each section, because some sections inevitably draw on the assumption of prior knowledge. This will enable such readers to identify where such methods might prove useful. In case the reader with little background knowledge of GLMs identifies a need to apply GLMs and GLMMs, and is horrified at the prospect of having to master textbooks on the subject, we would emphasise that fitting such models can often be achieved without an encyclopaedic knowledge of the topic. The final section of this chapter and sections of subsequent chapters will illustrate the application of these models and present the SAS code needed to implement them.

We will start by describing the GLM in Section 3.1 and then show how it is extended to the GLMM in Section 3.2. GLMMs are more complex than normal mixed models and there is therefore more potential for problems such as biased estimates and a failure to converge. These are considered in Section 3.3, which also gives some practical information on fitting GLMMs. A worked example is given in Section 3.4.

3.1 GENERALISED LINEAR MODELS

3.1.1 Introduction

GLMs can be used to fit fixed effects models to certain types of non-normal data: those with a distribution from the exponential family. Consider the following example. We wish to conduct a clinical trial to investigate the effect of a new treatment for epilepsy. A suitable variable for assessing efficacy is the number of seizures which occur during a pre-determined period of time. Thus, the response variable is a count. Such variables are often found to follow a Poisson distribution. This is a member of the exponential family and GLMs or GLMMs can be considered, depending on the details of how the trial is designed. Such an example is considered in Section 6.4. As a second example, consider the analysis of a particular adverse event in a clinical trial. In some situations, a simple contingency-table-based analysis will be sufficient. If the design of the trial is relatively complicated, and the adverse event is not sufficiently serious to cause withdrawal from the trial, contingency tables will be less attractive. In the multi-centre trial which we are regularly revisiting in this book, the occurrence of cold feet was such an adverse event, and could be reported at any of the follow-up visits. As a binary outcome, this is also from the exponential family, and in Section 3.4 we will show how GLMMs can be applied to this data.

As with the models we have met for normally distributed data, the models use a linear combination of variables to 'predict' the response. In the case of normally distributed data the fixed effects model is $\mathbf{y} = \mathbf{X}\boldsymbol{\alpha} + \mathbf{e}$. That is, the response is determined by the linear component, $\mathbf{X}\boldsymbol{\alpha}$, which gives the expected response, which we will denote by $\boldsymbol{\mu}$, and by a randomly determined error term. In a somewhat convoluted way we could write the model as

$$\mathbf{y} = \boldsymbol{\mu} + \mathbf{e},$$

$$\boldsymbol{\mu} = \mathbf{X}\boldsymbol{\alpha}.$$

The GLM can easily be derived from this artificial-looking model by allowing μ and $\mathbf{X}\boldsymbol{\alpha}$ to be related by a 'link function', g, so that

$$g(\boldsymbol{\mu}) = \mathbf{X}\boldsymbol{\alpha}.$$

Thus, normal models are a special case of GLMs in which the link function is the identity function. In general, the link function is not the identity function but takes a form suitable for the distribution of the data.

An alternative, less mathematical way, of familiarisation with the concept of the GLM is to think of the link function as a method of mapping the response data from their scale of observation to the real scale $(-\infty, +\infty)$. For example, binomial probabilities have a range $0-1$ and the logit link function, $\log(\mu/(1 - \mu))$, will translate this range to the real scale. This is necessary because fitting a linear model directly to the binomial parameter could lead to estimates of probabilities which were negative or greater than one. Use of the link function allows the model parameters to be included in the model linearly, just as in the models we have described for normal data. This often gives the GLM an advantage over contingency table methods, which are sometimes used to analyse binary data (e.g. chi-squared tests), because these methods cannot incorporate several fixed effects simultaneously.

Here, we will give only a brief introduction to GLMs. However, more detail can be found in McCullagh and Nelder (1989). Before defining the GLM, basic details of the binomial and Poisson distributions will be given for those who are not completely familiar with these distributions, and the general form for distributions from the exponential family will be specified. This general distributional form will be needed for setting a particular form of link function known as the 'canonical' link.

3.1.2   Distributions

We now define the Bernoulli, binomial and Poisson distributions. These can all be described as 'one-parameter' distributions, i.e. using a single parameter completely describes the distribution.

The Bernoulli distribution

This distribution is used to model binary data where observations have one of two possible outcomes, which can be thought of as 'success' or 'failure'. If one is used to denote success, and zero failure, the density function is

$$f(y) = \mu^y (1 - \mu)^{(1-y)}, \quad y = 0, 1.$$

Thus, μ corresponds to the probability of success and the mean and variance are given by

$$\text{mean}(y) = \mu,$$

$$\text{var}(y) = \mu(1 - \mu).$$

The Binomial distribution

This distribution is also suitable for binary data. However, observations are now recorded as the number of successes out of a number of 'tries'. The parameter of interest is the proportion of successes. If z and n are the observed numbers of successes and tries, then the proportion $y = z/n$ has a density function

$$f(y, n) = \frac{n!(\mu)^{ny}(1 - \mu)^{n-ny}}{(ny)!(n - ny)!}$$

and

$$\text{mean}(y) = \mu,$$

$$\text{var}(y) = \mu(1 - \mu)/n.$$

Note that when $n = 1$, the Bernoulli density function is obtained. Thus, the Bernoulli distribution is a special case of the binomial distribution.

The Poisson distribution

This distribution can be used to model 'count' data. The number of episodes of dizziness over a fixed period, or the number of abnormal heart beats on an ECG tape of a prescribed length, are examples of count data. Its density function is

$$f(y) = \mu^y e^{-\mu}/y!, \quad y = 0, 1, 2, \ldots$$

and

$$\text{mean}(y) = \mu,$$

$$\text{var}(y) = \mu.$$

The Poisson distribution with offset

Sometimes the underlying scale for count data varies with each observation. For example, observations may be made over varying time periods (e.g. number of epileptic seizures measured over different numbers of days for each patient). Alternatively, the underlying scale may relate to some other factor such as the size of a geographical region over which counts of subjects with a specific disease are taken. To take account of such a varying scale, the scale for each observation needs to be utilised in forming the distribution density. The scale variable is often referred to as the *offset*. The parameter of interest is then the number of counts per unit scale of the offset variable. If we denote the offset variable by t (even though it is not always time) and the observed number of counts by z, the distribution of $y = z/t$ has density function

$$f(y, t) = (\mu t)^{yt} e^{-\mu t}/(yt)!$$

and

$$\text{mean}(y) = \mu,$$

$$\text{var}(y) = \mu/t.$$

Note that when $t = 1$, the density function for the Poisson distribution without an offset variable above is obtained, confirming it as a special case of this distribution.

3.1.3   The general form for exponential distributions

To show how the GLM can be used for data with any exponential family distribution, we first need to define a general form in which all exponential family density functions can be expressed. This can be written as

$$f(y; \theta, \phi) = \exp\{[y\theta - b(\theta)]/a(\phi) + c(y, \phi)\},$$

where

$\theta = $ a location parameter (not necessarily the mean),

$\phi = $ a dispersion parameter (only appears in distributions which have two parameters such as the normal distribution).

The form of the functions a, b and c will be different for each distribution. Distributions that are not from the exponential family cannot be expressed in this way.

The one-parameter distributions considered in this book can be defined solely in terms of the location parameter, θ, and the general form above then simplifies to

$$f(y; \theta) = \exp\{[y\theta - b(\theta)]/a + c(y)\},$$

where a is now a constant. Expressions for a, $b(\theta)$ and $c(y)$ are listed below for the Bernoulli, binomial and Poisson distributions:

Distribution	a	$b(\theta)$	$c(y)$
Bernoulli	1	$\log(1 + \exp(\theta))$	1
Binomial	$1/n$	$\log(1 + \exp(\theta))$	$\log[n!/((ny)!(n - ny)!)]$
Poisson	1	$\exp(\theta)$	$-\log(y!)$
Poisson with offset	$1/t$	$\exp(\theta)$	$yt\log(yt) - \log(yt!)$

where n and t are the denominator and offset terms, respectively. Details of exactly how these expressions are obtained from their density functions will be given later in Section 3.1.7.

It can be shown that the mean and variance of a distribution can then be written in terms of the functions a and b as

$$\text{mean}(y) = \mu = b'(\theta),$$

$$\text{var}(y) = ab''(\theta).$$

Hence, $\theta = b'^{-1}(\mu)$ and we can alternatively write the variance in terms of μ as

$$\text{var}(y) = ab''(b'^{-1}(\mu)).$$

Using the above expressions, the means and variances for the distributions can then be written in terms of θ or μ as follows.

Distribution	Mean $\mu = b'(\theta)$	Variance in terms of θ, $ab''(\theta)$	Variance in terms of μ
Bernoulli	$(1 + \exp(-\theta))^{-1}$	$\exp(\theta)/(1 + \exp(\theta))^2$	$\mu(1 - \mu)$
Binomial	$(1 + \exp(-\theta))^{-1}$	$\exp(\theta)/(1 + \exp(\theta))^2/n$	$\mu(1 - \mu)/n$
Poisson	$\exp(\theta)$	$\exp(\theta)$	μ
Poisson with offset	$\exp(\theta)$	$\exp(\theta)/t$	μ/t

3.1.4   The GLM definition

In Section 3.1.1 we defined the GLM using the matrix notation used for normal models by

$$\mathbf{y} = \boldsymbol{\mu} + \mathbf{e}$$

and related $\boldsymbol{\mu}$ to a linear sum of the fixed effects, $\mathbf{X\alpha}$ (the linear component) by a 'link function', g, so that

$$g(\boldsymbol{\mu}) = \mathbf{X\alpha}.$$

To relate the GLM directly to the general exponential density function introduced in the previous section, we label the linear component $\boldsymbol{\theta}$ so that $\boldsymbol{\theta} = \mathbf{X\alpha}$. We will next consider how link functions can be constructed.

Canonical link functions

A type of link function known as the *canonical link function* is given by

$$g = b'^{-1},$$

where b is obtained from the general form for the density function for exponential distributions given above. For the distributions we have considered, the canonical link functions are given by

Distribution	$g(\mu) = b'^{-1}(\mu)$	Name
Bernoulli	$\log(\mu/(1 - \mu))$	Logit
Binomial	$\log(\mu/(1 - \mu))$	Logit
Poisson	$\log(\mu)$	Log
Poisson with offset	$\log(\mu)$	Log

In most situations, use of the canonical link function will lead to a satisfactory analysis model. However, we should also mention that there is not a strict requirement for canonical link functions to be used in the GLM and non-canonical link functions are also available. These are not derived from the density function but still map the data from its original scale onto the real scale. For example, the link function known as the probit function, $g(\mu) = \Phi^{-1}(\mu)$ (where Φ is the cumulative normal density function), is sometimes used for binary data recorded in toxicology experiments, since values of μ corresponding to specific probabilities can easily be obtained using the normal density function. Despite not being canonical, this link function does still map the original range of the data (0 to 1) to $-\infty$ to ∞ as required for the GLM. In this book we will not be considering non-canonical link functions. However, further information can be found in McCullagh and Nelder (1989).

We earlier specified a general formula for the variance in the GLM as $\text{var}(y) = ab''(\theta)$. Using the relationship $g = b'^{-1}$, for canonical link functions we can now equivalently write the variance in terms of μ and the function g as,

$$\text{var}(y) = ag'^{-1}(\mu).$$

The variance matrix, V

The variance matrix for the GLM may be written:

$$\text{var}(\mathbf{y}) = \text{var}(\mathbf{e}) = \mathbf{V}.$$

Since the GLM is a fixed effects model the observations are assumed to be uncorrelated and the variance matrix, $\mathbf{V}$, is therefore diagonal. The diagonal terms of this matrix are equal to the variances of each observation given the underlying distribution. So, for example, in an analysis of six Bernoulli observations, where $\boldsymbol{\mu} = (\mu_1, \mu_2, ..., \mu_6)'$, $\mathbf{V}$ can be written:

$$\mathbf{V} = \begin{pmatrix} \mu_1(1-\mu_1) & 0 & 0 & 0 & 0 & 0 \\ 0 & \mu_2(1-\mu_2) & 0 & 0 & 0 & 0 \\ 0 & 0 & \mu_3(1-\mu_3) & 0 & 0 & 0 \\ 0 & 0 & 0 & \mu_4(1-\mu_4) & 0 & 0 \\ 0 & 0 & 0 & 0 & \mu_5(1-\mu_5) & 0 \\ 0 & 0 & 0 & 0 & 0 & \mu_6(1-\mu_6) \end{pmatrix}.$$

Note that unlike the fixed effects model for normal data, the variances are different for each observation.

In the previous subsection we showed that the variance could be written in the general form

$$\text{var}(\mathbf{y}) = ab''(\boldsymbol{\theta}) = ag'^{-1}(\boldsymbol{\mu}).$$

We can use a and $b''(\theta)$ to express $\mathbf{V}$ in matrix form as a product of two diagonal matrices:

$$\mathbf{V} = \mathbf{AB},$$

where

$\mathbf{A} = \text{diag}\{a_i\},$
$\mathbf{B} = \text{diag}\{b''(\theta_i)\} = \text{diag}\{g'^{-1}(\mu_i)\}.$

For the binomial distribution, $\mathbf{A}$ is a diagonal matrix of inverses of the denominator terms (number of 'tries'). For example, for a dataset with six observations, $\mathbf{A}$ would be

$$\mathbf{A} = \begin{pmatrix} 1/n_1 & 0 & 0 & 0 & 0 & 0 \\ 0 & 1/n_2 & 0 & 0 & 0 & 0 \\ 0 & 0 & 1/n_3 & 0 & 0 & 0 \\ 0 & 0 & 0 & 1/n_4 & 0 & 0 \\ 0 & 0 & 0 & 0 & 1/n_5 & 0 \\ 0 & 0 & 0 & 0 & 0 & 1/n_6 \end{pmatrix}.$$

For a Poisson distribution with offset, the n_i are replaced by the offset variable values, t_i.

$\mathbf{B}$ is a diagonal matrix of variance terms. For the Bernoulli and binomial distributions it is

$$\mathbf{B} = \begin{pmatrix} \mu_1(1-\mu_1) & 0 & 0 & 0 & 0 & 0 \\ 0 & \mu_2(1-\mu_2) & 0 & 0 & 0 & 0 \\ 0 & 0 & \mu_3(1-\mu_3) & 0 & 0 & 0 \\ 0 & 0 & 0 & \mu_4(1-\mu_4) & 0 & 0 \\ 0 & 0 & 0 & 0 & \mu_5(1-\mu_5) & 0 \\ 0 & 0 & 0 & 0 & 0 & \mu_6(1-\mu_6) \end{pmatrix}.$$

Note that for the Bernoulli and Poisson distributions, $a_i = 1$ and thus $\mathbf{A} = \mathbf{I}$ (the identity matrix) and $\mathbf{V} = \mathbf{B}$.

The dispersion parameter

Variance in the model can be increased (or decreased) from the observation variances specified by the underlying distribution (i.e. $a_i b''(\theta_i)$), by multiplying the variance matrix by a dispersion parameter, ϕ:

$$\mathbf{V} = \phi \mathbf{AB}.$$

An alternative dispersion parameter is suggested by Williams (1982) for binary data. Here, ϕ_i is calculated using a formula that varies depending on the denominator of each observation and so adjusts for their differing variances:

$$\mathbf{V} = \phi_i \mathbf{AB}.$$

It is referred to as the 'Williams modification'. Before GLMMs were developed, dispersion parameters were frequently used as a limited facility to model variance at the residual level in one-parameter distributions. We will consider the implications of using a dispersion parameter further in Section 3.3.8.

3.1.5   Interpreting results from GLMs

In this section we look at how results from GLMs using logit and log link functions can be interpreted.

The logit link function

The logit link function, $\log(\mu/(1 - \mu))$, is the canonical link function for Bernoulli and binomial distributions. Analyses using this link function are often referred to as *logistic regression* analyses. Results are obtained on the logit scale and can be expressed in terms of *odds ratios (ORs)* when exponentiated. To see this, we consider a simple, single centre, parallel group trial to compare two treatments, with a binary outcome. The model could be written as

$$\log(\mu/(1 - \mu)) = a + bx,$$

where x is an indicator variable denoting treatment (say, one if the treatment is A and zero if the treatment is B).

Thus, using p_A and p_B to denote the probabilities of success,

$$\log(p_A/(1 - p_A)) = a + b,$$

$$\log(p_B/(1 - p_B)) = a,$$

and, by subtraction,

$$\log(p_A/(1 - p_A)) - \log(p_B/(1 - p_B)) = b.$$

Hence,

$$\log \frac{(p_A/(1 - p_A))}{(p_B/(1 - p_B))} = b,$$

and, on exponentiating,

$$\frac{p_A/(1 - p_A)}{p_B/(1 - p_B)} = e^b.$$

The numerator of this expression gives us the odds of success on treatment A (i.e. the probability of success divided by the probability of failure). Similarly, the denominator is the odds of success on treatment B, leading to e^b being the estimate of the OR.

Calculation of such an OR will be illustrated in the analysis of cold feet in the hypertension trial in Section 3.4.

The log link function

The log function is the canonical link function for Poisson distributions. Models using this link function can be described as *log-linear* models. Results are obtained on the log scale which, as for logistic regression, is not easy to interpret directly. In a similar way though, exponentiating the coefficients allows them to be expressed this time in terms of *relative rates (RRs)*. The RR is simply the ratio of two event rates; for example, the rate on treatment A divided by the rate on treatment B can be used to compare the two treatments.

$$RR = \frac{\text{Rate of event on treatment A}}{\text{Rate of event on treatment B}}.$$

The calculation of an RR will be illustrated in the analysis of epileptic seizure frequencies in Section 6.4.

If the log link function is used to analyse data which are binary, the rates then become risks (or probabilities) and RRs are then sometimes instead referred to as *relative risks*.

3.1.6 Fitting the GLM

The GLM can be fitted use PROC GENMOD in SAS. For readers who wish to understand the 'mechanics' of how the GLM is fitted, we now look at the numerical procedures involved.

The GLM is fitted by maximising the log likelihood function. Using the general form for a one-parameter exponential distribution given in Section 3.1.3:

$$f(y; \theta) = \exp([y\theta - b(\theta))/a + c(y)).$$

The log likelihood for a set of observations can be written:

$$\log(L) = \sum_i (y_i\theta_i - b(\theta_i))/a_i + K,$$

or, in matrix/vector notation as

$$\log(L) = \mathbf{y}'\mathbf{A}^{-1}\boldsymbol{\theta} - b(\boldsymbol{\theta})^{1/2'}\mathbf{A}^{-1}b(\boldsymbol{\theta})^{1/2} + K, \tag{A}$$

where

$\boldsymbol{\theta} = (\theta_1, \theta_2, \dots, \theta_n)'$,
$b(\boldsymbol{\theta}) = (b(\theta_1), b(\theta_2), \dots, b(\theta_n))'$,
$K = \text{constant}$.

In fixed effects models for normal data we saw that a solution for $\boldsymbol{\alpha}$ was easily obtained by differentiating the log likelihood with respect to $\boldsymbol{\alpha}$ and setting the resulting expression to zero (Section 2.2.2). However, in GLMs the differentiated log likelihood, $d\log(L)/d\boldsymbol{\alpha}$, is non-linear in $\boldsymbol{\alpha}$ and an expression giving a direct solution for $\boldsymbol{\alpha}$ cannot be formed. To obtain $d\log(L)/d\boldsymbol{\alpha}$ we differentiate (A) above after substituting $\boldsymbol{\theta} = \mathbf{X}\boldsymbol{\alpha}$:

$$d\log(L)/d\boldsymbol{\alpha} = d(\mathbf{y}'\mathbf{A}^{-1}\mathbf{X}\boldsymbol{\alpha})/d\boldsymbol{\alpha} - d(b(\mathbf{X}\boldsymbol{\alpha})^{1/2'}\mathbf{A}^{-1}b(\mathbf{X}\boldsymbol{\alpha})^{1/2})/d\boldsymbol{\alpha}$$

$$= \mathbf{X}'\mathbf{A}^{-1}\mathbf{y} - \mathbf{X}'\mathbf{A}^{-1}b'(\mathbf{X}\boldsymbol{\alpha})$$

$$= \mathbf{X}'\mathbf{A}^{-1}(\mathbf{y} - b'(\mathbf{X}\boldsymbol{\alpha})).$$

Setting this differential to zero leads to equations that can, in principle, be solved for $\boldsymbol{\alpha}$. However, because they are non-linear in $\boldsymbol{\alpha}$, one of the approaches below is usually used instead to obtain estimates of $\boldsymbol{\alpha}$.

The likelihood function can be maximised directly for $\boldsymbol{\alpha}$ using an iterative procedure such as Newton–Raphson (see Section 2.2.4). The variance of the resulting $\hat{\boldsymbol{\alpha}}$ can be calculated at the final iteration by (from McCullagh and Nelder, 1989, Chapter 9):

$$\text{var}(\hat{\boldsymbol{\alpha}}) = (\mathbf{BX}'\mathbf{V}^{-1}\mathbf{XB})^{-1},$$

where

$$\mathbf{B} = \text{diag}\{g'^{-1}(\boldsymbol{\mu})\} = \text{diag}\{b''(\boldsymbol{\theta})\}.$$

We see that $\mathbf{B}$ is a diagonal matrix containing the variances of the individual observations.

Alternatively, the likelihood can be maximised using an *iterative weighted least squares* method. This approach (defined by McCullagh and Nelder, 1989, Section 2.5) can be based on analysing the following pseudo variable, $\mathbf{z}$, which can be thought of as a linearised observation vector.

$$\mathbf{z} = g(\boldsymbol{\mu}) + (\mathbf{y} - \boldsymbol{\mu})g'(\boldsymbol{\mu})$$

$$= g(\boldsymbol{\mu}) + (\mathbf{y} - \boldsymbol{\mu})\mathbf{B}^{-1}.$$

$\mathbf{z}$ can, in fact, be defined as a first-order Taylor series expansion for $g(\mathbf{y})$ about $\boldsymbol{\mu}$.

Recalling that $g(\boldsymbol{\mu}) = \mathbf{X}\boldsymbol{\alpha}$ we see that $\mathbf{z}$ has variance

$$\mathbf{V}_{\mathbf{z}} = \text{var}(\mathbf{X}\boldsymbol{\alpha}) + \mathbf{B}^{-1}\text{var}(\mathbf{y} - \boldsymbol{\mu})\mathbf{B}^{-1}$$

$$= \mathbf{0} + \mathbf{B}^{-1}\mathbf{ABB}^{-1}$$

$$= \mathbf{AB}^{-1}.$$

A normal model can then be expressed in terms of the linearised pseudo variable, $\mathbf{z}$:

$$\mathbf{z} = \mathbf{X}\boldsymbol{\alpha} + \mathbf{e}.$$

$\mathbf{z}$ is then analysed iteratively using weighted least squares (see Section 2.2.1). Weights are taken as the inverse of the variance matrix $\mathbf{V}_{\mathbf{z}}$. Iteration is required because $\mathbf{z}$ and $\mathbf{V}_{\mathbf{z}}$ are dependent on $\boldsymbol{\alpha}$. The raw data, $\mathbf{y}$, can be taken as initial values for $\mathbf{z}$. Alternatively, $g(\mathbf{y})$ can be used for the initial values, although an adjustment may be necessary to prevent infinite values. The identity matrix can be used initially for $\mathbf{V}_{\mathbf{z}}$. The initial fixed effects solution is calculated using these values of $\mathbf{z}$ and $\mathbf{V}_{\mathbf{z}}$ as

$$\hat{\boldsymbol{\alpha}} = (\mathbf{X}'\mathbf{V}_{\mathbf{z}}^{-1}\mathbf{X})^{-1}\mathbf{X}'\mathbf{V}_{\mathbf{z}}^{-1}\mathbf{z},$$

as in Section 2.2.2. This forms the first iteration. $\hat{\boldsymbol{\alpha}}$ is then used to calculate new values for $\mathbf{z}$ and $\mathbf{V}_{\mathbf{z}}$. From these new values, $\hat{\boldsymbol{\alpha}}$ is recalculated to form the second iteration. The process is continued until $\hat{\boldsymbol{\alpha}}$ convergences. The asymptotic variance of $\hat{\boldsymbol{\alpha}}$ can be calculated at the final iteration by

$$\text{var}(\hat{\boldsymbol{\alpha}}) = (\mathbf{X}'\mathbf{V}_{\mathbf{z}}^{-1}\mathbf{X})^{-1}.$$

3.1.7 Expressing individual distributions in the general exponential form

In Section 3.1.3 we introduced the idea of expressing distributions in a general form for exponential distributions:

$$f(y; \theta, \phi) = \exp\{[y\theta - b(\theta)]/a(\phi) + c(y, \phi)\}.$$

Forms for a, b and c were given for the Bernoulli, binomial and Poisson distributions. We now show how these forms are obtained from the distribution densities.

The Bernoulli distribution

The density function

$$f(y) = \mu^y(1 - \mu)^{(1-y)}$$

is, by logging and then exponentiating the right-hand side of this equation, rearranged in the general exponential form as

$$f(y) = \exp[y \log(\mu/(1 - \mu)) + \log(1 - \mu)].$$

Thus, $\theta = \log(\mu/(1 - \mu))$ and we obtain the logit as the canonical link function. The mean, μ, can then be expressed as the inverse of the logit function, $\mu = \exp(\theta)/(1 + \exp(\theta)) = (1 + \exp(-\theta))^{-1}$. Writing the distribution in terms of θ we obtain

$$f(y) = \exp[y\theta + \log\{1 - \exp(\theta)/[1 + \exp(\theta)]\}]$$

$$= \exp[y\theta - \log\{1 + \exp(\theta)\}].$$

Therefore, $b(\theta) = \log(1 + \exp(\theta))$, $a = 1$ and $c(y) = 1$.

The Binomial distribution

The density function,

$$f(y, n) = \frac{n!(\mu)^{ny}(1 - \mu)^{n-ny}}{(ny)!(n - ny)!}$$

is rearranged in the general exponential form using the same trick as was used for the Bernoulli distribution, as

$$f(y, n) = \exp\{[[y \log[\mu/(1 - \mu)] + n \log(1 - \mu)]n + \log[n!/((ny)!(n - ny)!)]\}.$$

This again gives $\theta = \log[\mu/(1 - \mu)]$ and the logit as the canonical link function. Therefore, μ can again be expressed as $\exp(\theta)/(1 + \exp(\theta)) = (1 + \exp(-\theta))^{-1}$ and we can write,

$$f(y, n) = \exp\{[y\theta + \log\{1 - \exp(\theta)/[1 + \exp(\theta)]\}]n + \log[n!/((ny)!(n - ny)!)]\}$$

$$= \exp\{[y\theta - \log[1 + \exp(\theta)]n + \log[n!/((ny)!(n - ny)!)]\}.$$

Therefore, $b(\theta) = \log(1 + \exp(\theta))$, $a = 1/n$, and $c(y) = \log\{n!/((ny)!(n - ny)!)\}$.

The Poisson distribution

The density function

$$f(y) = \mu^y e^{-\mu} / y!$$

is rearranged in the general exponential form as

$$f(y) = \exp\{y \log(\mu) - \mu - \log(y!).$$

Therefore, $\theta = \log(\mu) = g(\mu)$ and we obtain the log as the canonical link function. Substituting for $\mu = \exp(\theta)$ gives

$$f(y) = \exp\{y\theta - \exp(\theta) - \log(y!)\}.$$

Thus, $b(\theta) = \exp(\theta)$, $a = 1$, and $c(y) = -\log(y!)$.

The Poisson distribution with offset

The density function

$$f(y, t) = (\mu t)^{yt} e^{-\mu t} / (yt)!$$

is rearranged in the general exponential form as

$$f(y, t) = \exp\{[y \log(\mu) - \mu]t + yt \log(t) - \log((yt)!)\}.$$

Thus, $\theta = \log(\mu)$ and again the log is the canonical link function. Substituting $\mu = \exp(\theta)$ we may write

$$f(y) = \exp\{[y\theta - \exp(\theta)]t + yt \log(t) - \log(yt!)\},$$

giving $b(\theta) = \exp(\theta)$, $a = 1/t$, and $c(y) = yt \log(yt) - \log(yt!)$.

The normal distribution

The normal distribution has both a location and a dispersion parameter. Although it is well known that a GLM is not necessary for analysing normal data, it is helpful to see this by showing that the canonical link function for the normal distribution is the identity function. The density function is

$$f(y) = \exp(-(y - \mu)^2 / 2\sigma^2) / \sqrt{(2\pi\sigma^2)},$$

which can be rearranged as

$$f(y) = \exp[(y\mu - \mu^2/2)/\sigma^2 - y^2/2\sigma^2 - \tfrac{1}{2} \log(2\pi\sigma^2)].$$

This is now in the general exponential form for two-parameter distributions. Thus, the canonical link is the identity function, $\theta = g(\mu) = \mu$, and $\phi = \sigma^2$, $a(\phi) = \phi$, $b(\theta) = \theta^2/2$ and $c(y, \phi) = -(y^2/\phi + \log(2\pi\phi))/2$.

3.1.8   Conditional logistic regression

This model does not, strictly speaking, form a GLM but it is mentioned here because it can be very useful for modelling binary data in datasets where there are only a few observations in each category of a fixed effect (e.g. there are only two observations per patient in a two-period, cross-over trial). We shall see later in Section 3.2.4 that fitting problems can arise in GLMs with fixed effects containing only a few observations per category. Using a conditional logistic regression analysis is one way in which these problems can be avoided. It works by omitting the 'problem' effect (e.g. patients in a cross-over trial) as a fixed effect in the model but instead the likelihood is 'conditioned' on this effect. Other effects are fitted as fixed just as in ordinary GLMs, and results can be interpreted as if the 'problem' effect had been fitted as fixed. A more complete description of the method can be found in Clayton and Hills (1993, Chapter 29) and Collett (1991, Section 7.7.1). The model can be fitted using PROC PHREG in SAS. An example of its use will be given in Section 8.4.

3.2   GENERALISED LINEAR MIXED MODELS

GLMMs are based on extending the fixed effects GLM to include random effects, random coefficients and covariance patterns. They are still a fairly new class of models and much research work is still being carried out. Here, in Section 3.2.1, we will specify a general form for the GLMM that encompasses all types of mixed model. In Section 3.2.2 we define the likelihood function for random effects and random coefficients GLMMs and introduce a similar function known as the quasi-likelihood, which is required for fitting covariance pattern models. Following this, in Section 3.2.3 fitting methods for GLMMs will be outlined. These last two sections can be omitted by readers who do not desire a detailed understanding of the more theoretical aspects of fitting GLMMs. In Section 3.2.4 we discuss some difficulties that can arise with fitting random effects models and suggest when these models are inadvisable. It is recommended that all readers should read this section.

3.2.1   The GLMM definition

The GLMM can be defined by

$$\mathbf{y} = \boldsymbol{\mu} + \mathbf{e}.$$

As in the GLM, $\boldsymbol{\mu}$ is the vector of expected means of the observations and is linked to the model parameters by a link function, g:

$$g(\boldsymbol{\mu}) = \mathbf{X}\boldsymbol{\alpha} + \mathbf{Z}\boldsymbol{\beta}.$$

X and Z are the fixed and random effects design matrices, and α and β are the vectors of fixed and random effects parameters as in the normal mixed model. The random effects, β, can again be assumed to follow a normal distribution:

$$\beta \sim N(\mathbf{0}, \mathbf{G})$$

and G is defined just as in Section 2.2. The variance matrix can be written

$$\text{var}(\mathbf{y}) = \mathbf{V} = \text{var}(\boldsymbol{\mu}) + \mathbf{R},$$

where R is the residual variance matrix, $\text{var}(\mathbf{e})$. However, V is not as easily specified as it was for normal data where $\mathbf{V} = \mathbf{ZGZ'} + \mathbf{R}$. This is because $\boldsymbol{\mu}$ is not now a linear function of β. A first-order approximation used by some fitting methods is

$$\mathbf{V} \approx \mathbf{BZGZ'B} + \mathbf{R},$$

where B is a diagonal matrix of variances determined by the underlying distribution as described in Section 3.1 (for example $\mathbf{B} = \text{diag}\{\mu_i(1 - \mu_i)\}$ for binary data). In random effects and random coefficients models the residual matrix, R, is diagonal since the residuals are assumed uncorrelated. The diagonal variance terms are equal to the expected variances given the underlying distribution, and thus $\mathbf{R} = \mathbf{AB}$ as in the GLMs. For random effects and random coefficients models, V can then be written

$$\mathbf{V} \approx \mathbf{BZGZ'B} + \mathbf{AB}.$$

In covariance pattern models, correlated residuals are allowed and R can be expressed as a product of a correlation matrix defined on the linear scale, P, and AB:

$$\mathbf{R} = \mathbf{A}^{1/2}\mathbf{B}^{1/2}\mathbf{PB}^{1/2}\mathbf{A}^{1/2}.$$

The reason for defining P on a linear scale is because it can then be parameterised to have a covariance pattern in the same way as normal data. We will be looking more closely at how to define covariance patterns in Section 6.2. The approximation to V then becomes

$$\mathbf{V} \approx \mathbf{BZGZ'B} + \mathbf{A}^{1/2}\mathbf{B}^{1/2}\mathbf{PB}^{1/2}\mathbf{A}^{1/2}.$$

The above formula can be considered a general form for V since by taking $\mathbf{P} = \mathbf{I}$ for random effects and random coefficients models, we obtain $\mathbf{V} \approx \mathbf{BZGZ'B} + \mathbf{AB}$ as above.

The dispersion parameter

As in the GLMs, variance at the residual level can be increased (or decreased) by using a dispersion parameter. The residual variance is multiplied by the dispersion parameter, ϕ, so that,

$$\mathbf{R} = \phi \mathbf{A}^{1/2}\mathbf{B}^{1/2}\mathbf{PB}^{1/2}\mathbf{A}^{1/2}.$$

If the observed residual variance were exactly equal to that predicted ($\mathbf{A}^{1/2}\mathbf{B}^{1/2}$ $\mathbf{PB}^{1/2}\mathbf{A}^{1/2}$), then the dispersion parameter would equal one. However, often this is not the case. The value of the dispersion parameter is influenced by several factors and these will be explained in more detail in Section 3.3.

3.2.2   The likelihood and quasi-likelihood functions

As in normal mixed models a popular way of fitting the GLMM is based on maximising the likelihood function for the model parameters. However, a difficulty with this is that true likelihood functions can only be defined for random effects and random coefficients models. A true likelihood function is not available for covariance pattern models since a general multivariate distributional form does not exist for non-normal data (for normal data the multivariate normal distribution was used). However, we will show how it is possible to get around this difficulty by defining an alternative function known as the *quasi-likelihood* function, which has very similar properties to the likelihood function. In this section we will specify the likelihood function for random effects and random coefficients models, define a quasi-likelihood function for covariance pattern models, and then give a general form of the quasi-likelihood function that is appropriate for all types of mixed model.

The likelihood function for random effects and random coefficients models

For these models we can obtain a true likelihood function from the product of the likelihoods based on $\mathbf{y}|\boldsymbol{\beta}$ and $\boldsymbol{\beta}$ (by $\mathbf{y}|\boldsymbol{\beta}$ we mean $\mathbf{y}$ conditional on $\boldsymbol{\beta}$ so that $\boldsymbol{\beta}$ is treated as constant when defining the variance of $\mathbf{y}|\boldsymbol{\beta}$). A true likelihood function is possible because the distributions of $\mathbf{y}|\boldsymbol{\beta}$ and $\boldsymbol{\beta}$ are known and hence likelihood functions can be formed from them. The likelihood for the fixed effects, $\boldsymbol{\alpha}$, and the variance parameters in the $\mathbf{G}$ matrix, $\boldsymbol{\gamma}_{\mathbf{G}}$, can be written

$$L(\boldsymbol{\alpha}, \boldsymbol{\gamma}_{\mathbf{G}}; \mathbf{y}) = L(\boldsymbol{\alpha}; \mathbf{y}|\boldsymbol{\beta})L(\boldsymbol{\gamma}_{\mathbf{G}}; \boldsymbol{\beta}). \tag{B}$$

Now $\boldsymbol{\beta}$ is assumed to have a multivariate normal distribution, $\boldsymbol{\beta} \sim N(\mathbf{0}, \mathbf{G})$, so substituting the multivariate normal density for $L(\boldsymbol{\gamma}_{\mathbf{G}}; \boldsymbol{\beta})$ we have

$$L(\boldsymbol{\alpha}, \boldsymbol{\gamma}_{\mathbf{G}}; \mathbf{y}) \propto L(\boldsymbol{\alpha}; \mathbf{y}|\boldsymbol{\beta}) \, |\mathbf{G}|^{-1/2} \exp(-1/2\boldsymbol{\beta}'\mathbf{G}^{-1}\boldsymbol{\beta}).$$

The $\mathbf{y}|\boldsymbol{\beta}$ are independent because we have assumed uncorrelated residuals ($\mathbf{R}$ is diagonal) and therefore $L(\boldsymbol{\alpha}; \mathbf{y}|\boldsymbol{\beta})$ is simply defined using the assumed distribution of $\mathbf{y}|\boldsymbol{\beta}$ (e.g. binomial, Poisson). This can be expressed using the same form obtained in Section 3.1.6 for the GLMs:

$$L(\boldsymbol{\alpha}; \mathbf{y}|\boldsymbol{\beta}) = \exp\{\mathbf{y}'\mathbf{A}^{-1}\boldsymbol{\theta} - b(\boldsymbol{\theta})^{1/2'}\mathbf{A}^{-1}b(\boldsymbol{\theta})^{1/2} + K\}$$

$$\propto \exp\{\mathbf{y}'\mathbf{A}^{-1}\boldsymbol{\theta} - b(\boldsymbol{\theta})^{1/2'}\mathbf{A}^{-1}b(\boldsymbol{\theta})^{1/2}\},$$

where

$$\theta = \mathbf{X}\alpha + \mathbf{Z}\beta,$$

$\mathbf{A} = \mathrm{diag}\{a_i\}$, where a_i are constant terms (see Section 3.1.4),

$b(\theta) = (b(\theta_1), b(\theta_2), \ldots, b(\theta_n))'$, where b is the function used in the general distributional form (see Section 3.1.3),

$K = $ constant.

The overall likelihood for α and $\gamma_\mathbf{G}$ can then be expressed as

$$L(\alpha, \gamma_\mathbf{G}; \mathbf{y}) \propto \exp\{\mathbf{y}'\mathbf{A}^{-1}\theta - b(\theta)^{1/2'}\mathbf{A}^{-1}b(\theta)^{1/2}\} \, |\mathbf{G}|^{-1/2} \exp(-1/2\beta'\mathbf{G}^{-1}\beta),$$

and the log likelihood as

$$\log\{L(\alpha, \gamma_\mathbf{G}; \mathbf{y})\} = \mathbf{y}'\mathbf{A}^{-1}\theta - b(\theta)^{1/2'}\mathbf{A}^{-1}b(\theta)^{1/2} - 1/2 \log |\mathbf{G}|$$
$$- 1/2\beta'\mathbf{G}^{-1}\beta + K \tag{C}$$

The quasi-likelihood function for covariance pattern models

In these models the observations are correlated and the model is parameterised by the fixed effects, α, and the variance parameters used in the $\mathbf{R}$ matrix, $\gamma_\mathbf{R}$. However, since a general multivariate distributional form is not available for non-normal data we cannot define a true likelihood function. This difficulty is overcome by instead specifying a quasi-likelihood function, $QL(\alpha, \gamma_\mathbf{R}; \mathbf{y})$, which has similar properties to a true likelihood. It is defined so that the differential of its log with respect to α has the same form as the differential of a true log likelihood with respect to α. The differential of the true log likelihood for random effects models (C) with respect to α can be shown to be

$$\delta \log\{L(\alpha, \gamma_\mathbf{G}; \mathbf{y})\}/\delta\alpha = \mathbf{X}'\mathbf{A}^{-1}(\mathbf{y} - \mu)$$

(it is obtained in a similar way to d $\log(L)/\mathrm{d}\alpha$ in Section 3.1.6).

Now since $\mathbf{R} = \mathbf{AB}$, this can be equivalently written by substituting $\mathbf{A}^{-1} = \mathbf{BR}^{-1}$ as

$$\delta \log\{L(\alpha; \mathbf{y}|\beta)\}/\delta\alpha = \mathbf{X}'\mathbf{BR}^{-1}(\mathbf{y} - \mu),$$

where $\mathbf{R}$ is the residual covariance matrix. This form can then be used to define the differentiated log quasi-likelihood function for covariance pattern models, so we may write

$$\delta \log\{QL(\alpha, \gamma_\mathbf{R}; \mathbf{y})\}/\delta\alpha = \mathbf{X}'\mathbf{BR}^{-1}(\mathbf{y} - \mu).$$

We note that some authors define the quasi-likelihood function as the log of the quasi-likelihood specified here. However, it would seem to make more sense to define it as we have so that it corresponds to the likelihood function.

The general quasi-likelihood function for all GLMMs

It is helpful to define a quasi-likelihood form that is appropriate for all types of GLMMs which may contain random effects, coefficients and covariance patterns. This is obtained by replacing $L(\boldsymbol{\alpha}; \mathbf{y}|\boldsymbol{\beta})$ in (B) by $QL(\boldsymbol{\alpha}, \boldsymbol{\gamma_R}; \mathbf{y}|\boldsymbol{\beta})$. Doing this we obtain

$$QL(\boldsymbol{\alpha}, \boldsymbol{\gamma}; \mathbf{y}) = QL(\boldsymbol{\alpha}, \boldsymbol{\gamma_R}; \mathbf{y}|\boldsymbol{\beta})L(\boldsymbol{\gamma_G}; \boldsymbol{\beta}),$$

and,

$$\log\{QL(\boldsymbol{\alpha}, \boldsymbol{\gamma}; \mathbf{y})\} = \log\{QL(\boldsymbol{\alpha}, \boldsymbol{\gamma_R}; \mathbf{y}|\boldsymbol{\beta})\} - 1/2\log|\mathbf{G}| - 1/2\boldsymbol{\beta}'\mathbf{G}^{-1}\boldsymbol{\beta} + K, \quad \text{(D)}$$

where

$$\boldsymbol{\gamma} = (\boldsymbol{\gamma_G}, \boldsymbol{\gamma_R}).$$

This function will correspond to a true likelihood function whenever the residuals are uncorrelated (i.e. no $\boldsymbol{\gamma_R}$ parameters are included). From now on the term 'quasi-likelihood' will be used to infer either a true likelihood (for models with uncorrelated residuals) or a quasi-likelihood (for models with correlated residuals).

3.2.3   Fitting the GLMM

The quasi-likelihood is less straightforward to maximise than the likelihood function defined for normal mixed models. This is mainly because it is not linear in $\boldsymbol{\alpha}$ and a solution for $\boldsymbol{\alpha}$ cannot be expressed in terms of the variance parameters. Many methods have been proposed for maximising the quasi-likelihood and it is likely that further improvements to methodology will become available over the next few years as more experience is gained. Here, we will describe two methods known as pseudo-likelihood and generalised estimating equations (GEEs) that are used within SAS procedures and macros. We also briefly mention a method known as marginal quasi-likelihood and show how the Bayesian approach can be used. For those who wish for a greater understanding of the area, Breslow and Clayton (1993) give an in-depth coverage of approaches to maximising the quasi-likelihood.

Pseudo-likelihood

This method was proposed by Wolfinger and O'Connell (1993) and is used within a macro provided by SAS called GLIMMIX (see Section 9.3). Pseudo-likelihood maximises the quasi-likelihood by iteratively analysing a linearised pseudo variable (i.e. a transformation of $\mathbf{y}$ onto the linear scale) using weighted normal mixed models. The method is referred to as 'pseudo-likelihood' because the likelihood function maximised at each iteration is that of the pseudo variable and not that of the original data. The pseudo variable introduced in Section 3.1.6

based on a first-order Taylor series expansion is again used:

$$\mathbf{z} = g(\boldsymbol{\mu}) + (\mathbf{y} - \boldsymbol{\mu})\mathbf{B}^{-1}$$
$$= \mathbf{X}\boldsymbol{\alpha} + \mathbf{Z}\boldsymbol{\beta} + (\mathbf{y} - \boldsymbol{\mu}))\mathbf{B}^{-1}.$$

$\mathbf{z}$ has variance

$$\mathbf{V_z} = \mathrm{var}(\mathbf{X}\boldsymbol{\alpha} + \mathbf{Z}\boldsymbol{\beta}) + \mathbf{B}^{-1}\,\mathrm{var}(\mathbf{y} - \boldsymbol{\mu})\mathbf{B}^{-1}$$
$$= \mathbf{Z}\mathbf{G}\mathbf{Z}' + \mathbf{B}^{-1}\mathbf{R}\mathbf{B}^{-1}.$$

By rewriting the residual matrix $\mathbf{R}$ as a product of a correlation matrix on the linear scale, $\mathbf{P}$, and $\mathbf{AB}^{-1}$, $\mathbf{V_z}$ can be re-expressed as

$$\mathbf{V_z} = \mathbf{Z}\mathbf{G}\mathbf{Z}' + \mathbf{A}^{1/2}\mathbf{B}^{-1/2}\mathbf{P}\mathbf{B}^{-1/2}\mathbf{A}^{1/2}.$$

In random effects and random coefficients models the residuals are uncorrelated and $\mathbf{P} = \mathbf{I}$, so $\mathbf{V_z}$ then simplifies to

$$\mathbf{V_z} = \mathbf{Z}\mathbf{G}\mathbf{Z}' + \mathbf{AB}^{-1}.$$

$\mathbf{z}|\boldsymbol{\beta}$ has the following multivariate normal distribution:

$$\mathbf{z}|\boldsymbol{\beta} \sim \mathrm{N}(\mathbf{X}\boldsymbol{\alpha} + \mathbf{Z}\boldsymbol{\beta}, \mathbf{A}^{1/2}\mathbf{B}^{-1/2}\mathbf{P}\mathbf{B}^{-1/2}\mathbf{A}^{1/2})$$

(conditioning the $\mathbf{z}$ on $\boldsymbol{\beta}$ allows $\mathbf{ZGZ'}$ to be omitted from the variance matrix formula).

$\mathbf{z}$ can now be analysed as a weighted normal mixed model with residual matrix $\mathbf{P}$ (defined on the linear scale), and diagonal weight matrix $\mathbf{A}^{-1}\mathbf{B}$ (inverse product of pre- and post-multipliers of $\mathbf{P}$, $\mathbf{A}^{1/2}\mathbf{B}^{-1/2}\mathbf{B}^{-1/2}\mathbf{A}^{1/2}$). Because $\mathbf{z}$ and $\mathbf{B}$ are dependent on the estimates of $\boldsymbol{\alpha}$ and $\boldsymbol{\beta}$, the normal mixed model needs to be fitted iteratively. Any of the methods described in Section 2.2 can be used to do this. However, again, variance parameters will be biased downwards to some extent if ML or IGLS are used, but this problem is not expected with REML and RIGLS. We will now define the iterative procedure more explicitly.

The iterative procedure   The raw data, $\mathbf{y}$, can be taken as initial values for $\mathbf{z}$ and the identity matrix, $\mathbf{I}$, can be used for the initial weight matrix. Alternatively, $g(\mathbf{y})$ can be taken as initial values for $\mathbf{z}$, with an adjustment if necessary to prevent infinite values. A weighted normal mixed model is then fitted using a method such as REML. This completes the first iteration and provides initial estimates for $\hat{\boldsymbol{\alpha}}$ and $\hat{\boldsymbol{\beta}}$. New values of $\mathbf{z}$ and $\mathbf{B}$ are calculated using $\hat{\boldsymbol{\alpha}}$ and $\hat{\boldsymbol{\beta}}$, and from these a second weight matrix, $\mathbf{A}^{-1}\mathbf{B}$. A second weighted mixed model is then fitted and the process is continued until the parameter estimates converge (i.e. until parameter values change very little between successive iterations). This method is computationally costly because normal mixed models, themselves requiring iterations, are fitted iteratively.

Fixed and random effects variance   Once the model has converged, the fixed and random effects and their variances can be calculated using the formula specified for normal mixed models in Sections 2.2.3 and 2.2.4:

$$\hat{\alpha} = (\mathbf{X}'\mathbf{V}_z^{-1}\mathbf{X})^{-1}\mathbf{X}'\mathbf{V}_z^{-1}\mathbf{y},$$

$$\text{var}(\hat{\alpha}) = (\mathbf{X}'\mathbf{V}_z^{-1}\mathbf{X})^{-1},$$

and, for random effects

$$\hat{\beta} = \mathbf{G}\mathbf{Z}'\mathbf{V}_z^{-1}(\mathbf{y} - \mathbf{X}\alpha),$$

$$\text{var}(\hat{\beta}) = \mathbf{G}\mathbf{Z}'\mathbf{V}_z^{-1}\mathbf{Z}\mathbf{G} - \mathbf{G}\mathbf{Z}'\mathbf{V}_z^{-1}\mathbf{X}(\mathbf{X}'\mathbf{V}_z^{-1}\mathbf{X})^{-1}\mathbf{X}'\mathbf{V}_z^{-1}\mathbf{Z}\mathbf{G}.$$

Just as in normal mixed models, because $\mathbf{V}_z$ is estimated and not known there can be a small amount of downward bias in $\text{var}(\hat{\alpha})$ and $\text{var}(\beta)$ (see Sections 2.4.3 and 3.3.4).

Generalised estimating equations (GEEs)

This method was first suggested by Liang and Zeger (1986) and was initially developed for analysing covariance pattern models. However, it can also be extended to fit random effects and random coefficients models. We will initially describe it in the context of fitting covariance pattern models and then show how random effects models can also be included. This method is based on alternating between solutions for the fixed effects and for the variance parameters. The iterative procedure can be defined as follows:

1. A solution for the fixed effects is obtained while holding the variance parameters constant.
2. A solution for the variance parameters is obtained while holding the fixed effects constant.
3. The variance parameters are fixed at the values obtained in (2) and a second solution for the fixed effects is obtained.
4. Steps 2 and 3 are repeated until the parameter estimates converge.

We now describe how solutions for the fixed effects, variance parameters and random effects are obtained at each step.

The fixed effects solution (steps (1) and (3))   The solution for the fixed effects can be obtained by differentiating the log quasi-likelihood with respect to α and setting the resulting expression to zero. Differentiating the log quasi-likelihood given by (D) above and setting to zero gives

$$\mathbf{X}'\mathbf{B}\mathbf{R}^{-1}(\mathbf{y} - \mu) = 0.$$

The above expression gives rise to n equations which are often referred to as *score equations* or *generalised estimating equations (GEEs)*. A solution for α cannot be obtained by rearranging this expression, as was the case in normal mixed models, because the equations are non-linear in $\boldsymbol{\alpha}$ (recall that $\boldsymbol{\mu} = g^{-1}(\mathbf{X}\boldsymbol{\alpha})$) and for this reason the solution needs to be found iteratively. There are various methods available for obtaining the solution. Often a linearised pseudo variable of the form described for the pseudo-likelihood above is used. However, we will not go into more detail on these methods in this book.

The variance parameters solution (step (2))   Having obtained estimates of the fixed effects, the variance parameters are estimated in a similar way to that used for IGLS (see Section 2.2). The matrix of products of the full residuals $(\mathbf{y} - \mathbf{X}\boldsymbol{\alpha})$ is set equal to the variance matrix, $\mathbf{V}$, specified in terms of the variance parameters. This gives

$$\mathbf{V} = (\mathbf{y} - \boldsymbol{\mu})(\mathbf{y} - \boldsymbol{\mu})',$$

which leads to a set of $n \times n$ simultaneous equations (one for each element in the $n \times n$ matrices) that can be solved for the variance parameters ($n =$ number of observations). The equations will have a similar form to those used in Section 2.2.4.

The equations do not take account of the fact that $\boldsymbol{\alpha}$ will be estimated and not known. However, as in normal mixed models it is possible to adapt them to provide the unbiased REML estimators (see Section 2.2.1).

As described above, steps 2 and 3 are used iteratively until convergence is obtained.

Random effects and random coefficients models   GEEs can also be used to fit random effects and random coefficients models by obtaining a solution for the random effects, $\boldsymbol{\beta}$, at each iterative step and calculating $\boldsymbol{\mu}$ as $g^{-1}(\mathbf{X}\boldsymbol{\alpha} + \mathbf{Z}\boldsymbol{\beta})$. The solution for $\boldsymbol{\beta}$ can be obtained by differentiating the log quasi-likelihood with respect to $\boldsymbol{\beta}$ and setting the resulting expression to zero. Differentiating the log quasi-likelihood given by (D) above and setting to zero gives

$$\mathbf{Z}'\mathbf{BR}^{-1}(\mathbf{y} - \boldsymbol{\mu}) - \mathbf{G}^{-1}\boldsymbol{\beta} = \mathbf{0}.$$

As with the fixed effects, a general expression for the solution for $\boldsymbol{\beta}$ cannot be obtained because the equations are non-linear in $\boldsymbol{\beta}$ and the equations need to be solved iteratively.

Marginal quasi-likelihood (MQL)

This method simplifies the computations in GEEs for random effects models but is identical to GEEs for covariance pattern models. The method avoids estimating the random effects at each iteration and hence saves on the amount of computation required. The random effects are omitted when calculating values of $\boldsymbol{\mu}$, and $\boldsymbol{\mu}$ is taken to be $g^{-1}(\mathbf{X}\boldsymbol{\alpha})$ rather than $g^{-1}(\mathbf{X}\boldsymbol{\alpha} + \mathbf{Z}\boldsymbol{\beta})$ at each iteration. However, the variance matrix, $\mathbf{V}$, is still parameterised in terms of the variance parameters, allowing these to be estimated. The results obtained from MQL can sometimes be satisfactory but it has been found that estimates of variance parameters are seriously biased in some circumstances. The problem has been addressed by Rodriguez and Goldman (1995) and Goldstein (1996). Although we do not advocate MQL for fitting random effects models, we mention it here because it has been frequently applied in practice.

In the literature MQL is sometimes referred to as a 'population averaged' method, whereas methods using the full $\mu = g^{-1}(\mathbf{X\alpha} + \mathbf{Z\beta})$ are referred to as 'population specific'.

Bayesian methods

Bayesian methods can be used to fit GLMMs in much the same way as normal mixed models. Again, they have been developed most fully for fitting random effects and random coefficients models where the true likelihood function is available. The posterior distribution of the parameters can be determined in the same way as defined in Section 2.3.2 except that a non-normal distribution is assumed for $\mathbf{y}$. The resulting posterior density surface can then be used to provide estimates, standard deviations and probability intervals for the model parameters. As in normal mixed models, some of the bias problems are overcome when a Bayesian approach is used, although we will see later that the potential problems associated with random effects shrinkage are not avoided.

There is not yet a SAS procedure that can be used to fit GLMMs using a Bayesian method. Instead, a package developed at the Medical Research Council (Cambridge) known as BUGS will be used to analyse the examples in this book. This package uses the Gibbs sampler (see Section 2.3.5) to sample the joint posterior distribution of the model parameters. It is currently available free of charge and information on how to obtain it can be found on Web page: www.mrc-bsu.cam.ac.uk/bugs.

3.2.4   Some flaws with GLMMs

In this section we will describe two technical flaws that can occur in GLMMs.

Uniform effect categories

We define a fixed or random effect category as 'uniform' if an infinite value is obtained when the link function is applied to the mean of observations in the category. It occurs when all observations within the category are zero in binary and Poisson data, or when all observations are n_i/n_i in binary data. For example, centre effects are fitted in some of the analysis models we will use in Section 3.4 to analyse an adverse event 'cold feet' in a multi-centre trial. In some centres no subjects were recorded as having cold feet and this caused these centre effects to be uniform.

For *uniform fixed effects* a corresponding effect estimate on the linear scale cannot then be estimated and will to tend towards plus or minus infinity. For example, if in a simple between patient trial the frequency of success for one of the treatments is 100%, then a model using a logit link function would attempt to estimate the treatment mean as $\log(1.00/0.00)$.

However, sensible estimates of *uniform random effects* (and their corresponding variance components) can still be obtained provided not all categories within a particular random effect are uniform (although in some circumstances the estimates of the variance components may be biased). This is possible because the random effect estimates are shrunken and thus information from all observations is used in forming the estimates. However, when all categories of a random effect are uniform the model will not converge. In other situations, whether or not convergence is achieved is less predictable and depends on the number of uniform categories, the number of fixed effects fitted and on how the fixed effects relate to random effects.

Uniform categories are most likely when the probability of success is very small or large in binary data, or when the event rate is very small in Poisson data. They are also likely when there are small numbers of observations within some of the fixed effect categories. In Sections 3.3.6 and 3.3.7 we will discuss how uniform effects should be handled.

Random effects shrinkage

The random effects GLMM assumes that $\mathbf{y}|\boldsymbol{\beta}$ has a distribution where the variance (in one-parameter distributions) is defined as a function of the expected means (e.g. $\boldsymbol{\mu}(1 - \boldsymbol{\mu})$ and $\boldsymbol{\mu}$ for binomial and Poisson data, respectively). However, this relationship between the mean and variance does not hold exactly because random effects estimates are shrunken compared with their raw means.

For binary data this causes the predicted residual variance, **AB**, to be greater than that observed (i.e. $\text{var}(\mathbf{y} - g^{-1}(\mathbf{X}\boldsymbol{\alpha} + \mathbf{Z}\boldsymbol{\beta})))$ whenever shrinkage occurs towards 0.5, and less than that observed otherwise. This can be seen for binomial data by considering observations within a particular random effect category. If their raw mean is μ and their shrunken mean is μ_s, their observed variance can be shown to be $\text{var}(y) = \Sigma_i(y_i - \mu_s)^2/n = \mu_s^2 - 2\mu\mu_s + \mu$. This is greater than the predicted variance, $\mu_s(1 - \mu_s)$, whenever μ_s is in the range $(\mu, 0.5)$ for $\mu < 0.5$, or $(0.5, \mu)$ for $\mu > 0.5$. For uniform categories shrinkage is always towards 0.5 and the predicted variance is therefore always greater than that observed. In datasets with several uniform random effect categories this discrepancy can cause appreciable bias in the variance component estimates and their values will be affected by whether a dispersion parameter is fitted. Variance parameter bias is likely to be small when no random effect categories are uniform because shrinkage will be smaller and it can occur in either direction (towards or away from 0.5).

For Poisson data it can be shown that $\text{var}(y) = \mu_s^2 - 2\mu\mu_s + \mu^2 + \mu$, where μ is the raw mean. This is greater than the predicted variance, μ_s, except when μ_s

is in the range $(\mu, \mu + 1)$. For uniform categories (where $\mu = 0$) the shrunken estimate may or may not be in the range (0,1) so the predicted variance can be either less than or greater than that predicted. For this reason the problem of biased variance components is less likely in analyses of Poisson data.

In Section 3.3.2 we will discuss under what circumstances random effects shrinkage is likely to cause bias problems and suggest when a random effects model is not advisable.

3.2.5   Reparameterising random effects models as covariance pattern models

One approach to overcoming the potential problems caused by random effects shrinkage is to reparameterise the random effects model as a covariance pattern model. For example, a cross-over trial analysis fitting patient effects as random could be reparameterised as a covariance pattern model with a constant correlation between observations on the same patient (see Section 2.4.1 for an illustration of such a reparameterisation). This reparameterised model still has the benefits of the random effects model (e.g. greater efficiency when there are missing data) but does not suffer from the fitting problems experienced when parameterised as a random effects model. A general form for a random effects model reparameterised as a covariance pattern model can be written,

$$\mathbf{y} = \boldsymbol{\mu} + \mathbf{e},$$

$$g(\boldsymbol{\mu}) = \mathbf{X}\boldsymbol{\alpha}.$$

The random effects are now not included in the linear part of the model but are assumed to be incorporated into the error term and lead to a variance matrix of the form:

$$\mathbf{V} = \mathbf{A}^{1/2}\mathbf{B}^{1/2}(\mathbf{ZGZ}' + \mathbf{P})\mathbf{B}^{1/2}\mathbf{A}^{1/2},$$

where $\mathbf{Z}$ defines the random effects levels and $\mathbf{G}$ is a matrix of variance parameters corresponding to the random effects. We have not had the opportunity to explore fully such models, but believe they may have the potential to avoid the problems with bias that can occur with random effects GLMMs.

3.3   PRACTICAL APPLICATION AND INTERPRETATION

In this section some points relating to the practical application of GLMMs and their interpretation will be covered. Experience with these models is still limited and therefore some of the issues are far from resolved. As well as the coverage in this chapter, additional points relating to covariance pattern and random coefficients models will be made in Sections 6.2 and 6.5. The worked example in Section 3.4 will illustrate many of the points made.

3.3.1   Specifying binary data

Binary data can be specified either as a series of zeros and ones (*Bernoulli form*) or as frequencies of 'success' out of a number of 'tries' (*binomial form*). If data are recorded in Bernoulli form (as is often the case in clinical trials), then it is usually most convenient to analyse them as such. This also has the advantage that other measurements made at the observation level (e.g. baseline effects) can be included in the model as covariates. If there are no baseline effects, then data can alternatively be aggregated to give frequencies (e.g. at the centre-treatment level) and analysed in binomial form. When the binomial form is used we would suggest that the dispersion parameter is fixed at one and a random effect is fitted at the observation level (e.g. an effect for each centre-treatment frequency). This approach might be appropriate for an analysis of centre-treatment frequencies of an adverse event from a multi-centre trial. Fitting centre-treatment effects as random with the dispersion parameter fixed at one will allow variation at the observation level to be modelled by the centre-treatment variance component. The meta analysis example given in Section 5.7 shows how binomial trial-treatment frequencies can be analysed in this way (Model 3a).

Analysing binomial frequencies is less intensive computationally than using Bernoulli data. However, results will differ to some extent between the two analyses because variance at the residual level is modelled separately from that at the random effects level in a Bernoulli model. The difference will be more noticeable in datasets where there are uniform random effect categories (see Section 3.3.7). In this situation we would prefer an analysis of the data in Bernoulli form because the dispersion parameter can then help overcome any biases caused by random effects shrinkage (see Section 3.3.8).

3.3.2   Difficulties with fitting random effects (and random coefficients) models

In practice, there can be difficulties with fitting random effects and random coefficients models and in some circumstances these models are not recommended at all; the variance parameters and fixed effects standard errors can be badly biased or convergence may not occur. These problems are most likely to occur in datasets where there are only a few observations within each random effect category (e.g. a two-period, cross-over trial has only two observations per patient), or when the probability of an outcome is low. The problems are largely caused by the effect of random effects shrinkage, particularly when there are uniform random categories (see Section 3.2.4). Also, the approximations used by the model fitting methods could contribute to some extent to the biases. For example, the Taylor series expansion used for the linearised pseudo variable is only to the first order, and more accurate approximations can be obtained by using higher order expansions.

In designs such as the cross-over trial where the number of observations per patient is small we would not recommend fitting a random effects model. Instead, we would suggest that the model is reparameterised as a covariance pattern model with a constant (compound symmetry) correlation between observations on the same patient (see Section 2.4.1 for an illustration of such a reparameterisation). This reparameterised model still has the benefits of the random effects model (e.g. greater efficiency when there are missing data) but does not suffer from the fitting problems experienced when it is parameterised as a random effects model.

In situations where there are a reasonable number of observations per random effects category, biases are much less likely. We will meet such an example in the meta-analysis data we will consider in Section 5.7, where there are over 75 patients per trial. Although the outcome probability is quite low (0.09) in this example, this is outweighed by the relatively large numbers of patients per trial and any bias in the variance parameter estimates is likely to be small. However, it can be difficult to decide when there are a 'reasonable number' of observations. For example, in the multi-centre trial we will consider in Section 3.4 the number of patients per centre varies between one and 39 and it is unclear whether significant bias problems are likely to occur. In general, respecifying the model as a covariance pattern model (see Section 3.2.4) would seem a good strategy whenever there is any doubt over the adequacy of the random effects model. However, this becomes less straightforward when there is more than one random effect (e.g. a model fitting centre and centre·treatment effects as random) because the required covariance patterns are not readily available in packages such as SAS. The best approach then is often to fit the random effects model and to examine the value of the dispersion parameter. If it is noticeably less than one it is possible that random effects shrinkage has caused bias problems. However, the dispersion parameter itself does help to some extent to overcome discrepancies between the observed and predicted residual variation caused by random effects shrinkage and/or approximation inaccuracy, although it is not completely clear to us in what circumstances the results will be reliable. However, a dispersion parameter of less than about 0.5 would seem too small for comfort. An alternative approach to avoid the problem might be to analyse the data as binomial frequencies, provided no covariates are modelled at the residual level.

3.3.3   Accuracy of variance parameters

It is important that variance parameters are estimated with a reasonable accuracy because of their effect on the calculation of fixed effects and their standard errors. We have already discussed how random effects shrinkage can cause biases in variance parameters, particularly when there are uniform random categories. Additionally, as in normal mixed models, the accuracy of the variance parameters is dependent on the number of DF used to estimate them. Although there are no hard and fast rules, it would seem inadvisable to fit an effect as random if less than about five DF were available for estimation (e.g. a multi-centre trial with five or less centres). When an insufficient number of DF are available to estimate a variance parameter accurately, the suggestions given for normal data in Section 2.4.2 can be considered.

3.3.4 Bias in fixed and random effects standard errors

Standard errors of fixed and random effects are calculated from variance parameters and therefore any bias in these parameters will also cause bias in the fixed effects standard errors. Additionally, bias can occur for the same reasons given for normal mixed models whenever an effect is estimated using information from several error strata (see Section 2.4.3).

The 'empirical' variance estimator suggested as a more robust estimator of the fixed effects variance for covariance pattern models (see Section 2.4.3) can also be used for GLMMs. This estimator takes into account the observed covariance in the data and may help alleviate some of the small sample bias. It does, however, cause the fixed effects variances to reflect observed covariances in the data rather than those specified by the covariance pattern modelled, so it may not always be the best choice. We discuss this further in Section 6.2. If the analysis is based on a linearised approximation (e.g. pseudo-likelihood, see Section 3.2.3), the empirical variance estimator can be based directly on the linearised data and their variance matrix at the last iteration. For example, in pseudo-likelihood the approximation $\mathbf{z} = g(\boldsymbol{\mu}) + (\mathbf{y} - \boldsymbol{\mu})\mathbf{B}^{-1}$ is used and the empirical estimator of var($\hat{\boldsymbol{\alpha}}$) can be calculated as

$$\text{var}(\hat{\boldsymbol{\alpha}}) = (\mathbf{X}'\mathbf{V}_{\mathbf{z}}^{-1}\mathbf{X})^{-1}(\mathbf{X}'\mathbf{V}_{\mathbf{z}}^{-1}\,\text{cov}(\mathbf{z})\mathbf{V}_{\mathbf{z}}^{-1}\mathbf{X})(\mathbf{X}'\mathbf{V}_{\mathbf{z}}^{-1}\mathbf{X})^{-1}.$$

Alternatively, it can be based on the raw data and its variance matrix by substituting a 'linked' design matrix, **BX**, for the **X** matrix in the formula given in Section 2.4.3 (i.e. the usual design matrix **X** is pre-multiplied by the diagonal matrix of expected observation variances, **B**, see Section 3.1.4) to give

$$\text{var}(\hat{\boldsymbol{\alpha}}) = (\mathbf{X}'\mathbf{BV}^{-1}\mathbf{BX})^{-1}\mathbf{X}'\mathbf{BV}^{-1}\,\text{cov}(\mathbf{y})\mathbf{V}^{-1}\mathbf{BX}(\mathbf{X}'\mathbf{BV}^{-1}\mathbf{BX})^{-1}.$$

The empirical estimator is calculated (using the second formula) by default when the REPEATED statement in PROC GENMOD in SAS is used (see Section 9.2).

3.3.5 Negative variance components

Negative estimates of variance components can occur in random effects and random coefficients GLMMs just as in the normal mixed model and the points made in Section 2.4.1 apply. An additional influence in the GLMM making a negative variance component estimate more likely is that bias is also possible due to the effects of random effects shrinkage (see Section 3.3.2).

3.3.6   Uniform fixed effect categories

Uniform fixed effect categories (see Section 3.2.4) are easily identified in the results of model fitting by estimates and standard errors that are extremely large. When this occurs, the results given for other effects are in fact equivalent to a reanalysis of the data with observations corresponding to the uniform category removed and thus they can still be used. If it is important to test the overall significance of a set of fixed effects containing a uniform category (e.g. treatments), this can be done by comparing the likelihoods (or quasi-likelihoods) between models that include and exclude the effects using a likelihood ratio test (see Section 2.4.4). However, we note that quasi-likelihood statistics are not always output by statistical software (e.g. SAS PROC GENMOD and the GLIMMIX macro) and it may not therefore always be possible to perform this test.

Alternatively, when a fixed effects model is being fitted to binary data, an exact logistic regression where all possible combinations of data values are considered can be used. This can be carried out using the package LOGEXACT.

3.3.7   Uniform random effect categories

A low dispersion parameter or lack of convergence is often an indicator of uniform random effects categories. In this situation biases in variance components may occur and the suggestions made in Section 3.3.2 should be followed.

3.3.8   The dispersion parameter

The dispersion parameter allows the residual variances to increase or decrease from their predicted values; the residual variance matrix is taken to be $\mathbf{R} = \phi\mathbf{AB}$ in GLMs and $\mathbf{R} = \phi\mathbf{A}^{1/2}\mathbf{B}^{1/2}\mathbf{PA}^{1/2}\mathbf{B}^{1/2}$ in GLMMs. It has different roles in the GLM and the GLMM.

The GLM

The dispersion parameter in the GLM can largely be interpreted as *over-* or *under-dispersion*. When binomial or Poisson frequencies are modelled the dispersion parameter has a role similar to a variance component at the residual level. Observed residual variation may be greater or less than that predicted by $\mathbf{R} = \mathbf{AB}$ and the dispersion parameter will take account of this to some extent. For example, if trial·treatment frequencies from a meta-analysis were in binomial form and trial and treatment effects were fitted, the dispersion parameter would reflect the trial·treatment variation. A value greater than one indicates more variation than expected by chance. This is equivalent to obtaining a positive trial·treatment variance component and can be referred to as 'over-dispersion'. Conversely, when the dispersion parameter is less than one, it is equivalent to obtaining a negative trial-treatment variance component estimate which can be referred to as 'under-dispersion'. In this situation a decision needs to be made as to whether or not this negative variance is genuine or whether it is most likely to be an underestimate of a zero or small positive variance component (see Section 2.4.1). If the latter appears most likely, then the dispersion parameter should be omitted from the model. This is equivalent to fixing a variance component at zero.

When data are in Bernoulli form the observed variation is in most situations almost exactly equal to the predicted variances given by **AB** and the dispersion parameter will be close to one. However, if there are uniform effect categories it will be less than one. If it is noticeably less than one it is usually preferable to remove observations corresponding to the uniform effects from the model.

The GLMM

In the GLMM the dispersion parameter can be influenced by over- or under-dispersion of the data and by the effects of random effects shrinkage (see Section 3.2.3). Random effects shrinkage can cause the predicted residual variance to be greater than that observed (particularly when uniform random effect categories are present in binary data) and the dispersion parameter will help to overcome this discrepancy. Thus, interpretation of the dispersion parameter in the GLMM can be difficult, since it is not always clear which factors have affected it.

In Bernoulli datasets a dispersion parameter differing from one is most likely to be due to the effects of random effects shrinkage. If it is considerably less than one the use of a GLMM may be inappropriate (see Section 3.3.2).

In binomial datasets (i.e. where observations are expressed as frequencies over denominators) with no uniform random effect categories the dispersion value can be interpreted largely in terms of over- or under-dispersion. However, in most situations we would recommend that binomial data is analysed with the dispersion parameter fixed at one and a random effect fitted at the observation level (see Section 3.3.1).

In binomial datasets where there are uniform random effect categories it is more difficult to discriminate between the influences of random effects shrinkage and of over- or under-dispersion. (For this reason we would usually recommend that such datasets are analysed in Bernoulli form.) A dispersion parameter of less than one is most likely to be due to random effects shrinkage. A dispersion parameter greater than one is likely to indicate over-dispersion, since random effects shrinkage is expected to cause the dispersion parameter to decrease.

In Poisson datasets the dispersion value can mainly be interpreted largely in terms of over- or under-dispersion. This is because uniform random categories do not usually affect its value greatly (see Section 3.2.3).

Should a dispersion parameter be fitted?   We believe that it is always helpful to include a dispersion parameter unless a variance component is being modelled at the residual level. Its role will be either to model genuine over- or under-dispersion in binomial or Poisson data, or to help overcome discrepancies between the observed and predicted residual variation caused by random effects shrinkage. It is also a useful diagnostic aid to help determine when the GLMM may be misspecified or inappropriate.

Covariance pattern models

The dispersion parameter in covariance pattern models fitting no random effects has the same role as in GLMs and the points made above for GLMs apply.

3.3.9   Significance testing

The GLMs

Fixed effects can be tested using chi-squared tests. These are based on asymptotic theory; that is, as sample sizes become larger, the distribution of fixed effects estimates conform more closely to normal distributions, the standard deviations of which are increasingly well estimated by fixed effects standard errors. If a dispersion parameter has been fitted it may be more appropriate to use instead Wald F tests to take account of the fact that a residual variance parameter is estimated. The Wald F statistic is calculated as described in Section 2.4.4. However, note that the F test still relies on the asymptotic normality of the fixed effects just as the chi-squared test does. Theory to underpin the analysis of datasets with small sample sizes is, however, remarkably sparse.

The GLMMs

Fixed and random effects   These effects can be tested using Wald F tests calculated on the linear scale as described in Section 2.4.4. These tests take into account the uncertainty in the variance parameters (including the dispersion parameter) and may therefore be preferable to Wald chi-squared tests. However, as stated for GLMs, both the chi-squared and F tests rely on the asymptotic normality of the fixed effects estimates. The Satterthwaite DF can be calculated for the F (and t) tests in just the same way as for normal mixed models. However, there is a situation where these DF would not be appropriate, although it is a situation which we have already recommended should be avoided. If neither a dispersion parameter, nor a variance component are modelled at the residual level a conservative estimate could be taken as the lowest DF of the error strata used for estimating the fixed effect. If a Bayesian method is used, then tests can be performed by calculating exact 'Bayesian' p-values from the marginal posterior distribution for each effect (see Section 2.3.3).

Variance parameters A variance parameter can be tested by comparing likelihoods (or quasi-likelihoods) between models fitting and not fitting the parameter using a likelihood ratio test as described in Section 2.4.4. The theoretical basis for this approach has only been proved for true likelihoods, although we believe that quasi-likelihood-based tests will still give good approximations. As we noted in Section 3.3.6, quasi-likelihood statistics are not always output by statistical software (e.g. SAS) and it may not therefore always be possible to perform such tests in practice. Note that Bayesian models are usually set up to sample only positive variance components and in that situation variance components cannot be tested for significance.

3.3.10 Confidence intervals

Confidence intervals for an effect are obtained using the mean and standard error estimates given on the linear (linked) scale. These confidence intervals can often be converted to a more interpretable scale, usually by exponentiation. For example, odds ratios can be obtained when the logit link function is used, and relative risks for the log link function.

The GLMs

Confidence intervals are calculated using the usual formula for normal data. Note that t statistics are not used because the GLMs do not estimate variance parameters:

$$\text{lower } 95\% \text{ confidence limit} = \text{mean effect} - 1.96 \times \text{SE},$$
$$\text{upper } 95\% \text{ confidence limit} = \text{mean effect} + 1.96 \times \text{SE}.$$

The GLMMs

Here, percentage points from the t distribution with the DF calculated as described in Section 2.4.4 can be used to take into account the fact that the variance and dispersion parameters are estimated:

$$\text{lower } 95\% \text{ confidence limit} = \text{mean effect} - t_{DF,0.975} \times \text{SE},$$
$$\text{upper } 95\% \text{ confidence limit} = \text{mean effect} + t_{DF,0.975} \times \text{SE}.$$

3.3.11 Model checking

In the GLMM it is assumed that the random effects are normally distributed and uncorrelated. Residual plots can be used to visually check normality of these effects and to identify any outlying effect categories as described in Section 2.4.6.

The residuals $\mathbf{y} - \hat{\boldsymbol{\mu}}$ do not need to be checked for Bernoulli data. However, if the data are binomial (i.e. expressed as frequencies with denominators) or Poisson, then residual plots can be used to identify outlying observations. It should be borne in mind that normal plots will not necessarily produce straight lines. The residuals, $\mathbf{y} - \hat{\boldsymbol{\mu}}$, do not have equal variances and therefore they should first be linearised by dividing by their predicted standard errors, $\mathbf{A}^{1/2}\mathbf{B}^{1/2}$ (e.g. $(\mu_i(1 - \mu_i))^{1/2}$ for binary data) to give what are known as the 'Pearson' residuals. Note that when data are analysed using a linearised pseudo variable (as in pseudo-likelihood) the 'pseudo' residuals will already be on a linear scale and can be checked directly.

Residuals are correlated in covariance pattern models and random coefficients models and model checking needs to reflect this. Model checking methods for these models are considered in Sections 6.2 and 6.5.

3.4   EXAMPLE

3.4.1   Introduction and models fitted

The multi-centre trial of treatments for lowering blood pressure introduced in Section 1.3 is used again here. An adverse event, 'cold feet', is analysed as a binary variable and observations at the final or last attended visit are used. Cold feet was, in fact, recorded on a scale of $1 - 5$: $1 =$ none, $2 =$ occasionally, $3 =$ on most days, $4 =$ most of the time, $5 =$ all of the time. A binary 'cold feet' variable was created by taking categories 1 and 2 as negative and categories $3 - 5$ as positive. The frequencies of cold feet by treatment and centre are shown in Table 3.1. In this trial 'cold feet' was recorded at baseline so, in order to include a baseline covariate in the model (and so reduce between-patient variation), we will analyse the data in Bernoulli form.

Table 3.1 indicates that there are several zero frequencies of cold feet and these will lead to uniform centre and centre·treatment categories. This in turn may cause variance component bias (see Section 3.3.2) and it is not clear whether a random effects model will be satisfactory. Here, we will fit a variety of models (see Table 3.2) and discuss their strengths and weaknesses. In practice, only Model 1 is likely to be considered as a fixed effects model since the large number of uniform categories will cause problems in estimating satisfactory treatment effects in Models 2 and 3 (as discussed in Section 3.4.2). In Model 4, centre effects are fitted as random and in Model 5 both centre and centre·treatment effects are fitted as random. Model 5 takes into account the random variation in the treatment effect between centres and results can be related with more confidence to the 'population' of potential centres.

Models 6 and 7 are the same as Models 4 and 5 except that they are fitted using a Bayesian model with non-informative priors to obtain a joint (posterior) distribution of the model parameters. The Bayesian models are set up in a similar way to the normal example described in Section 2.5 except that Bernoulli distributions are now assumed for the observations. Again, normal distributions with zero

Example **135**

Table 3.1 Frequencies of cold feet by treatment and centre.

Centre	Treatment			Total
	A	B	C	
1	3/13	5/14	1/12	9/39
2	2/3	0/4	0/3	2/10
3	0/3	0/3	0/2	0/8
4	1/4	1/4	0/4	2/12
5	1/4	3/5	0/2	4/11
6	0/2	1/1	1/2	2/5
7	0/6	1/6	0/6	1/18
8	1/2	0/1	1/2	2/5
9	—	—	0/1	0/1
11	0/4	1/4	0/4	1/12
12	0/3	1/3	0/4	1/10
13	1/1	0/1	0/2	1/4
14	0/8	2/8	1/8	3/24
15	1/4	0/4	0/3	1/11
18	0/2	0/2	0/2	0/6
23	1/1	—	0/2	1/3
24	—	—	0/1	0/1
25	0/3	0/2	0/2	0/7
26	0/3	1/4	0/3	1/10
27	—	1/1	0/1	1/2
29	1/1	—	0/1	1/2
30	0/1	0/2	0/2	0/5
31	0/12	0/12	0/12	0/36
32	1/2	0/1	0/1	1/4
35	0/2	0/1	—	0/3
36	0/9	5/6	0/8	5/23
37	0/2	0/1	1/2	1/5
40	0/1	1/1	—	1/2
41	0/2	0/1	0/1	0/4
Total	13/98	23/92	5/93	41/283

Table 3.2 Models used to analyse 'cold feet' in a multicentre trial.

Model	Fixed effects	Random effects	Method
1	Baseline, treatment	—	GLM
2	Baseline, treatment, centre	—	GLM
3	Baseline, treatment, centre·treatment	—	GLM
4	Baseline, treatment	Centre	P-L[1]
5	Baseline, treatment	Centre, centre·treatment	P-L[1]
6	Baseline, treatment	Centre	Bayes
7	Baseline, treatment	Centre, centre·treatment	Bayes

[1]P-L = pseudo-likelihood.

means and very large variances (of 1000) were used as non-informative priors for the fixed effects (baseline and treatment), and inverse gamma distributions with very small parameters (of 0.0001) were used as non-informative prior distributions for the centre and centre-treatment variance components. Note that this prior specification for the variance components ensures that negative variance component samples cannot be obtained. Ten thousand samples were taken using the Gibbs sampler using the package BUGS (see Section 2.3) to repeatedly sample conditionally from the posterior distribution of the model parameters. The values sampled were then used directly to obtain parameters estimates, standard deviations, probability intervals and 'Bayesian' *p*-values in exactly the same way as described for the example given in Section 2.5.

3.4.2   Results

Estimates of the variance components and fixed effects for each model are shown in Table 3.3.

Fixed effects models (1–3)

In Model 2 the treatment effect estimates differ from those in Model 1. This is because information on treatments is lost from all centres with no instances of

Table 3.3   Estimates of variance components and fixed effects (on the logit scale).

| Model | Variance components | | | |
	Centre	Treatment·centre	Dispersion parameter	$-2\log(L)$
1	—	—	1.00^1 (1.02)	169.11
2	—	—	1.00^1 (0.79)	141.09
3	—	—	1.00^1 (0.52)	95.94
4	0.06	—	0.79	—
5	0.00	1.79	0.54	—
6	0.02^2	—	1.00^1	—
7	0.03^2	0.14^2	1.00^1	—

| Model | Treatment effects (SEs) | | | |
	Baseline	A – B	A – C	B – C
1	3.05 (0.49)	−0.81 (0.45)	0.83 (0.61)	1.64 (0.58)
2	2.87 (0.58)	−0.97 (0.51)	0.93 (0.66)	1.90 (0.63)
3	3.10 (0.80)	—	—	—
4	3.02 (0.49)	−0.81 (0.45)	0.83 (0.60)	1.63 (0.57)
5	3.10 (0.46)	−0.70 (0.57)	1.04 (0.66)	1.74 (0.64)
6	3.05 (0.51)	−0.81 (0.45)	1.00 (0.61)	1.80 (0.59)
7	3.17 (0.56)	−0.68 (0.53)	1.12 (0.69)	1.80 (0.65)

[1] Dispersion parameter is fixed at one (value in brackets is its estimate);
[2] Estimates are median values from the marginal posterior distributions.

Example   **137**

cold feet (3, 9, 18, 24, 25, 30, 31, 35 and 41) and from any centres using only one treatment (9 and 24). This causes the treatment estimates to be equivalent to an analysis with patients from these centres omitted and also explains why the treatment standard errors are higher than those in Model 1. Of course, these smaller centres could be combined for the purposes of analysis as 'other centres'. However, the results below indicate that centre effects are not important and therefore Model 1 is likely to be satisfactory. In Model 3 the treatment effects are non-estimable because all treatments are not used at every centre (at least one observation is required in each category of a containing effect). However, the baseline effect is estimable because its containment stratum is that of the residual.

Although Models 2 and 3 are not recommended for estimating treatment effects, they can still be used to test the overall significance of the fixed centre and centre·treatment effects by using likelihood ratio tests. For example, to test centre effects in Model 2 we calculate twice the difference in the log likelihoods between Models 1 and 2, $169.11 - 141.09 = 28.02 \sim \chi^2_{28}$ ($p = 0.46$). This indicates that centre effects are non-significant. To test centre·treatment effects, twice the difference in the log likelihoods between Models 2 and 3 is taken, $141.09 - 95.94 = 45.16 \sim \chi^2_{47}$ ($p = 0.55$). This is also non-significant. However, it should be borne in mind that these tests have low power for detecting small centre or centre·treatment effects.

Several uniform effect categories (see Sections 3.2.4) occur in Models 2 and 3. These categories are easily identified in the results by estimates and standard errors that are extremely large. The fixed effect estimates resulting from Model 2 are listed in Table 3.4.

The uniform centre categories can be identified as centres 3, 9, 18, 24, 25, 30, 31 and 35. These are the centres where no patients had cold feet. Centre 40 has no DF. This has occurred because centre 41 (the reference category) is also uniform. Standard errors for the other centre effects are based on comparisons with centre 40. However, their estimates in this output are still based on comparisons to centre 41! Thus, SAS does not produce useful centre estimates and standard errors when the reference category is uniform. Usually, centre estimates will not be of interest. However, if required they can be obtained by renumbering the centres so that the last one is non-uniform. Although we are clearly getting estimates and standard errors that are unstable, the likelihood still converges, since the uniform categories have little effect on it.

The dispersion parameter was fixed at 1 in Models 1–3. However, its estimated value is also given in brackets in Table 3.3. In Model 1 this is very close to one and fixing it at one has made very little difference to the fixed effects standard errors. In Models 2 and 3 the dispersion parameter is well below one. This is largely due to the influence of the uniform categories and thus the dispersion parameter estimate can be considered as another indicator of their presence.

Table 3.4   Fixed effect estimates resulting from Model 2.

Parameter		DF	Estimate	SE	Chi-square	Pr>Chi
INTERCEPT		1	−26.4696	2.2172	142.5189	0.0000
CF1		1	2.8658	0.5754	24.8066	0.0000
TREAT	1	1	0.9327	0.6595	2.0003	0.1573
TREAT	2	1	1.8976	0.6298	9.0786	0.0026
TREAT	3	0	0.0000	0.0000	.	.
CENTRE	1	1	23.4489	2.1750	116.2291	0.0000
CENTRE	2	1	23.1337	2.2932	101.7682	0.0000
CENTRE	3	1	−0.1433	110410.898	0.0000	1.0000
CENTRE	4	1	23.2694	2.3248	100.1831	0.0000
CENTRE	5	1	24.1458	2.2649	113.6513	0.0000
CENTRE	6	1	25.2717	2.3524	115.4058	0.0000
CENTRE	7	1	21.7817	2.4299	80.3548	0.0000
CENTRE	8	1	23.6757	2.4809	91.0715	0.0000
CENTRE	9	1	1.1042	322114.211	0.0000	1.0000
CENTRE	11	1	22.5774	2.3855	89.5721	0.0000
CENTRE	12	1	21.9982	2.4695	79.3543	0.0000
CENTRE	13	1	24.5172	2.4706	98.4813	0.0000
CENTRE	14	1	23.0918	2.2415	106.1272	0.0000
CENTRE	15	1	23.0020	2.3927	92.4187	0.0000
CENTRE	18	1	−0.0504	126966.182	0.0000	1.0000
CENTRE	23	1	24.1708	2.7840	75.3765	0.0000
CENTRE	24	1	1.1042	322114.211	0.0000	1.0000
CENTRE	25	1	−0.0220	118043.191	0.0000	1.0000
CENTRE	26	1	23.0020	2.3927	92.4187	0.0000
CENTRE	27	1	25.5208	2.6736	91.1194	0.0000
CENTRE	29	1	24.5703	2.9971	67.2077	0.0000
CENTRE	30	1	−0.0888	138318.641	0.0000	1.0000
CENTRE	31	1	−0.0721	52615.0639	0.0000	1.0000
CENTRE	32	1	23.2847	2.5285	84.8046	0.0000
CENTRE	35	1	−0.2314	183222.204	0.0000	1.0000
CENTRE	36	1	22.7532	2.2297	104.1306	0.0000
CENTRE	37	1	23.4874	2.4492	91.9644	0.0000
CENTRE	40	0	23.6215	0.0000	.	.
CENTRE	41	0	0.0000	0.0000	.	.

Random effects models fitted using pseudo-likelihood (4 and 5)

In Model 4 there is a small positive centre variance component indicating that some recovery of treatment information from between the centres has occurred. However, it is not possible to assess the extent of this since there is no satisfactory equivalent fixed effects model (Model 2 has uniform centre effects). The treatment effect standard errors are very similar to those obtained in Model 1, indicating that little appears to have been gained by fitting centre effects as random in this example.

Example   **139**

In Model 5 the centre-treatment variance component is positive, indicating that the treatment effect varies randomly between centres. This is reflected in the treatment effects standard errors which are increased over those for Model 4 to allow for the additional variation occurring between centres. Since treatment effects are assumed to vary randomly between the centres, results can be related with more confidence to the population of centres. The centre variance component is zero, so there is no overall variability in the incidence of cold feet between centres.

The dispersion parameters in Model 4 and 5 are both notably below one due to the shrinkage of the uniform centre effect categories. This parameter is helping to overcome the discrepancy in the mean/variance relationship caused by the random effects estimates being shrunken compared with their raw means. If it were omitted, then it is likely that a downward bias in the centre and centre-treatment variance component would have occurred. However, it is difficult to tell how adequately the dispersion parameter has overcome this potential problem. Our own view is that the results are likely to be satisfactory. We can draw some comfort from the fact that the results from Model 4 are similar to those from Model 1, which does not suffer from the problems associated with uniform effect categories.

Bayesian models (6 and 7)

The variance component estimates in these models are on the whole smaller than those obtained by pseudo-likelihood (Models 4 and 5). However, they are taken as medians of their posterior distributions and are also based on using a prior that is constrained to be positive (pseudo-likelihood is equivalent to using flat priors). For these reasons they cannot be compared directly. The treatment standard errors in Model 7 are still noticeably increased over those in Model 6 despite the fact that the centre-treatment variance component median is small. This is likely to be because the prior distribution constrains all the samples of variance component parameters to be positive. It is difficult to decide whether the pseudo-likelihood or Bayesian models are preferable for this example.

3.4.3   Discussion of points from Section 3.3

Bias in fixed and random effects standard errors (Section 3.3.4)

The treatment standard errors in Models 4 and 5 will be affected by any bias in the variance component estimates. In this example there is a possibility that some bias in variance components (and hence fixed effects standard errors) has occurred, although to some extent this may have been overcome by fitting a dispersion parameter. Additionally, some downward bias may occur because information is combined across several error strata (see Section 2.4.3). However, since there are 29 centres, any bias occurring for this reason is likely to be small.

Negative variance components (Section 3.3.5)

A negative variance component estimate would have been obtained for centre effects in Model 5 had it not been constrained at zero. It would almost certainly have been an underestimate of a zero or small positive variance component. Centre effects have been retained in the analysis here because they form part of the study design and a centre-treatment interaction has been fitted. Identical fixed effects estimates and standard errors would have been obtained if the data had been reanalysed with centre effects omitted; however, the denominator DF used for F tests will be smaller when centre effects are retained. The problem does not arise in Models 6 and 7, since negative variance component samples cannot be obtained when an inverse gamma distribution is assumed for their prior.

Significance testing (Section 3.3.9)

Significance tests are illustrated for Model 5.

Tests of treatment effects were made using Wald F tests using Satterthwaite's approximation to the denominator DF. An $F_{2,65}$ value of 3.67 was obtained for the composite test of treatment equality. This gave a significant p-value of 0.03. Wald t tests were used to perform pairwise treatment comparisons:

$$A - B \quad t_{54} = 1.23, \quad p = 0.22,$$

$$A - C \quad t_{79} = 1.58, \quad p = 0.12,$$

$$B - C \quad t_{71} = 2.71, \quad p = 0.009.$$

Thus, cold feet are significantly less likely on treatment C than on treatment B. Treatment A is intermediate in its effect. The t statistic for baseline cold feet was 45.52 on 266 DF. This was highly significant ($p \ll 0.0001$) indicating that fitting baseline has greatly increased the sensitivity of the analysis.

Confidence intervals (Section 3.3.10)

95% confidence intervals for treatment effects were calculated from the mean treatment differences and SEs. The confidence intervals on a linear scale for $A - B$ are:

$$95\% \text{ CI} = -0.704 \pm t_{54,0.975} \times 0.570, \quad t_{54,0.975} = 2.01$$

so we obtain:

$$95\% \text{ CI} = -0.704 \pm 2.01 \times 0.570 = (-1.850, 0.442).$$

A comparison of treatments A and B in terms of an odds ratio is obtained by exponentiating the effect estimate:

$$\text{OR} = \frac{P(\text{cold feet on A})/(1 - P(\text{cold feet on A}))}{P(\text{cold feet on B})/(1 - P(\text{cold feet on B}))} = \exp(-0.704) = 0.49.$$

Example **141**

Table 3.5 Effect estimates and 95% confidence interval.

Effect	Linear scale	Odds ratio
A – B	−0.704 (−1.846, 0.446)	0.49 (0.16, 1.56)
A – C	1.038 (−0.273, 2.348)	2.82 (0.72–10.48)
B – C	1.742 (0.460, 3.024)	5.70 (1.58, 20.56)

Confidence intervals for the odds ratio are calculated by exponentiating the confidence intervals calculated on the linear scale, $\exp(-1.850, 0.442) = (0.16, 1.56)$.

Odds ratios and confidence intervals were calculated in the same way for the other treatment effects and are given in Table 3.5.

Checking model assumptions (Section 3.3.11)

Plots of the centre-treatment effects against their predicted values and normal plots are used to check assumptions for Model 5 (Figure 3.1). Note that the centre

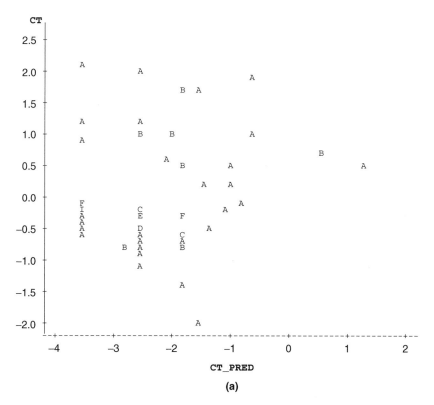

(a)

Figure 3.1 Plots of centre-treatment effects: (a) Centre-treatment effects against their predicted values. (b) Centre-treatment effects — normal plot. A = 1 obs, B = 2 obs, etc.

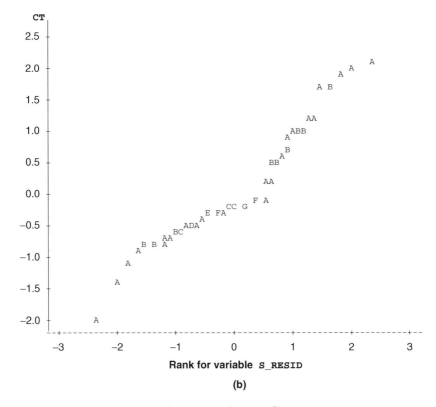

Figure 3.1   (*continued*).

effects do not need to be checked since the centre variance component estimate was zero. Since the data are in Bernoulli form, plots of the residuals are not useful for identifying outliers. The centre·treatment effect plots show no strong evidence of outliers or lack of symmetry. However, the normal plot indicates some deviation from normality. This is not of the systematic kind that we see through asymmetry and therefore would not give us cause to reanalyse the data under different assumptions.

SAS code and output

Variables

```
centre = centre,
treat  = treatment,
cf     = post-treatment cold feet (1 = yes, 0 = no),
cf1    = pre-treatment cold feet (1 = yes, 0 = no),
one    = 1, all observations.
```

Example   **143**

SAS code is provided for Models 1–5. However, output is only given for Models 2 and 5, which should be sufficient to illustrate the use of PROC GENMOD and the GLIMMIX macro in this example. Details on how to obtain the GLIMMIX macro are given in Section 9.1. The Bayesian models were fitted using the software package BUGS.

Model 1

```
PROC GENMOD; CLASS centre treat;
MODEL cf/one=cf1 treat/ DIST=B COVB;
```

Model 2

```
PROC GENMOD; CLASS centre treat;
MODEL cf/one=cf1 treat centre/ DIST=B COVB;
```

```
                    The GENMOD Procedure
                      Model Information
         Description                     Value
         Data Set                        WORK.A
         Distribution                    BINOMIAL
         Link Function                   LOGIT
         Dependent Variable              CF
         Dependent Variable              ONE
         Observations Used               279
         Number Of Events                39
         Number Of Trials                279
         Missing Values                  4

                  Class Level Information
      Class     Levels  Values
      CENTRE        29  1 2 3 4 5 6 7 8 9 11 12 13 14
                        15 18 23 24 25 26 27 29 30 31
                        32 35 36 37 40 41
      TREAT          3  1 2 3

                    Parameter Information
         Parameter     Effect     CENTRE  TREAT
         PRM1          INTERCEPT
         PRM2          CF1
         PRM3          TREAT              1
         PRM4          TREAT              2
         PRM5          TREAT              3
         PRM6          CENTRE     1
         PRM7          CENTRE     2
         ETC

          Criteria For Assessing Goodness Of Fit
        Criterion           DF      Value    Value/DF
        Deviance           247   141.0937     0.5712
        Scaled Deviance    247   141.0937     0.5712
        Pearson Chi-Square 247   194.5548     0.7877
```

```
              Scaled Pearson X2    247   194.5548      0.7877
              Log Likelihood         .   -70.5469           .
```

```
                    Analysis Of Parameter Estimates
         Parameter     DF Estimate     Std Err ChiSquare    Pr>Chi
         INTERCEPT      1 -26.4696      2.2172  142.5189     0.0001
         CF1            1   2.8658      0.5754   24.8066     0.0001
         TREAT      1   1   0.9327      0.6595    2.0003     0.1573
         TREAT      2   1   1.8976      0.6298    9.0786     0.0026
         TREAT      3   0   0.0000      0.0000         .          .
         CENTRE     1   1  23.4489      2.1750  116.2291     0.0001
         CENTRE     2   1  23.1337      2.2932  101.7682     0.0001
         CENTRE     3   1  -0.1433 110410.898    0.0000     1.0000
       ETC
```

(The full table appears within the main text.)

```
                        Estimated Covariance Matrix
      Parameter
      Number     PRM1       PRM2       PRM3       PRM4      PRM6       PRM7

      PRM1     4.91610   -0.17914   -0.36437   -0.35226  -4.56407   -4.57692
      PRM2    -0.17914    0.33107    0.006572   0.02065   0.09524    0.04791
      PRM3    -0.36437    0.006572   0.43490    0.28726   0.05976    0.09728
      PRM4    -0.35226    0.02065    0.28726    0.39662   0.03671    0.07221
      PRM6    -4.56407    0.09524    0.05976    0.03671   4.73076    4.49461
      PRM7    -4.57692    0.04791    0.09728    0.07221   4.49461    5.25868
      ETC
```

Model 3

```
PROC GENMOD; CLASS centre treat;
MODEL cf/one=cf1 treat centre centre*treat/ DIST=B COVB;
```

Model 4

```
%GLIMMIX(DATA=a,
STMTS=%STR(CLASS centre treat;
MODEL cf/one=cf1 treat/ DDFM=SATTERTH;
LSMEANS treat/ diff pdiff;
RANDOM centre;),
ERROR=B,
MAXIT=30);
```

Model 5

```
%GLIMMIX(DATA=a,
STMTS=%STR(CLASS centre treat;
MODEL cf/one=cf1 treat/ DDFM=SATTERTH;
LSMEANS treat/ diff pdiff;
RANDOM centre centre*treat;),
ERROR=B,
MAXIT=30);
```

Example **145**

```
                        Class Level Information
Class      Levels
CENTRE       29
TREAT         3
Values
1 2 3 4 5 6 7 8 9 11 12 13 14 15 18 23 24 25 26 27 29 30 31 32 35 36 37 40 41
1 2 3
```

```
                   Covariance Parameter Estimates
                   Cov Parm              Estimate
                   CENTRE               0.00000000
                   CENTRE*TREAT         1.78894392
```

```
                   GLIMMIX Model Statistics
            Description                     Value
            Deviance                      120.3983
            Scaled Deviance               221.9315
            Pearson Chi-Square            134.2882
            Scaled Pearson Chi-Square     247.5349
            Extra-Dispersion Scale          0.5425
```

```
                      Parameter Estimates
    Parameter  Estimate Std Error    DDF        T   Pr>|T|
    INTERCEPT   -3.5836    0.5230  100.561   -6.85  0.0001
    CF1          3.0990    0.4593  266.888    6.75  0.0001
    TREAT 1      1.0380    0.6587   79.427    1.58  0.1190
    TREAT 2      1.7420    0.6435   71.323    2.71  0.0085
    TREAT 3      0.0000        .        .        .      .
```

```
                   Tests of Fixed Effects
            Source   NDF     DDF    Type III F      Pr>F
            CF1        1   266.888      45.52      0.0001
            TREA       2    65.346       3.67      0.0308
```

```
                   Least Squares Means
    Level      LSMEAN Std Error    DDF       T  Pr>|T|    Mu
    TREAT 1 -2.2457    0.4182 64.7447  -5.37  0.0001 0.0957
    TREAT 2 -1.5417    0.3898 46.9868  -3.96  0.0003 0.1763
    TREAT 3 -3.2837    0.5147 96.1580  -6.38  0.0001 0.0361
```

```
               Differences of Least Squares Means
   Level 1   Level 2  Difference  Std Error    DDF       T  Pr>|T|
   TREAT 1   TREAT 2    -0.7041     0.5704  54.8040  -1.23  0.2223
   TREAT 1   TREAT 3     1.0380     0.6587  79.4271   1.58  0.1190
   TREAT 2   TREAT 3     1.7420     0.6435  71.3228   2.71  0.0085
```

Model checking for Model 5

An ascii file solut containing the random effects estimates first needs to be created by running the following GLIMMIX code and using the output produced by the SOLUTION option.

```
%GLIMMIX(DATA=a,
STMTS=%STR(CLASS centre treat;
MODEL cf/one=cf1 treat/ DDFM=SATTERTH;
RANDOM centre centre*treat/ SOLUTION;
LSMEANS treat/ DIFF PDIFF;
),
ERROR=B,
CONVERGE=1.0e-4,MAXIT=20);
```

The SOLUTION option will produce the following output:

```
                      Random Effects Estimates

Effect            CENTRE    TREAT  Estimate SE Pred    DF         t Pr>|t|

CENTRE              1                0.0000       .      .         .    .
CENTRE              2                0.0000       .      .         .    .
.
.
.
CENTRE *TREAT       1         1      0.5583  0.6450 110.535    0.87 0.3885
CENTRE *TREAT       1         2      0.2317  0.5959 107.143    0.39 0.6982
CENTRE *TREAT       1         3      0.8603  0.8226  75.286    1.05 0.2990
CENTRE *TREAT       2         1      1.6741  0.8602  69.999    1.95 0.0556
CENTRE *TREAT       2         2     -0.8430  1.0151  41.330   -0.83 0.4110
CENTRE *TREAT       2         3     -0.7978  1.0688  33.753   -0.75 0.4605
CENTRE *TREAT       3         1     -0.4647  1.1213  28.418   -0.41 0.6816
CENTRE *TREAT       3         2     -0.7135  1.0495  36.599   -0.68 0.5009
CENTRE *TREATx      3         3     -0.1535  1.2494  18.326   -0.12 0.9035
.
.
.
```

This output was manipulated manually to retain only the centre·treatment effects and delete all headings to give:

```
1    1     0.5582     0.64491311     200      0.87      0.3878
1    2     0.2316     0.59587671     200      0.39      0.6979
1    3     0.8600     0.82255895     200      1.05      0.2970
2    1     1.6737     0.86010530     200      1.95      0.0531
2    2    -0.8428     1.01496509     200     -0.83      0.4073
2    3    -0.7976     1.06857113     200     -0.75      0.4563
3    1    -0.4646     1.12108522     200     -0.41      0.6790
3    2    -0.7133     1.04931977     200     -0.68      0.4975
3    3    -0.1535     1.24911492     200     -0.12      0.9023
```

The centre, treatment and the centre·treatment estimates can now be read from columns 1–3.

A separate program is then required to produce the residual plots. This program (below) first reruns the analysis to produce a dataset resid containing the residuals. This is merged to dataset a containing the patient, centre and treatment values. Next the solut file containing the centre·treatment effects is read in

Example   **147**

and merged with dataset `resid` containing the residuals. The centre·treatment effects are deducted from the predicted values, $\mathbf{X}\hat{\boldsymbol{\alpha}} + \mathbf{Z}\hat{\boldsymbol{\beta}}$, to give the expected values, $\mathbf{X}\hat{\boldsymbol{\alpha}}$, for each observation (on the linear scale). The predicted values for each centre·treatment are then calculated as means of the expected values by centre·treatment. The dataset containing the centre·treatment effects `ct` is then merged with the dataset containing the centre·treatment predicted values `ct_pred` allowing residual plots and normal plots to be produced.

```
%GLIMMIX(DATA=a,
STMTS=%STR(CLASS centre treat;
MODEL cf/one=cf1 treat/ DDFM=SATTERTH;
RANDOM centre centre*treat/ SOLUTION;
LSMEANS treat/ DIFF PDIFF;
),
ERROR=B,
OUTPUT=OUT=resid PRED=pred RESCHI=resid,
CONVERGE=1.0e-4,MAXIT=20);

DATA A; SET A;
n=_n_;
KEEP pat centre treat;

DATA resid; SET resid;
n=_n_;

DATA resid; MERGE resid a;
KEEP pred resid pat centre treat;
PROC SORT; BY centre treat;

DATA ct(KEEP=centre treat ct); INFILE 'solut';
INPUT centre treat ct;
PROC SORT DATA=ct; BY centre treat;

DATA resid; MERGE resid ct; BY centre treat;
pred=pred-ct;
PROC MEANS NOPRINT DATA=resid; BY centre treat;
 VAR pred; OUTPUT OUT=ct_pred MEAN=ct_pred N=freq;

DATA ct; MERGE ct_pred ct; BY centre treat;
PROC PLOT; PLOT ct*ct_pred;
TITLE 'CENTRE.TREATMENT RESIDUALS AGAINST THEIR PREDICTED
VALUES';
PROC RANK OUT=norm NORMAL=tukey; VAR ct; RANKS rank;

PROC PLOT DATA=norm; PLOT ct*rank;
TITLE 'CENTRE.TREATMENT RESIDUALS - NORMAL PLOT';
```

Plots are given in the main text.

4

Mixed Models for Categorical Data

Categorical data often occur in clinical trials. For example, adverse events may be classified on an ordinal scale as mild, moderate or severe. In this chapter we will primarily consider the analysis of measurements made on ordered categorical scales; however, we also describe how unordered categorical data can be analysed. A fixed effects method for analysing ordinal data known as 'ordinal logistic regression' was first suggested by McCullagh (1980) and has been widely applied. The mixed categorical model is far less well established. The model we will define will be based on extending ordinal logistic regression to include random effects and covariance patterns. As we suggested in Chapter 3 for GLMMs, some readers with a less statistical background may wish to read only the introductory paragraphs of each section which will enable them to identify where these methods might prove useful. The final section of this chapter and sections of subsequent chapters will illustrate the application of mixed categorical models.

Ordinal logistic regression will be described in Section 4.1. It is extended to a mixed ordinal logistic regression model in Section 4.2. In Section 4.3 we describe how the model can be adapted to analyse unordered categorical data. In Section 4.4 some practical issues related to model fitting and interpretation are considered, and a worked example is given in Section 4.5.

4.1 ORDINAL LOGISTIC REGRESSION (FIXED EFFECTS MODEL)

Ordinal logistic regression is a fixed effects method for analysing ordinal data. It is often preferable to contingency table methods such as the chi-squared 'test for trend' because several fixed effects can be included in the model. The method works by

- Assuming observations have a multinomial distribution which can be expressed:

$$p(y_i) = \prod_{j=1}^{c} \mu_{ij}^{z_{ij}},$$

where

i = observation number,
j = category number,
c = number of categories,
$z_{ij} = 1$ if $y_i = j$
$\quad = 0$ otherwise,
$\mu_{ij} = p(y_i = j)$.

- Taking the ordered nature of the data into account by defining a model for the cumulative categorical probabilities. The cumulative probabilities, $\mu_{ij}^{[c]} = \sum_{k=1}^{j} \mu_{ik}$, correspond to the probability that observation i is in a category less than or equal to j. They can be thought of as partitioning the categories in every possible place. For example, if the response variable had categories labelled 1, 2, 3 and 4, three partitions would be possible: $1/2-4$, $1-2/3-4$ and $1-3/4$. The cumulative probabilities are linked to the model parameters using the logit link function (see Section 3.1.4):

$$\log(\mu_{ij}^{[c]}/(1 - \mu_{ij}^{[c]})) = I_j + \mathbf{x}_i\boldsymbol{\alpha}, \quad j = 1, 2, \ldots, c - 1,$$

where

I_j = intercept parameter for each partition j,
$\mu_{ij}^{[c]} = p(y_i <= j) = \sum_{k=1}^{j} \mu_{ik}$,
$\mathbf{x}_i$ = ith row of fixed effects design matrix $\mathbf{X}$,
$\boldsymbol{\alpha}$ = vector of fixed effect parameters.

Note that there is a separate equation for each partition.

- Maximising the multinomial likelihood function for the model parameters (I_j's and $\boldsymbol{\alpha}$). The likelihood function can be expressed:

$$L = \prod_i \prod_j \mu_{ij}^{z_{ij}}.$$

This can be defined in terms of the model parameters by first inverting the logit functions to give expressions for the cumulative probabilities:

$$\mu_{ij}^{[c]} = (1 + \exp(-I_j - \mathbf{x}_i\boldsymbol{\alpha}))^{-1}, \quad j = 1, 2, \ldots, c - 1,$$

then obtaining the multinomial probabilities by taking differences between adjacent cumulative probabilities as

$$\mu_{i1} = (1 + \exp(-I_1 - \mathbf{x}_i\boldsymbol{\alpha}))^{-1},$$

$$\mu_{ij} = (1 + \exp(-I_j - \mathbf{x}_i\boldsymbol{\alpha}))^{-1}$$
$$\quad - (1 + \exp(-I_{j-1} - \mathbf{x}_i\boldsymbol{\alpha}))^{-1}, \quad j = 2, \ldots, c - 1,$$

$$\mu_{ic} = 1 - (1 + \exp(-I_{c-1} - \mathbf{x}_i\boldsymbol{\alpha}))^{-1}.$$

These expressions for μ_{ij} are inserted into the likelihood function which is maximised for the model parameters. An iterative method such as Newton–Raphson is usually required to do this. In SAS an ordinal logistic regression can be performed using PROC LOGISTIC.

4.2 MIXED ORDINAL LOGISTIC REGRESSION

Extending the fixed effect ordinal logistic model to a mixed model is not as straightforward as extending the GLM to a GLMM. This is because the multinomial distribution is not from the exponential family and cannot be linked to the real scale using a single link function. A separate link function is now used for each partition of the categories. One way to get around this problem is to re-express the data in binary form so that it can be modelled as a GLMM and this is the approach we will be concentrating on.

In Section 4.2.1 the ordinal mixed model will be specified in a general form that can encompass random effects, random coefficients or covariance pattern models. The residual matrix for mixed categorical models has a more complex form than for GLMMs and will be defined in Section 4.2.2. As in GLMMs, there can be benefits in reparameterising random effects models as covariance pattern models and this will be discussed in Section 4.2.3. A quasi-likelihood function for the model is defined in Section 4.2.4 and in Section 4.2.5 model fitting methods are discussed.

4.2.1 Definition of the mixed ordinal logistic regression model

As noted above, the multinomial distribution is not a member of the exponential family and we will consider overcoming this hurdle by re-expressing the data in binary form so that a GLMM can then be fitted. To re-express the data in binary form we allow each observation to become a vector of $c - 1$ correlated binary observations (c = number of categories). For example, if there are four categories, then we could let: $y = 1$ become $(1, 0, 0)$, $y = 2$ become $(0, 1, 0)$, $y = 3$ become $(0, 0, 1)$ and $y = 4$ become $(0, 0, 0)$. Thus, the three terms correspond to the presence or absence of the first three categories, while the presence of the fourth category is implied by absence of the first three. The vector, $\mathbf{y}$, containing the n extended observations can then be defined:

$$\mathbf{y} = (y_{11}, y_{12}, y_{13}, \ y_{21}, y_{22}, y_{23}, \ \ldots, \ y_{n1}, y_{n2}, y_{n3})'.$$

To illustrate this redefinition consider the following data constituting the first five observations from a repeated measures trial in which $\mathbf{y}$ has range $1-4$.

Patient	Visit	Treatment	y
1	1	A	2
1	2	A	1
1	3	A	4
2	1	B	3
2	2	B	1

When **y** is expressed in its extended binary form it becomes

$$\mathbf{y} = (0, 1, 0, \;\; 1, 0, 0, \;\; 0, 0, 0, \;\; 0, 0, 1, \;\; 1, 0, 0)'.$$

Model definition

The ordinal mixed model can now be specified in a form of a GLMM using the cumulative probabilities that result from partitioning the categories in each possible place. A cumulative probability, $\mu_{ij}^{[c]}$, is defined as the probability that observation i is in a category less than or equal to j:

$$\mathbf{y} = \boldsymbol{\mu} + \mathbf{e},$$
$$\log(\boldsymbol{\mu}^{[c]}/(1 - \boldsymbol{\mu}^{[c]})) = \mathbf{X}\boldsymbol{\alpha} + \mathbf{Z}\boldsymbol{\beta},$$
$$\boldsymbol{\beta} \sim N(\mathbf{0}, \mathbf{G}),$$
$$\text{var}(\mathbf{e}) = \mathbf{R}.$$

$\boldsymbol{\mu}$ is the vector of expected multinomial probabilities corresponding to the n extended observations. If there are four categories then we may write

$$\boldsymbol{\mu} = (\mu_{11}, \mu_{12}, \mu_{13}, \;\; \mu_{21}, \mu_{22}, \mu_{23}, \;\; \ldots, \;\; \mu_{n1}, \mu_{n2}, \mu_{n3})',$$

where

$$\mu_{ij} = \text{probability observation } i \text{ is in category } j.$$

$\boldsymbol{\mu}^{[c]}$ is a vector containing the cumulative probabilities obtained by partitioning the four categories in the three possible places:

$$\boldsymbol{\mu}^{[c]} = \left(\mu_{11}^{[c]}, \mu_{12}^{[c]}, \mu_{13}^{[c]}, \;\; \mu_{21}^{[c]}, \mu_{22}^{[c]}, \mu_{23}^{[c]}, \;\; \ldots, \;\; \mu_{n1}^{[c]}, \mu_{n2}^{[c]}, \mu_{n3}^{[c]} \right)',$$

where

$$\mu_{ij}^{[c]} = \text{probability } (y_i <= j) = \sum_{k=1}^{j} \mu_{ik}.$$

So we may equivalently write

$$\boldsymbol{\mu}^{[c]} = (\mu_{11}, \mu_{11} + \mu_{12}, \mu_{11} + \mu_{12} + \mu_{13}, \mu_{21}, \mu_{21} + \mu_{22},$$
$$\mu_{21} + \mu_{22} + \mu_{23}, \ldots, \mu_{n1}, \mu_{n1} + \mu_{n2}, \mu_{n1} + \mu_{n2} + \mu_{n3})'.$$

$\boldsymbol{\alpha}$ and $\boldsymbol{\beta}$ are again vectors containing the fixed and random effects. They have the same form as given in Section 2.1 except that $\boldsymbol{\alpha}$ now additionally includes $c - 1$ intercept terms corresponding to each of the $c - 1$ partitions of the data. Thus, if a model fitting two treatments and three visits as fixed and patients as random were fitted to the example data above, we could write

$$\boldsymbol{\alpha} = (I_1, I_2, I_3, T_A, T_B, V_1, V_2, V_3)',$$
$$\boldsymbol{\beta} = (P_1, P_2)'.$$

$\mathbf{X}$ and $\mathbf{Z}$ are again design matrices for the fixed and random effects. However, they now have more rows than previously to correspond to the extended number of observations. For our data $\mathbf{X}$ and $\mathbf{Z}$ would be

$$
\mathbf{X} =
\begin{array}{cccccccc}
I_1 & I_2 & I_3 & T_A & T_B & V_1 & V_2 & V_3 \\
\end{array}
$$

$$
\mathbf{X} = \begin{pmatrix}
1 & 0 & 0 & 1 & 0 & 1 & 0 & 0 \\
0 & 1 & 0 & 1 & 0 & 1 & 0 & 0 \\
0 & 0 & 1 & 1 & 0 & 1 & 0 & 0 \\
1 & 0 & 0 & 1 & 0 & 0 & 1 & 0 \\
0 & 1 & 0 & 1 & 0 & 0 & 1 & 0 \\
0 & 0 & 1 & 1 & 0 & 0 & 1 & 0 \\
1 & 0 & 0 & 1 & 0 & 0 & 0 & 1 \\
0 & 1 & 0 & 1 & 0 & 0 & 0 & 1 \\
0 & 0 & 1 & 1 & 0 & 0 & 0 & 1 \\
1 & 0 & 0 & 0 & 1 & 1 & 0 & 0 \\
0 & 1 & 0 & 0 & 1 & 1 & 0 & 0 \\
0 & 0 & 1 & 0 & 1 & 1 & 0 & 0 \\
1 & 0 & 0 & 0 & 1 & 0 & 1 & 0 \\
0 & 1 & 0 & 0 & 1 & 0 & 1 & 0 \\
0 & 0 & 1 & 0 & 1 & 0 & 1 & 0 \\
\end{pmatrix},
$$

$$
\mathbf{Z} = \begin{array}{cc} P_1 & P_2 \end{array} \begin{pmatrix} 1 & 0 \\ 1 & 0 \\ 1 & 0 \\ 1 & 0 \\ 1 & 0 \\ 1 & 0 \\ 1 & 0 \\ 1 & 0 \\ 1 & 0 \\ 0 & 1 \\ 0 & 1 \\ 0 & 1 \\ 0 & 1 \\ 0 & 1 \\ 0 & 1 \end{pmatrix}.
$$

The G matrix

G is again a matrix of variance parameters corresponding to the random effects and coefficients and has the same form as given in Section 2.1. In the model considered above fitting one random effect (patient) **G** would have the form:

$$
\mathbf{G} = \begin{pmatrix} \sigma_{\mathrm{p}}^2 & 0 \\ 0 & \sigma_{\mathrm{p}}^2 \end{pmatrix},
$$

where σ_{p}^2 = patient variance component.

4.2.2 Residual variance matrix, R

The residual variance matrix needs to take into account firstly, the multinomial correlations that occur within the binary vectors used for each observation, and secondly any covariance patterns defined at the residual level.

Multinomial correlations

From the multinomial distribution it is known that covariances for the observation vectors, $(y_{i1}, y_{i2}, \ldots, y_{i,c-1})'$, are

$$\mathrm{cov}(y_{ij}, y_{ik}) = \mathrm{E}(y_{ij} - \mu_{ij})(y_{ik} - \mu_{ik})$$

$$= \mu_{ij}(1 - \mu_{ij}), \qquad\qquad j = k,$$

$$= \mathrm{E}(y_{ij}y_{ik}) - \mathrm{E}(y_{ij})\mathrm{E}(y_{ik}) = -\mu_{ij}\mu_{ik}, \quad j \neq k.$$

($\mathrm{E}(y_{ij}y_{ik}) = 0$ when $j \neq k$, because either y_{ij} or y_{ik} has to be zero.)

Thus, within-observation covariance matrices, $\mathbf{R}_i$, can be defined for each of the n original observations. If $c = 4$ we can write the covariance terms corresponding to each pair of partitions for patient i as

$$\mathbf{R}_i = \begin{pmatrix} \mu_{i1}(1 - \mu_{i1}) & -\mu_{i1}\mu_{i2} & -\mu_{i1}\mu_{i3} \\ -\mu_{i1}\mu_{i2} & \mu_{i2}(1 - \mu_{i2}) & -\mu_{i2}\mu_{i3} \\ -\mu_{i1}\mu_{i3} & -\mu_{i2}\mu_{i3} & \mu_{i3}(1 - \mu_{i3}) \end{pmatrix}.$$

Random effects models

In a random effects model (i.e. where observations are uncorrelated at the residual level) the $\mathbf{R}_i$ matrices form blocks along the diagonal of the full residual matrix, $\mathbf{R}$. For example, in the random effects model considered above $\mathbf{R}$ is

$$\mathbf{R} = \begin{pmatrix} \mathbf{R}_1 & \mathbf{0} & \mathbf{0} & \mathbf{0} & \mathbf{0} \\ \mathbf{0} & \mathbf{R}_2 & \mathbf{0} & \mathbf{0} & \mathbf{0} \\ \mathbf{0} & \mathbf{0} & \mathbf{R}_3 & \mathbf{0} & \mathbf{0} \\ \mathbf{0} & \mathbf{0} & \mathbf{0} & \mathbf{R}_4 & \mathbf{0} \\ \mathbf{0} & \mathbf{0} & \mathbf{0} & \mathbf{0} & \mathbf{R}_5 \end{pmatrix},$$

where

$$\mathbf{0} = \begin{pmatrix} 0 & 0 & 0 \\ 0 & 0 & 0 \\ 0 & 0 & 0 \end{pmatrix}.$$

As in the GLMM definition (Section 3.2) $\mathbf{R}$ can be arranged as a product of a correlation matrix, $\mathbf{P}$, and the matrix of expected Bernoulli variances, $\mathbf{B} = \mathrm{diag}\{\mu_{ij}(1 - \mu_{ij})\}$.

$$\mathbf{R} = \mathbf{B}^{1/2}\mathbf{P}\mathbf{B}^{1/2}.$$

Note that the $\mathbf{A}$ matrix of constants used for the GLMM is not now required, since the data are in Bernoulli form and $\mathbf{A} = \mathbf{I}$. For our example data $\mathbf{P}$ may be written

$$\mathbf{P} = \begin{pmatrix} \mathbf{P}_{11} & \mathbf{0} & \mathbf{0} & \mathbf{0} & \mathbf{0} \\ \mathbf{0} & \mathbf{P}_{22} & \mathbf{0} & \mathbf{0} & \mathbf{0} \\ \mathbf{0} & \mathbf{0} & \mathbf{P}_{33} & \mathbf{0} & \mathbf{0} \\ \mathbf{0} & \mathbf{0} & \mathbf{0} & \mathbf{P}_{44} & \mathbf{0} \\ \mathbf{0} & \mathbf{0} & \mathbf{0} & \mathbf{0} & \mathbf{P}_{55} \end{pmatrix},$$

where

$$\mathbf{P}_{ii} = \text{matrix blocks of 'within-observation' correlations}$$

$$= \begin{pmatrix} 1 & -\mu_{i1}\mu_{i2}/b_{i12} & -\mu_{i1}\mu_{i3}/b_{i13} \\ -\mu_{i1}\mu_{i2}/b_{i12} & 1 & -\mu_{i2}\mu_{i3}/b_{i23} \\ -\mu_{i1}\mu_{i3}/b_{i13} & -\mu_{i2}\mu_{i3}/b_{i23} & 1 \end{pmatrix},$$

$$b_{ikl} = [\mu_{ik}(1 - \mu_{ik})\mu_{il}(1 - \mu_{il})]^{1/2}.$$

As in GLMMs variance at the residual level can be increased (or decreased) by using a dispersion parameter. The residual variance is multiplied by the dispersion parameter, ϕ, so that

$$\mathbf{R} = \phi\mathbf{B}^{1/2}\mathbf{P}\mathbf{B}^{1/2}.$$

We suggest that it is usually beneficial to fit a dispersion parameter as in GLMMs (see Section 3.3.1).

Covariance pattern models

The $\mathbf{R}$ matrix specified above assumes that no correlation occurs between the repeated measurements. When a covariance pattern model is used to allow for such correlations, $\mathbf{P}$ will also include off-diagonal blocks of correlation parameters. For example, if a general covariance pattern were used to model our example data, then we would require a separate block of correlations for each pair of time points and $\mathbf{P}$ would have the form

$$\mathbf{P} = \begin{pmatrix} \mathbf{P}_{11} & \mathbf{Q}_{12} & \mathbf{Q}_{13} & \mathbf{0} & \mathbf{0} \\ \mathbf{Q}_{12} & \mathbf{P}_{22} & \mathbf{Q}_{23} & \mathbf{0} & \mathbf{0} \\ \mathbf{Q}_{13} & \mathbf{Q}_{23} & \mathbf{P}_{33} & \mathbf{0} & \mathbf{0} \\ \mathbf{0} & \mathbf{0} & \mathbf{0} & \mathbf{P}_{44} & \mathbf{Q}_{12} \\ \mathbf{0} & \mathbf{0} & \mathbf{0} & \mathbf{Q}_{12} & \mathbf{P}_{55} \end{pmatrix}.$$

The $\mathbf{Q}_{mn}$ are $(c - 1) \times (c - 1) = 3 \times 3$ sub-matrices of parameters corresponding to correlation between observations at visits m and n on the same patient. They replace the single correlation values used in GLMMs and here we assume they take the form

$$\mathbf{Q}_{mn} = \begin{pmatrix} q_{mn,11} & q_{mn,12} & q_{mn,13} \\ q_{mn,12} & q_{mn,22} & q_{mn,23} \\ q_{mn,13} & q_{mn,23} & q_{mn,33} \end{pmatrix},$$

with a separate correlation parameter used for each pair of partitions. Thus, six parameters are used for each $\mathbf{Q}_{mn}$ matrix here. This is the parameterisation used by Lipsitz *et al.* (1994) who has written an accompanying SAS macro for fitting the model in this form. However, because this model requires more covariance parameters than GLMMs, particularly when the number of categories is high (increased by a factor of $c(c - 1)/2$), more complex covariance patterns should be used with caution.

Simpler parameterisation for covariance pattern models

Here, we suggest an alternative simpler parameterisation with just one parameter corresponding to each of the parameters in the original covariance pattern (i.e. one for compound symmetry, three here for a general pattern). The correlation matrix for our example using a general covariance pattern would be

$$
\mathbf{P} = \begin{pmatrix}
\mathbf{P}_{11} & \theta_{12}\mathbf{P}_{12} & \theta_{13}\mathbf{P}_{13} & \mathbf{0} & \mathbf{0} \\
\theta_{12}\mathbf{P}_{12} & \mathbf{P}_{22} & \theta_{23}\mathbf{P}_{23} & \mathbf{0} & \mathbf{0} \\
\theta_{13}\mathbf{P}_{13} & \theta_{23}\mathbf{P}_{23} & \mathbf{P}_{33} & \mathbf{0} & \mathbf{0} \\
\mathbf{0} & \mathbf{0} & \mathbf{0} & \mathbf{P}_{44} & \theta_{12}\mathbf{P}_{12} \\
\mathbf{0} & \mathbf{0} & \mathbf{0} & \theta_{12}\mathbf{P}_{12} & \mathbf{P}_{55}
\end{pmatrix}.
$$

The $\mathbf{P}_{ij}$ sub-matrices have a similar form to that given earlier for the $\mathbf{P}_{ii}$'s except that they now use the expected values corresponding to observations i and j:

$$
\mathbf{P}_{ij} = \begin{pmatrix}
1 & -\mu_{i1}\mu_{j2}/b_{ij,12} & -\mu_{i1}\mu_{j3}/b_{ij,13} \\
-\mu_{i1}\mu_{j2}/b_{ij,12} & 1 & -\mu_{i2}\mu_{j3}/b_{ij,23} \\
-\mu_{i1}\mu_{j3}/b_{ij,13} & -\mu_{i2}\mu_{j3}/b_{ij,23} & 1
\end{pmatrix},
$$

$$
b_{ij,kl} = [\mu_{ik}(1 - \mu_{ik})\mu_{jl}(1 - \mu_{jl})]^{1/2}.
$$

The θ_{mn}'s define the covariance pattern and can be parameterised to fit most of the covariance patterns described in Section 6.2. For example, for a compound symmetry pattern the θ_{mn}'s would all have the same value and we could write,

$$
\mathbf{P} = \begin{pmatrix}
\mathbf{P}_{11} & \theta\mathbf{P}_{12} & \theta\mathbf{P}_{13} & \mathbf{0} & \mathbf{0} \\
\theta\mathbf{P}_{12} & \mathbf{P}_{22} & \theta\mathbf{P}_{23} & \mathbf{0} & \mathbf{0} \\
\theta\mathbf{P}_{13} & \theta\mathbf{P}_{23} & \mathbf{P}_{33} & \mathbf{0} & \mathbf{0} \\
\mathbf{0} & \mathbf{0} & \mathbf{0} & \mathbf{P}_{44} & \theta\mathbf{P}_{12} \\
\mathbf{0} & \mathbf{0} & \mathbf{0} & \theta\mathbf{P}_{12} & \mathbf{P}_{55}
\end{pmatrix}.
$$

We have not explored the use of this simpler parameterisation in our worked examples as it is not easily fitted using SAS. However, it may be preferable to the correlation matrix suggested by Lipsitz, since fewer parameters are used.

4.2.3 Reparameterising random effects models as covariance pattern models

In Section 3.2.4 we described how random effects shrinkage could cause non-convergence or variance parameter bias in GLMMs, particularly when there were uniform random effect categories present. These problems can occur for the same reason in mixed categorical models and we again suggest that they can sometimes be avoided by reparameterising random effects models as covariance pattern models. As in the GLMM, a general form for a random effects model reparameterised as a covariance pattern model can be written

$$\mathbf{y} = \boldsymbol{\mu} + \mathbf{e},$$

$$g(\boldsymbol{\mu}) = \mathbf{X}\boldsymbol{\alpha}.$$

The random effects are now not included in the linear part of the model but are assumed to be incorporated into the error term and lead to a variance matrix of the form

$$\mathbf{R} = \phi \mathbf{B}^{1/2}(\mathbf{Z}\mathbf{G}\mathbf{Z}' + \mathbf{P})\mathbf{B}^{1/2},$$

where $\mathbf{Z}$ defines the random effects levels and $\mathbf{G}$ is a matrix of variance parameters corresponding to the random effects.

4.2.4   Likelihood and quasi-likelihood functions

The model we have defined, based on binary observations, is now in the form of a GLMM and a quasi-likelihood function can be defined in the same way as described in Section 3.2.2. A general form for the log quasi-likelihood for a GLMM which may contain random effects, coefficients and/or covariance patterns is again

$$\log\{QL(\boldsymbol{\alpha}, \boldsymbol{\gamma}; \mathbf{y})\} = \log\{QL(\boldsymbol{\alpha}, \boldsymbol{\gamma_R}; \mathbf{y}|\boldsymbol{\beta})\} - 1/2 \log |\mathbf{G}| - 1/2\boldsymbol{\beta}'\mathbf{G}^{-1}\boldsymbol{\beta} + K,$$

where

$$\boldsymbol{\gamma} = (\boldsymbol{\gamma_G}, \boldsymbol{\gamma_R}).$$

This function will correspond to a true log likelihood function whenever the residuals are uncorrelated (i.e. no $\boldsymbol{\gamma_R}$ parameters are included) since $QL(\boldsymbol{\alpha}, \boldsymbol{\gamma_R}; \mathbf{y}|\boldsymbol{\beta})$ will then follow a multinomial distribution.

4.2.5   Model fitting methods

Now that the model is in the form of a GLMM, it can be fitted using the approaches suggested in Section 3.2.3. However, it is now necessary to accommodate the multinomial within-observation covariances and this adds a further degree of complexity to the computation. Most published examples have been applied to covariance pattern models and are based on using the generalised estimating equations approach (e.g. Lipsitz *et al.*, 1994; Liang *et al.*, 1992; Kenward *et al.*, 1994). Lipsitz *et al.* provide an SAS macro and for this reason we have used their approach to analysis the examples in this book. We are not aware of any SAS software which can be used to fit all types of categorical mixed model. For models with a single random effect (e.g. cross-over trial analysis) one possibility would be to reparameterise the model as a covariance pattern model (see Sections 3.2.4 and 4.2.3). For those who would like to pursue fitting random effects and random coefficients models further, we mention two approaches for which (non-SAS) software is freely available. Hedeker and Gibbons (1994) and Goldstein (1995) have each suggested approaches for fitting random effects (and coefficients) models. Heddeker and Gibbons (1994) have made available `Fortran`-based software to implement their method (see Section 9.1). Goldstein's method can be implemented with a macro for use with the package `MLWin` (see Section 9.1).

Alternatively, the Bayesian approach (see Sections 2.3 and 3.2) can be used for analysing random effects and coefficients models. For this approach it is not necessary formally to redefine the observations in the extended binary form. A method such as the Gibbs sampler (available in the package `BUGS`, see Section 9.1) can be used to simulate the posterior distribution from the categorical mixed model defined as follows:

$$\mathbf{y}_i \sim \text{multinomial } (\mu_{i1}, \mu_{i2}, \ldots, \mu_{ic}),$$

$$g(\mu_{ij}^{[c]}) = I_j + \mathbf{x}_i \boldsymbol{\alpha} + \mathbf{z}_i \boldsymbol{\beta},$$

$$\boldsymbol{\beta} \sim \text{N}(\mathbf{0}, \mathbf{G}),$$

where

$\mathbf{y}_i = (y_{1j}, y_{i2}, \ldots, y_{in})$,
$\mu_{ij} = $ probability observation i is in category j,
$\mu_{ij}^{[c]} = $ probability $(y_i <= j) = \sum_{k=1}^{j} \mu_{ik}$,
$I_j = $ intercept term for category j,
$\mathbf{x}_i = $ the ith row of fixed effects design matrix $\mathbf{X}$,
$\mathbf{z}_i = $ the ith row of random effects design matrix $\mathbf{Z}$,
$\mathbf{G} = $ covariance matrix.

Non-informative prior distributions can again be used for all parameters. For example, normal distributions with very large variances for fixed effects and inverse gamma distributions with very small parameters for variance components.

4.3   MIXED MODELS FOR UNORDERED CATEGORICAL DATA

So far in this chapter we have only considered models for ordered categorical data. Although less frequent, unordered categorical variables are sometimes encountered in medicine. Blood group and colour are examples, since there is no natural ordering of their categories. A mixed model for this type of data can be defined in a very similar way to the ordinal mixed model. Again, the data can be re-expressed in extended binary form so that they are in the form of a GLMM. The main difference from the ordinal mixed model is that the multinomial probabilities, $\boldsymbol{\mu}$, are now linked to the model parameters using 'generalised logits' rather than the logits of the cumulative probabilities used for ordinal data. The generalised logits can be calculated as the logs of the ratios of the probabilities of being in each category to that of being in the last category, i.e. by $\log(\boldsymbol{\mu}/\boldsymbol{\mu}_L)$, where $\boldsymbol{\mu}_L$ is a vector containing the multinomial probabilities of each observation being in the last category. The model can be specified by

$$\mathbf{y} = \boldsymbol{\mu} + \mathbf{e},$$

$$\log(\boldsymbol{\mu}/\boldsymbol{\mu}_L) = \mathbf{X}\boldsymbol{\alpha} + \mathbf{Z}\boldsymbol{\beta},$$

$$\boldsymbol{\beta} \sim N(\mathbf{0}, \mathbf{G}),$$

$$\text{var}(\mathbf{e}) = \mathbf{R}.$$

If there were four categories we could write

$$\boldsymbol{\mu}_4 = (\mu_{14}, \mu_{14}, \mu_{14}, \ \mu_{24}, \mu_{24}, \mu_{24}, \ \ldots)',$$

and the vector of generalised logits as

$$\log(\boldsymbol{\mu}/\boldsymbol{\mu}_4) = (\mu_{11}/\mu_{14}, \mu_{12}/\mu_{14}, \mu_{13}/\mu_{14}, \ \mu_{21}/\mu_{24}, \mu_{22}/\mu_{24}, \mu_{23}/\mu_{24}, \ \ldots)'.$$

The choice of the last category for the denominator is arbitrary. Any of the categories can, in fact, be used and sometimes convergence will be more likely if the largest category is chosen. $\boldsymbol{\alpha}$ and $\boldsymbol{\beta}$ are again vectors containing the fixed and random effects. However, a separate parameter is now needed for each category (except the last) because the proportional odds assumption used for ordinal data does not hold. We illustrate this using the following hypothetical dataset which contains the first five observations from a repeated measures trial in which $\mathbf{y}$ in an unordered categorical variable.

Patient	Visit	Treatment	y
1	1	A	2
1	2	A	1
1	3	A	4
2	1	B	3
2	2	B	1

In a model fitting treatments as fixed and patients as random, we could write

$$\boldsymbol{\alpha} = (I_1, I_2, I_3, T_{A,1}, T_{A,2}, T_{A,3}, T_{B,1}, T_{B,2}, T_{B,3})',$$

$$\boldsymbol{\beta} = (P_{1,1}, P_{1,2}, P_{1,3}, P_{2,1}, P_{2,2}, P_{2,3})',$$

where

$I_j = $ intercept for the jth category,
$T_{k,j} = $ effect for treatment k, category j,
$P_{i,j} = $ effect for patient i, category j.

Each treatment and patient effect now has a separate parameter corresponding to each category of the data (except the last). The $\mathbf{X}$ and $\mathbf{Z}$ design matrices also have extra columns corresponding to the extra parameters and have the form

$$\mathbf{X} = \begin{array}{c} \begin{array}{ccccccccc} I_1 & I_2 & I_3 & T_{A,1} & T_{A,2} & T_{A,3} & T_{B,1} & T_{B,2} & T_{B,3} \end{array} \\ \left(\begin{array}{ccccccccc} 1 & 0 & 0 & 1 & 0 & 0 & 0 & 0 & 0 \\ 0 & 1 & 0 & 0 & 1 & 0 & 0 & 0 & 0 \\ 0 & 0 & 1 & 0 & 0 & 1 & 0 & 0 & 0 \\ 1 & 0 & 0 & 1 & 0 & 0 & 0 & 0 & 0 \\ 0 & 1 & 0 & 0 & 1 & 0 & 0 & 0 & 0 \\ 0 & 0 & 1 & 0 & 0 & 1 & 0 & 0 & 0 \\ 1 & 0 & 0 & 1 & 0 & 0 & 0 & 0 & 0 \\ 0 & 1 & 0 & 0 & 1 & 0 & 0 & 0 & 0 \\ 0 & 0 & 1 & 0 & 0 & 1 & 0 & 0 & 0 \\ 1 & 0 & 0 & 0 & 0 & 0 & 1 & 0 & 0 \\ 0 & 1 & 0 & 0 & 0 & 0 & 0 & 1 & 0 \\ 0 & 0 & 1 & 0 & 0 & 0 & 0 & 0 & 1 \\ 1 & 0 & 0 & 0 & 0 & 0 & 1 & 0 & 0 \\ 0 & 1 & 0 & 0 & 0 & 0 & 0 & 1 & 0 \\ 0 & 0 & 1 & 0 & 0 & 0 & 0 & 0 & 1 \end{array} \right) \end{array},$$

$$\mathbf{Z} = \begin{array}{c} \begin{array}{cccccc} P_{1,1} & P_{1,2} & P_{1,3} & P_{2,1} & P_{2,2} & P_{2,3} \end{array} \\ \left(\begin{array}{cccccc} 1 & 0 & 0 & 0 & 0 & 0 \\ 0 & 1 & 0 & 0 & 0 & 0 \\ 0 & 0 & 1 & 0 & 0 & 0 \\ 1 & 0 & 0 & 0 & 0 & 0 \\ 0 & 1 & 0 & 0 & 0 & 0 \\ 0 & 0 & 1 & 0 & 0 & 0 \\ 1 & 0 & 0 & 0 & 0 & 0 \\ 0 & 1 & 0 & 0 & 0 & 0 \\ 0 & 0 & 1 & 0 & 0 & 0 \\ 0 & 0 & 0 & 1 & 0 & 0 \\ 0 & 0 & 0 & 0 & 1 & 0 \\ 0 & 0 & 0 & 0 & 0 & 1 \\ 0 & 0 & 0 & 1 & 0 & 0 \\ 0 & 0 & 0 & 0 & 1 & 0 \\ 0 & 0 & 0 & 0 & 0 & 1 \end{array} \right) \end{array}.$$

4.3.1 The G matrix

G now contains blocks of variance parameters to allow for the fact that separate effects are specified for each partition and that these effects are correlated within each patient. A separate covariance parameter can be specified for each of the parameters corresponding to each pair of partitions. For our example data we may write

$$\mathbf{G} = \begin{pmatrix} \mathbf{G}_p & \mathbf{0} \\ \mathbf{0} & \mathbf{G}_p \end{pmatrix},$$

where

$$\mathbf{G}_p = \begin{pmatrix} \sigma^2_{p,11} & \theta_{p,12} & \theta_{p,13} \\ \theta_{p,12} & \sigma^2_{p,22} & \theta_{p,23} \\ \theta_{p,13} & \theta_{p,23} & \sigma^2_{p,33} \end{pmatrix},$$

$\sigma^2_{p,jj}$ = patient variance component for category j,

$\theta_{p,jk}$ = patient covariance parameter for the category pair j, k.

Alternatively, a model with a simpler parameterisation could be proposed. For example, we could make the assumption that each category had the same variance component and that the random effects for the different partitions were uncorrelated within patients. We could then write

$$\mathbf{G}_p = \sigma^2_p \begin{pmatrix} 1 & 0 & 0 \\ 0 & 1 & 0 \\ 0 & 0 & 1 \end{pmatrix},$$

where

σ^2_p = patient variance component.

4.3.2 The R matrix

This matrix has the same form as the ordinal mixed model (see Section 4.2.2).

4.3.3 Fitting the model

The model can be fitted using similar techniques to those described for ordinal mixed models (Section 4.2.5). The method defined by Lipsitz *et al.* (1994) can also be used with unordered categorical data and their SAS macro contains an option for specifying that the data are unordered.

4.4 PRACTICAL APPLICATION AND INTERPRETATION

In this section some points relating to the practical application and interpretation of categorical mixed models will be considered. We should point out that experience with these models is very limited and therefore some of the issues are still far from resolved.

4.4.1 The proportional odds assumption

The fixed and random effects, α and β are assumed to be the same at each partition in models for ordinal data. For example, if there are four categories, an equal α and β are assumed whether a $1/2-4$, $1-2/3-4$ or $1-3/4$ partition is made. This means that the odds for effects are proportional across all partitions. This assumption could be tested by fitting a separate set of fixed and random effects at each partition, α_j and β_j, and testing the equality of the effects at different partitions. When there is a significant difference between fixed effect estimates for each partition (i.e. the proportional odds assumption does not hold) it may be informative to analyse each partition separately using binary GLMMs. However, often the 'average' α and β over all partitions will be of greatest interpretational value even if the results differ significantly between each partition.

4.4.2 Number of covariance parameters

The number of covariance parameters required by a covariance pattern model is increased by a factor of $(c-1) \times (c-2)/2$ over an equivalent GLMM (unless the alternative simpler parameterisation is used, see Section 4.2.2). Thus, the model can use a large number of covariance parameters and this can sometimes lead to inaccurate estimates or convergence problems. Therefore, the more complex patterns should be used only cautiously, particularly in small datasets or when the number of categories is high. It may also be worth considering combining any categories which have small frequencies with neighbouring categories. Alternatively, if there are a large number of categories, say about five or more, it might be worth trying a normal mixed model and checking the resulting residuals. If the model assumptions are approximately satisfied, this type of model would also have the advantage of being simpler to interpret.

4.4.3 Choosing a covariance pattern

As in GLMMs, approximate likelihood ratio tests based on the quasi-likelihood values could be used to compare models fitting different covariance parameters. However, we are not aware of a SAS macro for fitting categorical mixed models that outputs a quasi-likelihood value. For those without access to software providing a quasi-likelihood value, we suggest favouring the most simple patterns (e.g. compound symmetry or first-order autoregressive) and using more complex patterns only with larger datasets where there is a strong suggestion that the covariance pattern deviates from a simpler pattern.

4.4.4 Interpreting covariance parameters

It is difficult to interpret the size of covariance parameters fitted in the **R** matrix, since blocks containing $(c-1) \times (c-2)/2$ parameters are estimated rather than individual parameters. The diagonal terms representing correlation between observations in the same categories are perhaps the most helpful. However, if the alternative simpler parameterisation is used (see Section 4.2.2), there is only one parameter per block and the parameters can be interpreted in the same way as those from ordinary repeated measures analyses (see Section 6.2).

4.4.5   Fixed and random effects estimates

The fixed and random effects estimates are given in terms of logits which become more interpretable when exponentiated to give odds ratios (see Section 3.1.5).

4.4.6   Checking model assumptions

As in the GLMM, it is assumed that the random effect residuals are normally distributed and uncorrelated. Residual plots can be used to visually check normality and identify any outlying effect categories as described for normal data in Section 2.4.6.

4.4.7   The dispersion parameter

The same considerations apply as for Bernoulli data described in Section 3.3.8.

4.4.8   Other points

The points relating to: difficulties with fitting random effects models (3.3.2); bias in fixed and random effects standard errors (3.3.4); negative variance components (3.3.5); uniform fixed and random effect categories (3.3.6, 3.3.7); significance testing (3.3.9); and confidence intervals (3.3.10) also apply to categorical mixed models.

4.5   EXAMPLE

In this example we will consider the analysis of an adverse event, 'cold feet', which was recorded at each visit in the hypertension study introduced in Section 1.3. Previously (see Section 3.4) we considered this as a binary variable but cold feet were actually recorded on a scale of $1-5$: $1 =$ none, $2 =$ occasionally, $3 =$ on most days, $4 =$ most of the time, $5 =$ all of the time. Although the data are recorded from different centres, here for simplicity we will ignore the effect of centres and perform only repeated measures analyses. The frequencies of each category by treatment and visit are shown in Table 4.1. Some of the frequencies in categories 4 and 5 are low and therefore it may later be necessary to consider combining these categories in the analysis.

Example **165**

Table 4.1 Frequencies of 'cold feet' severity by treatment and visit.

Treatment	Visit	Category				
		1	2	3	4	5
A	3	83	4	6	0	4
	4	72	5	6	3	3
	5	70	3	5	2	3
	6	63	5	3	3	2
B	3	69	9	5	2	6
	4	65	7	10	3	3
	5	54	10	8	6	8
	6	55	5	8	4	7
C	3	79	2	7	1	1
	4	85	1	4	0	1
	5	82	3	3	2	1
	6	78	4	3	1	1
Total		855	58	68	27	40

We will consider analysing the data using a variety of covariance patterns (Model 1 — uncorrelated; Model 2 — compound symmetry; Model 3 — Toeplitz; Model 4 — general) using the SAS macro written by Lipsitz *et al.* (see Section 9.1) Each model will fit baseline cold feet, treatment and visit effects as fixed effects. Treatment.visit effects were found to be non-significant on initial analysis and therefore have been excluded from each model. Model 4 using a general pattern did not converge even when categories 4 and 5 were combined. This is likely to be due to the large number of covariance parameters that needed to be estimated by the model (60).

We will first consider the correlation parameter estimates arising from the models. These are rather more difficult to interpret than parameter estimates from covariance patterns in normal mixed models or GLMMs, because there is now a 4×4 matrix block of parameters representing the correlation between a pair of visits. Recall from Section 4.2.2 that

$$\mathbf{R} = \mathbf{B}^{1/2}\mathbf{P}\mathbf{B}^{1/2}.$$

In this example the correlation matrix, $\mathbf{P}$, has the block diagonal form

$$\mathbf{P} = \begin{pmatrix} \mathbf{P}_{11} & \mathbf{Q}_{12} & \mathbf{Q}_{13} & \mathbf{Q}_{14} & \mathbf{0} & \mathbf{0} & . & . \\ \mathbf{Q}_{12} & \mathbf{P}_{22} & \mathbf{Q}_{23} & \mathbf{Q}_{24} & \mathbf{0} & \mathbf{0} & . & . \\ \mathbf{Q}_{13} & \mathbf{Q}_{23} & \mathbf{P}_{33} & \mathbf{Q}_{34} & \mathbf{0} & \mathbf{0} & . & . \\ \mathbf{Q}_{14} & \mathbf{Q}_{24} & \mathbf{Q}_{34} & \mathbf{P}_{44} & \mathbf{0} & \mathbf{0} & . & . \\ \mathbf{0} & \mathbf{0} & \mathbf{0} & \mathbf{0} & \mathbf{P}_{55} & \mathbf{Q}_{12} & . & . \\ \mathbf{0} & \mathbf{0} & \mathbf{0} & \mathbf{0} & \mathbf{Q}_{12} & \mathbf{P}_{66} & . & . \\ . & . & . & . & . & . & . & . \\ . & . & . & . & . & . & . & . \end{pmatrix},$$

where the $\mathbf{P}_{ii}$'s are the multinomial 'within-observation' correlation matrices and the $\mathbf{Q}_{mn}$'s give the correlations between observations on the same patient between visits m and n. The $\mathbf{Q}_{mn}$ matrices obtained from Models 1–3 are shown in Table 4.2.

Statistical comparisons between the models using likelihood ratio tests were not readily available because quasi-likelihood values were not produced by the SAS macro. On informal examination the positive correlations in Models 2 and 3 indicate that the repeated observations on the same patient are correlated. Therefore, Model 1, which has zero correlations and assumes that the observations are independent, should be rejected. The banded pattern (Model 3) does not show marked differences between the correlations depending on visit separation. Therefore, we might choose to base our conclusions on Model 2 with a simpler covariance pattern.

Fixed effects estimates obtained from Models 1–3 are shown in Table 4.3. The standard errors are 'model based' and are thus calculated directly from the covariance pattern parameters estimated by the model. The four intercept terms (arising from the four possible partitions of the five categories), and the visit terms are of little interest. The large size of the baseline terms relative to their standard errors indicates that the model has benefited from their inclusion. Models 2 and 3 show very similar results.

Treatment effect estimates are shown with both 'model-based' standard errors and the 'empirical' standard errors in Table 4.4. In Models 2 and 3, where the covariance pattern of the data has been modelled, the empirical estimates

Table 4.2 $\mathbf{Q}_{mn}$ correlation matrices.

Model														
1	Parti-tion	**All visit pairs (i.e. all $\mathbf{Q}_{lm}$'s)**												
		1	**2**	**3**	**4**									
	1	0.00												
	2	0.00	0.00											
	3	0.00	0.00	0.00										
	4	0.00	0.00	0.00	0.00									
2	Parti-tion	**All visit pairs (i.e. all $\mathbf{Q}_{mn}$'s)**												
		1	**2**	**3**	**4**									
	1	0.61												
	2	0.11	0.39											
	3	0.04	0.09	0.27										
	4	0.07	0.05	0.15	0.21									
3		**Visit separation = 1 (i.e. $\mathbf{Q}_{12}$, $\mathbf{Q}_{23}$, $\mathbf{Q}_{34}$)**				**Visit separation = 2 (i.e. $\mathbf{Q}_{13}$, $\mathbf{Q}_{24}$)**				**Visit separation = 3 (i.e. $\mathbf{Q}_{14}$)**				
	Parti-tion	**1**	**2**	**3**	**4**	**1**	**2**	**3**	**4**	**1**	**2**	**3**	**4**	
	1	0.67				0.56				0.56				
	2	0.11	0.45			0.12	0.32			0.08	0.37			
	3	0.00	0.07	0.39		0.02	0.10	0.19		0.14	0.14	0.08		
	4	0.09	0.05	0.22	0.11	0.10	0.07	0.13	0.17	−0.04	0.00	−0.02	0.56	

Example   167

Table 4.3 Fixed effects estimates.

Effect		1 Uncorrelated	2 Compound symmetry	3 Toeplitz
		Model		
Intercepts	1	0.64(0.30)	0.65(0.41)	0.62(0.40)
	2	1.12(0.30)	1.11(0.41)	1.08(0.41)
	3	1.97(0.31)	1.96(0.42)	1.89(0.42)
	4	2.55(0.32)	2.53(0.46)	2.46(0.46)
Baseline	1 vs 5	1.92(0.23)	1.92(0.36)	1.89(0.36)
cold feet	2 vs 5	1.29(0.28)	1.20(0.44)	1.13(0.43)
	3 vs 5	1.14(0.35)	1.07(0.55)	0.99(0.54)
	4 vs 5	0.85(0.40)	0.52(0.60)	0.55(0.61)
Treatment	A − B	0.79(0.19)	0.76(0.30)	0.75(0.30)
	A − C	−0.59(0.24)	−0.62(0.37)	−0.60(0.36)
	B − C	−1.34(0.22)	−1.38(0.34)	−1.34(0.34)
Visit	1 vs 4	0.29(0.24)	0.35(0.16)	0.44(0.16)
	2 vs 4	0.24(0.24)	0.31(0.15)	0.34(0.16)
	3 vs 4	−0.11(0.23)	−0.03(0.14)	0.02(0.13)

Table 4.4 Treatment effect estimates with model-based and empirical estimates of standard errors.

Effect	Difference	Mean difference	Model-based SE	Empirical SE
Model 1	A − B	0.79	0.19	0.31
(Uncorrelated)	A − C	−0.66	0.24	0.37
	B − C	−1.44	0.22	0.35
Model 2	A − B	0.76	0.30	0.30
(Compound	A − C	−0.62	0.37	0.36
symmetry)	B − C	−1.38	0.34	0.35
Model 3	A − B	0.75	0.30	0.31
(Toeplitz)	A − C	−0.60	0.36	0.36
	B − C	−1.34	0.34	0.34

are similar to the model-based estimates. They are closest in Model 3, because this model uses more parameters and hence more closely reflects the observed covariance pattern of the data. However, in Model 1, which naively assumes that the repeated observations are independent, the standard error estimates are noticeably different. The empirical estimates reflect the covariances between the repeated observations even though the model has assumed there is no covariance. Although the empirical estimates are based on the observed covariance of the data, note that they differ between the models. This is because the models differ in their estimates of the fixed effects, with resultant differences in the covariances.

The overall treatment effects in Model 2 were highly significant ($p = 0.0002$). Cold feet were significantly more likely on treatment B than on treatment A ($p = 0.01$), and on treatment B than on treatment C ($p = 0.0007$). The coefficients for the treatment effects are difficult to interpret directly, but, as before, by exponentiation we can calculate odds ratios and 95% confidence intervals:

A − B   2.14 (1.19 - 3.85),
A − C   0.54 (0.26 - 1.11),
B − C   0.25 (0.13 - 0.49).

These intervals are calculated using the 'model-based' standard errors and the $z_{0.975}$ statistic. The exact DF for t statistics were not available from the SAS macro. However, the patient DF of over 300 can be taken as a conservative estimate and the t statistic is then well approximated by the $z_{0.975}$ statistic. The confidence intervals estimate the odds ratio for the probability of a 'favourable' outcome on the first treatment compared with the second; for example, the odds ratio corresponding to the difference A − B is based on the probability of a 'favourable' outcome on treatment A compared with treatment B. In this context, favourable could be considered as category 1 (none) versus the other categories; as categories 1 or 2 (occasionally or less) versus the rest; as categories 1, 2 or 3 (less than 'most of the time') versus the rest; or as categories 1 to 4 versus category 5 (all of the time). Note that it is an inherent assumption of this model that the same odds ratio applies to every partition between the categories.

SAS code and output

Variables

        cf = post-treatment cold feet (1−5),
cf11-cf14 = pre-treatment cold feet, four dummy binary variables such
            that cf1i = 1 if cold feet = i, = 0 otherwise,
    t1-t2 = treatment, two dummy binary variables for three treatments,
    v1-v3 = visit, three dummy binary variables for three visits.

Model 1

```
PROC LOGISTIC;
MODEL cf = cf11 cf12 cf13 cf14 t1 t2 v1 v2 v3;
```

                    The LOGISTIC Procedure

  Data Set: WORK.A
  Response Variable: CF
  Response Levels: 5
  Number of Observations: 1048
  Link Function: Logit

Example   169

```
               Response Profile

          Ordered
          Value     CF      Count

            1        1        855
            2        2         58
            3        3         68
            4        4         27
            5        5         40
```

```
    Score Test for the Proportional Odds Assumption

       Chi-Square = 39.8740 with 27 DF (p=0.0526)
```

```
Model Fitting Information and Testing Global Null Hypothesis BETA=0

                          Intercept
              Intercept      and
  Criterion     Only      Covariates   Chi-Square for Covariates

  AIC          1522.590    1425.478          .
  SC           1542.408    1489.889          .
  -2 LOG L     1514.590    1399.478     115.111 with 9 DF (p=0.0001)
  Score           .           .        118.914 with 9 DF (p=0.0001)
```

```
            Analysis of Maximum Likelihood Estimates

              Parameter Standard    Wald       Pr>     Standardized Odds
Variable DF   Estimate  Error   Chi-Square Chi-Square   Estimate    Ratio

INTERCP1 1     0.6437   0.2953    4.7521    0.0293        .           .
INTERCP2 1     1.1175   0.2970   14.1619    0.0002        .           .
INTERCP3 1     1.9708   0.3063   41.4080    0.0001        .           .
INTERCP4 1     2.5518   0.3211   63.1440    0.0001        .           .
CF11     1     1.9273   0.2333   68.2635    0.0001      0.502429    6.871
CF12     1     1.2850   0.2839   20.4849    0.0001      0.247530    3.615
CF13     1     1.1422   0.3508   10.6035    0.0011      0.161399    3.134
CF14     1     0.8522   0.3951    4.6530    0.0310      0.086750    2.345
T1       1    -0.5936   0.2370    6.2727    0.0123     -0.153864    0.552
T2       1    -1.3443   0.2197   37.4267    0.0001     -0.348186    0.261
V1       1     0.2887   0.2389    1.4606    0.2268      0.070300    1.335
V2       1     0.2431   0.2401    1.0250    0.3113      0.058502    1.275
V3       1    -0.1098   0.2309    0.2261    0.6344     -0.026160    0.896
```

Estimates for the treatment difference A − B were obtained by reparameterising the model to fit variables $t1$ and $t3$ (representing treatments A and C) so that treatment B was the redundant category.

Models 2 and 3   The SAS macro written by Lipsitz *et al.* (1994) was used to perform these analyses (see Section 9.1). This macro can be obtained from Web page [www.med.ed.ac.uk/phs/mixed].

5

Multi-Centre Trials and Meta-Analyses

In this chapter we consider the analysis of data that are collected from several centres or trials. Such datasets in which observations have a natural grouping can be described as *hierarchical*. Use of a random effects model to analyse a hierarchical dataset often leads to results that can be generalised more widely. Section 5.1 provides an introduction to multi-centre trials; the implications of fitting different models are considered in Section 5.2; a worked example is given in Section 5.3; some general points specific to hierarchical datasets are made in Section 5.4 and sample size estimation methods are introduced in Section 5.5. Meta-analysis is considered in Section 5.6 and an example follows in Section 5.7.

5.1 INTRODUCTION TO MULTI-CENTRE TRIALS

5.1.1 What is a multi-centre trial?

A multi-centre trial is carried out at several centres either because insufficient patients are available for the study at any one centre, or with the deliberate intention of assessing the effectiveness of treatments in several settings. Sometimes there will be extra variability in treatment effect estimates which can be due to differences between the centres (e.g. different investigators, types of patients, climates). This extra variation can be taken into account in the analysis by including centre and centre·treatment effects as random effects in the model. Such variation is likely to be most noticeable in trials that do not compare drugs. For example, in a trial to compare surgical procedures there may be varying levels of experience available at each centre with the different procedures. This will lead us to expect a positive variance component for the centre·treatment effects.

5.1.2 Why use mixed models to analyse multi-centre data?

When centre and centre·treatment effects are fitted as random, allowance is made for variability in the magnitude of the treatment effects between centres. However, deciding whether centre and centre·treatment effects should be fixed or random is often the subject of debate. In practice, the choice will depend on whether treatment estimates are to relate only to the set of centres used in the study or, more widely, to the circumstances and locations of which the trial centres can be regarded as a sample. In the former case, local treatment estimates for the sampled set of individual centres are obtained by fitting centre and centre·treatment effects as fixed. To obtain global treatment estimates, centre and centre·treatment effects should be fitted as random. When this is done the standard error of treatment differences is increased to reflect the heterogeneity of the treatment effects across centres.

If the centre·treatment term is omitted, there is a choice of whether to fit centre effects as fixed or random. Taking centre effects as random can increase the accuracy of treatment estimates, since information from the centre error stratum is used in addition to that from the residual stratum. Thus, it is nearly always beneficial to fit centres as random, regardless of whether a local or global interpretation is required. The amount of extra information will depend on the degree of treatment imbalance within the centres and the relative sizes of the variance components.

In the analysis of multi-centre trials it is important to check whether results from any particular centre are outlying. If this occurs it may be an indication that a centre has not followed the protocol correctly. In the fixed effects model, spurious outlying estimates caused by random variation may occur, particularly in small centres. In contrast, the shrunken estimates of centre and centre·treatment effects obtained by the random effects model do not have this problem.

5.2   THE IMPLICATIONS OF USING DIFFERENT ANALYSIS MODELS

In this section we look more closely at the implications of fitting centre and centre·treatment effects as fixed or random. We will consider four different models and in each of them treatment effects and baseline effects (if available) are fitted as fixed.

5.2.1   Centre and centre·treatment effects fixed

Treatments effects

These are estimated with equal weight given to results from each centre regardless of size. If centre sizes very greatly, this can cause results to differ markedly from analyses not fitting centre·treatment effects as fixed. Another potential difficulty with this method is that treatment effects cannot be estimated at all unless all treatments are received at every centre. For example, if no patients received treatment A at one centre then all comparisons involving treatment A would be non-estimable. In practice, we note, however, that this problem could be resolved by the amalgamation of centres with small numbers of patients.

Treatment standard errors

These are based on the residual (within-centre) variation. The variance (SE^2) of a treatment difference is given by

$$\text{var}(t_i - t_j) = \sigma^2(1/n_i + 1/n_j),$$

where σ^2 is the residual variance and n_i and n_j are the numbers of patients receiving treatments i and j. When equal numbers of patients, r, receive each treatment at each of c centres, we have $n_i = rc$ and the variance can be written

$$\text{var}(t_i - t_j) = 2\sigma^2/rc.$$

This variance will always be less than or equal to that for the model fitting centre·treatment effects as random (see Subsection 5.2.3).

Outlying centres

These can be determined from the centre and centre·treatment effect means. However, these can be misleading for small centres which could appear outlying due only to random variation.

Inference

This strictly applies only to those centres which were included in the trial.

5.2.2 Centre effects fixed, centre·treatment effects omitted

This fixed effects model is often used when the centre·treatment effects in the previous model are non-significant. However, in practice, there is often a lack of power to detect small centre·treatment effects and hence variability in the treatment effects across centres is often ignored.

Treatment estimates

These take account of differing centre sizes and are not estimated with equal weight given to results from each centre as in the previous model.

Treatment standard errors

These are based on the residual (within-centre) variation. The variances of treatment differences are given by the formula used in the previous model.

Outlying centres

These can be determined from the centre means. However, again, these can be misleading for small centres which could be apparently outlying due only to random variation.

Inference

This strictly applies only to those centres which were included in the trial. However, extrapolation of the inferences regarding the effect of treatment to other centres seems more reasonable when an assumption has been made that the treatment effect does not depend on the centre in which it has been applied.

5.2.3   Centre and centre·treatment effects random

Unlike the commonly used fixed effects approach of dropping a non-significant centre·treatment interaction, centre·treatment effects are retained in the random effects model provided the centre·treatment variance component, σ_{ct}^2, is positive. Thus, variation in treatment effects across centres is allowed even though its existence may not have been 'proven' by a significance test.

Treatment estimates

These take account of differing centre sizes. They are estimated using information from the centre·treatment error stratum, and also from the centre stratum if treatment effects are not balanced across centres. They are estimable even if some treatments are not received at every centre.

Treatment standard errors

These are based on the centre·treatment variation. If an equal number of patients receives each treatment at every centre the variance of treatment effect differences can be obtained as

$$\text{var}(t_i - t_j) = 2(\sigma^2/rc + \sigma_{ct}^2/c).$$

Thus, the variance is directly related to the size of the centre·treatment variance component, σ_{ct}^2, and the number of centres sampled, c. It is always expected to be greater than the fixed effects model variance, $2\sigma^2/rc$, unless σ_{ct}^2 is negative.

The variance is not as easily specified when centre sizes are unequal. However, if the proportion of patients allocated to each treatment is the same across centres, the variance is of the form

$$\text{var}(t_i - t_j) = \text{var}_{\text{FE}}(t_i - t_j) + k\sigma_{ct}^2,$$

where $\text{var}_{\text{FE}}(t_i - t_j)$ is the fixed effects model variance and k is a positive constant. Thus, the variance is again always expected to be greater or equal to that in the fixed effects model.

In most situations though, the proportions of patients receiving each treatment will differ to some extent from centre to centre. When this is the case, a general formula for the variance cannot be specified. The effect of the centre·treatment variance component will, however, still cause an increase in the variance of the treatment differences.

Outlying centres

These can be determined using the shrunken centre and centre·treatment estimates. Shrinkage is greater for small centres and therefore spurious outlying estimates will not be obtained for small centres.

Inference

This relates to the 'population' of possible centres from which those in the trial can be regarded as a random sample.

5.2.4 Centre effects random, centre·treatment effects omitted

This model is useful when local estimates are required, since smaller treatment standard errors are often obtained compared with a fixed effects model by recovering extra information from the centre error stratum.

Treatment estimates

These take account of differing centre sizes. Additional information on treatments is recovered from the centre error stratum.

Treatment standard errors

These are based on the residual (within-centre) variation. When the proportion of patients allocated to each treatment is the same across centres, the variances of treatment differences are given by the formula used in the first fixed effects model. When it is not, the variance is expected to be less, since extra information is recovered on treatment effects from the centre error stratum.

Outlying centres

These can be determined using the shrunken centre estimates. Shrinkage is greater for small centres and therefore spurious outlying estimates are not obtained for small centres.

Inference

This relates to the centres sampled only. Only under the strong assumption that there is no centre·treatment interaction can inferences be applied to the population of centres from which those in the trial can be considered a sample.

5.3 EXAMPLE: A MULTI-CENTRE TRIAL

We have already considered in some detail the analysis of a multi-centre trial of treatments for hypertension in Sections 1.3 and 2.5. Here, we discuss more fully the interpretation of the results from these analyses and consider estimates of treatment effects by centre.

Results from fixed and random effects analyses of DBP are summarised in Table 5.1. An initial fixed effects model including centre·treatment effects was also fitted. However, overall treatment effects were not estimable in this model, because all treatments were not received at every centre (see Table 1.1 in Section 1.3). This model gave a non-significant p-value for centre·treatment effects ($p = 0.19$) and thus the usual fixed effects approach would have been to remove centre·treatment effects to give Model 1.

The centre·treatment variance component is positive in Model 3 and this leads to an increase in the treatment standard errors over the fixed effects model (as indicated by the variance formulae given in Section 5.2). However, note that the baseline standard error is similar between the models. This is because baseline effects are estimated at the residual error level and not at the centre·treatment level.

Table 5.1   Results from fixed and random effects analyses of diastolic blood pressure.

Model	Fixed effects	Random effects	Method
1	Treatment, baseline, centre	—	OLS
2	Treatment, baseline	Centre	REML
3	Treatment, baseline	Centre, treatment·centre	REML

	Variance components (SEs)		
Model	Centre	Treatment·centre	Patient
1	—	—	71.9 (−)
2	7.82 (3.99)	—	70.9 (6.2)
3	6.46 (4.35)	4.10 (5.33)	68.4 (6.5)

	Treatment effects (SEs)			
Model	Baseline	A − B	A − C	B − C
1	0.22 (0.11)	1.20 (1.24)	2.99 (1.23)	1.79 (1.27)
2	0.30 (0.11)	1.03 (1.22)	2.98 (1.21)	1.95 (1.24)
3	0.28 (0.11)	1.29 (1.39)	2.93 (1.37)	1.64 (1.40)

Results from Model 3 can be related to the potential population of centres. Since there are 29 centres, there are no problems arising from an inadequate number of DF for the variance components, and we can be confident in presenting these results if global inference is required.

The centre variance component is positive in Models 2 and 3 and therefore some information on treatments will be recovered from the centre error stratum. However, the standard errors in Model 2 are only slightly smaller than in Model 1, indicating that only a small amount of information has been recovered. Also, some of the improvement in the standard error may be due to the smaller residual variance which has resulted from the use of this model.

Plots of the centre and centre-treatment effects from Model 3 were used in Section 2.5 to assess the normality of the random effects and to check whether any centres were outlying. Additionally, we can now take differences between the centre-treatment effects within each centre to calculate treatment effect estimates for each centre. In Model 3 these will be shrunken towards the overall treatment mean. We illustrate this by calculating the treatment difference $A - C$ for just the first eight centres in the study. Unshrunken fixed effects estimates are also calculated for comparison using results from the initial model (fitting treatment, centre and centre-treatment effects as fixed).

		Number of treatment estimates (SE)	
Centre	**Patients**	**Fixed model**	**Random model**
1	39	3.81 (3.34)	3.19 (2.30)
2	10	−5.67 (6.80)	1.56 (2.88)
3	8	25.66 (7.63)	6.05 (2.93)
4	12	−0.12 (5.90)	2.42 (2.80)
5	11	14.12 (7.21)	4.56 (2.89)
6	5	2.89 (8.37)	3.03 (2.97)
7	18	7.38 (4.82)	4.22 (2.66)
8	6	−4.68 (8.33)	2.15 (2.97)

It can be seen that, in general, shrinkage is towards the overall treatment difference of 2.92 (although this is not the case for all centres, because the models make different adjustments for baseline effects). The relative shrinkage (i.e. (fixed estimate − random estimate)/(fixed estimate)) is usually greatest for the smaller centres. The standard errors of the random effect estimates are smaller than those of the fixed effects estimates, because the random effects model utilises information on the treatment effects in the full sample as well as information from the centre of interest. By contrast, the fixed effects standard errors do not utilise the full sample information and are larger, because they are calculated using only information from the centre of interest. This also causes the fixed effects standard errors to vary greatly between the centres, because they are directly related to the centre sizes. It is difficult to determine whether any of the centres are outlying using the fixed effects estimates, because they need to be considered bearing in mind centre size. For example, at centre 3 a very large treatment difference is given by the fixed estimate, but the shrunken random estimate appears acceptable.

SAS code and output

Variables

```
centre  = centre number,
treat   = treatment (A, B, C),
patient = patient number,
dbp     = diastolic blood pressure at last attended visit,
dbp1    = baseline diastolic blood pressure.
```

The SAS code to produce the main results is given at the end of Section 2.5. Here, we give the code for obtaining the shrunken and unshrunken treatment effects at the first eight centres. PROC MIXED is used first to fit Model 3. ESTIMATE statements are included to calculate the shrunken treatment differences at the first eight centres. Next, a fixed effects model is fitted, which again uses ESTI-MATE statements to calculate (unshrunken) treatment differences. Two datasets, random and fixed, are then created to abstract and label the random and fixed effects estimates. The rest of the code listed is concerned with merging and printing the two sets of estimates.

```
PROC MIXED; CLASS centre treat;
TITLE 'RANDOM EFFECTS MODEL';
MODEL dbp = dbp1 treat;
RANDOM centre centre*treat;
ESTIMATE 'A-C,1' treat 1 0 -1| centre*treat 1 0 -1;
ESTIMATE 'A-C,2' treat 1 0 -1| centre*treat 0 0 0  1 0 -1;
ESTIMATE 'A-C,3' treat 1 0 -1| centre*treat 0 0 0  0 0 0  1 0 -1;
ESTIMATE 'A-C,4' treat 1 0 -1| centre*treat 0 0 0  0 0 0  0 0 0  1 0 -1;
ESTIMATE 'A-C,5' treat 1 0 -1| centre*treat 0 0 0  0 0 0  0 0 0  0 0 0  1 0
-1;
ESTIMATE 'A-C,6' treat 1 0 -1|
 centre*treat 0 0 0  0 0 0  0 0 0  0 0 0  0 0 0 1 0 -1;
ESTIMATE 'A-C,7' treat 1 0 -1|
 centre*treat 0 0 0  0 0 0  0 0 0  0 0 0  0 0 0  1 0 -1;
ESTIMATE 'A-C,8' treat 1 0 -1|
 centre*treat 0 0 0  0 0 0  0 0 0  0 0 0  0 0 0  0 0 0  1 0 -1;
MAKE 'estimate' OUT=random;

PROC MIXED data=a1; CLASS centre treat;
TITLE 'FIXED EFFECTS MODEL';
MODEL dbp = dbp1 treat centre centre*treat;
ESTIMATE 'A-C,1' treat 1 0 -1 centre*treat 1 0 -1;
ESTIMATE 'A-C,2' treat 1 0 -1 centre*treat 0 0 0  1 0 -1;
ESTIMATE 'A-C,3' treat 1 0 -1 centre*treat 0 0 0  0 0 0  1 0 -1;
ESTIMATE 'A-C,4' treat 1 0 -1 centre*treat 0 0 0  0 0 0  0 0 0  1 0 -1;
ESTIMATE 'A-C,5' treat 1 0 -1 centre*treat 0 0 0  0 0 0  0 0 0  0 0 0  1 0
-1;
```

```
ESTIMATE 'A-C,6' treat 1 0 -1
 centre*treat 0 0 0  0 0 0  0 0 0  0 0 0  0 0 0  1 0 -1;
ESTIMATE 'A-C,7' treat 1 0 -1
 centre*treat 0 0 0  0 0 0  0 0 0  0 0 0  0 0 0  0 0 0  1 0 -1;
ESTIMATE 'A-C,8' treat 1 0 -1
 centre*treat 0 0 0  0 0 0  0 0 0  0 0 0  0 0 0  0 0 0  0 0 0  1 0 -1;

MAKE 'estimate' OUT=fixed;
DATA random; SET random;
centre=substr(parm,5,2)*1;
type=1;
KEEP centre est se;

DATA fixed; SET fixed;
centre=substr(parm,5,2)*1;
type=2;
estf=est;
sef=se;
KEEP centre estf sef;

PROC SORT DATA=a; BY centre;
PROC MEANS NOPRINT DATA=a; BY centre; VAR dbp; OUTPUT OUT=freq N=freq;

DATA a; MERGE fixed random freq; BY centre;
IF centre<=8;
shrink=abs(estf-est);
PROC PRINT SPLIT='*' noobs; VAR centre freq estf sef est se;
LABEL centre='**CENTRE'
freq='PATIENTS*AT*CENTRE'
estf='FIXED*MODEL*ESTIMATE'

sef='FIXED*MODEL*SE'
est='RANDOM*MODEL*ESTIMATE'
se='RANDOM*MODEL*SE'
shrink='SHRINKAGE';
FORMAT estf est se sef 8.2;
```

RANDOM EFFECTS MODEL

The MIXED Procedure

Class Level Information

Class	Levels	Values
CENTRE	29	1 2 3 4 5 6 7 8 9 11 12 13 14
		15 18 23 24 25 26 27 29 30 31
		32 35 36 37 40 41
TREAT	3	A B C

REML Estimation Iteration History

Iteration	Evaluations	Objective	Criterion
0	1	1550.3451721	
1	3	1533.6847949	0.00000322
2	1	1533.6822800	0.00000000

Convergence criteria met.

Covariance Parameter Estimates (REML)

Cov Parm	Ratio	Estimate	Std Error	Z	Pr>|Z|
CENTRE	0.09453026	6.46282024	4.35101647	1.49	0.1374
CENTRE *TREAT	0.05991350	4.09615054	5.33114055	0.77	0.4423
Residual	1.00000000	68.36773938	6.52122280	10.48	0.0001

Model Fitting Information for DBP

Description	Value
Observations	288.0000
Variance Estimate	68.3677
Standard Deviation Estimate	8.2685
REML Log Likelihood	-1027.82
Akaike's Information Criterion	-1030.82
Schwarz's Bayesian Criterion	-1036.29
-2 REML Log Likelihood	2055.639

Tests of Fixed Effects

Source	NDF	DDF	Type III F	Pr>F
DBP1	1	208	6.31	0.0128
TREAT	2	48	2.28	0.1131

ESTIMATE Statement Results

Parameter	Estimate	Std Error	DDF	T	Pr > |T|
A-C,1	3.19143079	2.30383126	48	1.39	0.1724
A-C,2	1.55956887	2.88100385	48	0.54	0.5908
A-C,3	6.04931256	2.92885838	48	2.07	0.0443
A-C,4	2.42075237	2.79968665	48	0.86	0.3915
A-C,5	4.56090946	2.89200757	48	1.58	0.1213
A-C,6	3.02957362	2.97056871	48	1.02	0.3129
A-C,7	4.21760605	2.65563569	48	1.59	0.1188
A-C,8	2.15105586	2.96967610	48	0.72	0.4724

FIXED EFFECTS MODEL

The MIXED Procedure

Class Level Information

Class	Levels	Values
CENTRE	29	1 2 3 4 5 6 7 8 9 11 12 13 14 15 18 23 24 25 26 27 29 30 31 32 35 36 37 40 41
TREAT	3	A B C

Model Fitting Information for DBP

Description	Value
Observations	288.0000
Variance Estimate	69.2614
Standard Deviation Estimate	8.3223
REML Log Likelihood	-779.059

```
Akaike's Information Criterion    -780.059
Schwarz's Bayesian Criterion      -781.727
-2 REML Log Likelihood            1558.117
```

Tests of Fixed Effects

Source	NDF	DDF	Type III F	Pr>F
DBP1	1	208	0.99	0.3198
REAT	2	208	1.24	0.2905
CENTRE	28	208	1.98	0.0038
CENTRE*TREAT	48	208	1.20	0.1884

ESTIMATE Statement Results

Parameter	Estimate	Std Error	DDF	T	Pr > |T|
A-C,1	3.80805091	3.33644322	208	1.14	0.2550
A-C,2	-5.67204919	6.80364977	208	-0.83	0.4054
A-C,3	25.65590162	7.62753885	208	3.36	0.0009
A-C,4	-0.11894466	5.89717973	208	-0.02	0.9839
A-C,5	14.12298155	7.20848683	208	1.96	0.0514
A-C,6	2.88912913	8.36997182	208	0.35	0.7303
A-C,7	7.38105534	4.82007888	208	1.53	0.1272
A-C,8	-4.68245388	8.32843415	208	-0.56	0.5746

The table of shrunken and unshrunken treatment estimates is given within the main text.

5.4 PRACTICAL APPLICATION AND INTERPRETATION

In this section we consider some general points relating specifically to analyses of multi-centre data.

5.4.1 Plausibility of a centre·treatment interaction

One approach to analysing multi-centre trials of drug treatments works from the premise that it is not plausible for a treatment effect to vary across centres. If it does, it is deemed a fault with the study design. If a significant centre·treatment interaction is not detected, then the design is assumed to be sound and global inference is made from a model not allowing for any variation in treatment effects between centres. However, a drug effect can sometimes vary due to differences in the centre populations even when they are defined within the constraints of the protocol. For example, one drug may work better on severely ill patients than another but less well on moderately ill patients. Thus, a centre containing more severely ill patients could produce larger treatment effects than centres containing a more even mixture of patients. For this reason it is our belief that centre·treatment effects are always plausible and that if global inference is required from a multi-centre trial, then the random effects model is likely to be the most appropriate.

Interactions are often even more plausible in trials not involving drugs. For example, in a trial of surgical techniques, one centre may have much more expertise with one technique than another. In this type of trial a random effects model should almost always be used in order to provide global inference.

5.4.2   Generalisation

Consideration of different interpretations of results from fixed effects and random effects analyses brings the issue of generalisation to the fore. Results are often generalised from the situation in which they were sampled to other situations. However, strictly speaking, results should only be generalised when the study sample has been taken at random from the whole population of interest. Since centres are rarely sampled at random, even the global results from a multi-centre trial cannot be formally generalised to the population of possible centres.

There are analogies though with a single-centre study. In such studies, patients are not usually selected at random from the potential population of patients available at the centre. However, results are usually seen as some indication of those expected in the future both in the same centre and elsewhere.

Thus, in practice, generalisation needs to be by degree and will always to some extent involve subjective judgements, e.g. of how well the centres (or patients) sampled represent their full potential populations. Multi-centre studies would usually be considered more generalisable than single-centre studies even though the centres are not randomly sampled, and even if a fixed effects model is employed.

5.4.3   Number of centres

The accuracy with which variance components are estimated is dependent upon the number of centres used. If the centre-treatment variance component is inaccurate, then this will have a direct effect on the accuracy of the treatment standard errors (calculated as $2(\sigma^2/rc + \sigma_{ct}^2/c)$ when data are balanced). Thus, if a study uses only a few centres (say less than about five), σ_{ct}^2 and hence the treatment effect standard error may not be accurate. In this situation a random effects analysis may be inadvisable, although it could used to provide a rough idea of the global treatment estimates. However, the main conclusions from the study should be based on local results obtained from a fixed effects model, usually omitting the centre-treatment interaction term.

5.4.4   Centre size

Sometimes a trial contains several centres that stop participating in the trial after recruiting only a few patients. Since little information is available for measuring effects at such centres, a strategy which is often adopted is to combine such centres into one centre in the analysis. This has most use when a fixed effects model is used, because the centre-treatment interaction can then be assessed more effectively. However, in a mixed model it usually makes little difference whether such centres are combined together or fitted separately.

5.4.5   Negative variance components

If centre-treatment effects are retained but the centre variance component estimate is negative, then centre effects could either be removed altogether from the model or retained with their variance component constrained to zero. Although the same effect estimates will be obtained using either approach, the DF used by the significance tests will differ. The latter option of retaining the centre effect DF is perhaps the more satisfactory because centre-treatment effects are retained. If both centre and centre-treatment variance components are negative, then they can be removed from the model and the data analysed as a simple between-patient trial. Inference can still be made globally to the population of centres, since there is no variation in the treatment effect across centres.

5.4.6   Balance

In models including treatment-centre effects balance will only be achieved in the unlikely situation where there are an equal number of patients per treatment per centre (this condition is also required to achieve balance across random effects). Treatment mean estimates from either a fixed or random effects model will then equal the raw treatment means. In the more usual situations where there are unequal numbers of patients per treatment per centre, treatment means will differ between fixed effects and random effects models.

In models omitting centre-treatment effects, balance across random effects is achieved when treatments are allocated in equal proportions at each centre (even if the overall centre sizes vary). Treatment estimates are then the same regardless of whether centre effects are taken as fixed or random. However, if treatments are not allocated in equal proportions at each centre, treatment estimates will differ between the models because information is combined from both the centre and residual error strata in the random effects model.

5.5   SAMPLE SIZE ESTIMATION

When designing a multi-centre trial with the intention of estimating global treatment effects, sample size estimates can be calculated in a way which takes into account variation of treatment effects between centres. Here, we will obtain sample sizes that can be used for trials, based on considering differences between pairs of treatments.

5.5.1 Normal data

In Section 5.2 the variance of the difference between a pair of treatments in a balanced dataset was given as

$$\text{var}(t_i - t_j) = 2(\sigma^2/rc + \sigma_{ct}^2/c),$$

where,

r = number of patients per treatment per centre (replicates),
t = number of treatments,
c = number of centres,
t_i = the ith treatment effect,
σ^2 = residual variance,
σ_{ct}^2 = centre·treatment variance component.

Estimates for the number of centres (c) and number of patients per treatment per centre (r) can be obtained from the usual sample size estimation equation:

$$\Delta = (t_{\text{DF}, 1-\alpha/2} + t_{\text{DF}, \beta}) \times \text{SE}(t_i - t_j),$$

where

α = significance level,
β = power,
Δ = difference to be detected,
$\text{DF} = (c - 1) \times (t - 1)$, the centres·treatment DF.

One difficulty is that estimates of both the patients and centres·treatment variance components are required. Unless multi-centre data are available from a previous study it is likely that only an estimate of the between-patient residual variance will be available. However, it may still be preferable to use the above formulae with a guessed value for the treatment·centre variance component, rather than assuming it is zero.

In this section we consider the situation where an equal number of patients will be used per treatment per centre. However, the calculations are also appropriate when there will be varying numbers per centre, as long as patients are evenly allocated to treatments. In this case, r is taken to be the average number of replicates, $\Sigma r_i/c$. There are three ways in which a sample size can be calculated:

1. *Number of centres (c) specified* This approach would be applicable if a decision had been made to use a specific number of centres. After substitution of the formula for $\text{SE}(t_i - t_j)$ in the sample size estimation equation, with some

reorganisation we find that the number of patients per replicate (i.e. per treatment per centre) required is given by

$$r = \frac{2(t_{DF,1-\alpha/2} + t_{DF,\beta})^2 \sigma^2}{c\Delta^2 - 2(t_{DF,1-\alpha/2} + t_{DF,\beta})^2 \sigma_{ct}^2}.$$

Therefore, $t \times r \times c$ patients are required in total. If this formula gives a negative value for r, then it is not possible to detect the specified difference with the required power unless more centres are used. Either c should be increased or, alternatively, the power could be decreased or Δ increased.

2. *Number of patients per centre $(t \times r)$ specified* This approach might be appropriate if the duration of the trial is limited and there is only time to recruit a specified number of patients per centre. The number of centres required is given by

$$c = \frac{2(t_{DF,1-\alpha/2} + t_{DF,\beta})^2 (\sigma^2 + r\sigma_{ct}^2)}{r\Delta^2}.$$

Obviously, $DF = (c-1) \times (t-1)$ will not be known in advance. z-values from the normal distribution can be used instead of values from the t distribution to obtain an initial estimate of c. A more accurate value can then be calculated by using the DF obtained for this value of c in the above formula, and re-estimating c. This can be repeated until convergence is obtained, but changes are usually minimal after the first iteration.

3. *Neither number of centres nor average patients per centre specified* In this situation an optimal sample size can only be calculated by specifying the relative cost of sampling centres compared with sampling patients. The cost of sampling centres will depend on the type of centre being used. For example, the cost of a centre in an international study would be extremely expensive, but centres would be much cheaper in a study using local practitioners. The cost of sampling patients relates to the amount to be paid to the investigator per patient plus the cost of monitoring, validating and processing each patient's data. If we denote the relative cost by g, then the total cost is proportional to $c \times r \times t + c \times g$. This is minimised when

$$r = \sqrt{\frac{g\sigma^2}{t\sigma_{ct}^2}}.$$

c is then obtained by substituting r into the formula given earlier:

$$c = \frac{2(t_{DF,1-\alpha/2} + t_{DF,\beta})^2 (\sigma^2 + r\sigma_{ct}^2)}{r\Delta^2}.$$

Sometimes the values calculated might appear impracticable. For example, if the relative cost, g, of sampling a centre were set to be not much higher than that of sampling a patient (i.e. g close to one), then the number of centres estimated is likely to be very high. In this situation, g has clearly been set too low and should be increased.

Example

We will calculate sample sizes for a new hypertension study to compare three treatments. Assuming that DBP is again the primary endpoint, the variance components obtained in Section 2.5 for centre·treatment effects ($\sigma_{ct}^2 = 4.10$) and the residual ($\sigma^2 = 68.4$) can be used to estimate sample sizes. A difference of 5 mmHg is to be detected at the 5% significance level with 90% power.

1. *Number of centres specified*   It has been decided that the new study will use four centres. Therefore, the DF for the t distribution is $(4 - 1) \times (3 - 1) = 6$ and we have

    $$r = \frac{(2 \times (t_{6,0.975} + t_{6,0.90})^2 \times 68.4)}{(4 \times 5^2 - 2 \times (t_{6,0.975} + t_{6,0.90})^2 \times 4.10)}$$

    $$= \frac{(2 \times (2.45 + 1.44)^2 \times 68.4)}{(4 \times 5^2 - 2 \times (2.45 + 1.44)^2 \times 4.10)}$$

    $$= -86.0.$$

    Since r is negative it is not possible to obtain the required power when only four centres are used. If the number of centres is increased to six, then DF $= (6 - 1) \times (3 - 1) = 10$ and

    $$r = \frac{(2 \times (t_{10,0.975} + t_{10,0.90})^2 \times 68.4)}{(6 \times 5^2 - 2 \times (t_{10,0.975} + t_{10,0.90})^2 \times 4.10)}$$

    $$= \frac{(2 \times (2.23 + 1.37)^2 \times 68.4)}{(6 \times 5^2 - 2 \times (2.23 + 1.37)^2 \times 4.10)}$$

    $$= 40.5.$$

    Thus, $41 \times 3 = 123$ patients would be required in each of the six centres and the total number of patients is $123 \times 6 = 738$.

2. *Number of patients per centre specified*   It has been decided that an average of only 15 patients per centre will be used so that the study can be completed quickly. Using $r = 5$ we obtain an initial estimate of c using z statistics as

    $$c = 2 \times (1.96 + 1.28)^2 \times (68.4 + 5 \times 4.10)/(5 \times 5^2) = 14.9.$$

    With 15 centres the t distribution DF for use in the original formula would be $(15 - 1) \times (3 - 1) = 28$. On recalculating using t_{28} statistics c becomes 16.1, which should be rounded up to 17. Therefore, 17 centres should be used with five patients per treatment group ($17 \times 5 \times 3 = 255$ patients in total).

3. *Neither number of centres n or average patients per centre specified*   The study will use centres from different countries and therefore the cost of sampling centres compared with patients is high. If we set the relative cost of sampling centres compared with sampling patients at $g = 100$, then

$$r = \sqrt{[100 \times 68.4/(3 \times 4.10)]} = 23.6.$$

Rounding r to 24 we obtain an initial estimate of c:

$$c = 2 \times (1.96 + 1.28)^2 \times (68.4 + 24 \times 4.10)/(24 \times 5^2) = 5.8.$$

This value of c is rounded to six. As c is small, it should be recalculated more accurately using DF $= (6 - 1) \times (3 - 1) = 10$ as

$$c = 2 \times (t_{10,0.975} + t_{10,0.90})^2 \times (68.4 + 24 \times 4.10)/(24 \times 5^2)$$
$$= 2 \times (2.23 + 1.37)^2 \times (68.4 + 24 \times 4.10)/(24 \times 5^2) = 7.21.$$

However, we can be more precise, because the necessity to have an integer number of centres means that in moving from 7.21 centres to eight centres, our value for r can be recalculated. In this instance it is reduced appreciably from 23.6 to 16.6. We can therefore obtain our desired power from 17 patients per treatment per centre, using eight centres. We could also consider the alternative of using seven centres and recalculating r. This gives $r = 23.6$. We can therefore compare the cost of seven centres with 24 patients per treatment per centre, with the design with eight centres.

5.5.2 Non-normal data

When the variable of primary interest is non-normal the formulae above can be adapted if the variable is suitable for analysis using a GLMM. The variance of the difference between a pair of treatments can be written

$$\mathrm{var}(t_i - t_j) = 2(v/rc + \sigma_{\mathrm{ct}}^2/c),$$

where

$v = \phi\, a/b$,

$a =$ denominator term for binomial data or offset term for count data (see Section 3.1.3),

$b =$ expected variance, e.g. $\mu(1 - \mu)$ for binary data or μ for count data (see Section 3.1.3)

$\phi =$ dispersion parameter.

a will be one in most situations and an average value can be used for μ when setting b (for example, if the expected proportions for two treatments were 0.4 and 0.2, μ could be taken as 0.3 and b as $0.3 \times 0.7 = 0.21$). The sample size formulae given above for normal data can then be used with $\phi a/b$ substituted for σ^2, and Δ taken as the required treatment difference defined on the 'linked' scale (e.g. using logits for binary data or logs for count data).

Example: binary data

We will again calculate sample sizes for a new hypertension study to compare three treatments. However, now we assume (unrealistically!) that the incidence of the adverse event, cold feet, is the primary endpoint. A doubling in the proportion of cold feet is to be detected from 0.1 to 0.2 and thus the required difference in logits $(\log(\text{odds}(0.2)) - \log(\text{odds}(0.1)))$ is 0.81. Taking the average of 0.1 and 0.2 we obtain $\mu = 0.15$ and $b = 0.15(1 - 0.15) = 0.128$. a is taken as one, since the data are in Bernoulli form. From Section 3.4 we have $\sigma_{ct}^2 = 1.79$ and $\phi = 0.97$. Detection at the 5% significance level is required with 90% power.

We assume that there are no stipulations for the number of centres or patients and seek to design the cheapest trial. If we set the relative cost as $g = 50$, then

$$r = \sqrt{\{g \times \phi/(b \times t \times \sigma_{ct}^2)\}}$$
$$= \sqrt{\{50 \times 0.97/(0.128 \times 3 \times 1.79)\}}$$
$$= 8.4.$$

Rounding this value of r up to nine we obtain as an initial estimate of c:

$$c = 2(1.96 + 1.28)^2 \times (\phi/b + r\sigma_{ct}^2)/(r \times \Delta^2)$$
$$= 2(1.96 + 1.28)^2 \times (0.97/0.128 + 9 \times 1.79)/(9 \times 0.81^2)$$
$$= 84.2.$$

A very large number of centres is required because the centre-treatment variability, σ_{ct}^2, is high compared with the residual. Obtaining 85 centres is likely to be unrealistic, and we could either increase the relative cost of sampling centres to more than 50 or, alternatively, assume that $\sigma_{ct}^2 = 1.97$ may be an overestimate of the variability of treatment effects across centres and substitute a lower value.

Categorical data

Sample size estimation is always difficult when the variable of interest is categorical. If there are greater than about five categories the formula for continuous data is likely to provide a reasonable approximation. In other situations the best approach might be to partition the categories and use the formula for binary data.

Precision of sample size estimates

The use of sample size calculations gives very precise answers, but they are based on assumed values for variance components which may be quite imprecise, and specified differences which may be somewhat arbitrary. We do not therefore recommend slavish adherence to the precise numbers obtained from the formulae. It is sensible to undertake a kind of sensitivity analysis to see the extent to which the sample size depends on the assumptions made. We believe that the correct use of sample size calculations is to obtain reliable ball park figures.

5.6 META ANALYSIS

This type of analysis is increasingly used to combine results from several clinical trials which assess the same treatments in order to provide a more precise overall estimate of the treatment effects. When the original data are available, an identical hierarchical structure to the multi-centre trial arises, with trials replacing centres. The implications of fitting trial and trial-treatment effects as fixed or random are then the same as in multi-centre analyses (see Section 5.2).

If treatment estimates are to relate only to the trials included, then *local* treatment estimates are obtained by fitting trial and treatment-trial effects as fixed (although in practice the trial-treatment interaction is usually removed if non-significant). If they are to relate more widely to the circumstances and locations sampled by the trials, *global* estimates can be obtained by fitting trial and treatment-trial effects as random. When this is done the standard errors of treatment estimates are increased to reflect the heterogeneity across trials. Trial-treatment variance components are often larger than centre-treatment variance components, because different protocols are used by the different trials. Thus, there are often more noticeable increases in the treatment standard errors in meta-analyses than in multi-centre trials.

As with multi-centre trials, taking trial effects as random has the advantage of increasing the accuracy of treatment estimates. This is because information from the trial error stratum is used in addition to that from the residual stratum. Sometimes there are factors that differ at the trial level that can help to explain differences in results between trials. For example, race or type of clinic may affect the treatment effect size. These can be included as covariates in a mixed model and may reduce the trial-treatment variability (and hence lead to more precise treatment estimates).

Outlying trials can be checked for, using the shrunken estimates of trial and trial-treatment effects. Since shrinkage is greater when there are fewer observations per trial, spurious outlying estimates caused by random variation are less likely to occur. However, this would not be the case in a fixed effects model where there is no shrinkage of the trial and trial-treatment estimates. Often, the estimates of treatment effects from individual trials are themselves of interest. The shrunken estimates which utilise information from all trials are more robust than estimates from a fixed effects model, although it has to be recognised that there may be difficulty in conveying the concept of shrunken estimates to a medical researcher!

Meta-analysis tends to be carried out most frequently with binomial data and most published work has related to this. Although meta-analysis can be used with normal data, in practice individual trials are often adequate to achieve the desired power. However, in a pharmaceutical company it may be advantageous to undertake a meta-analysis on normal data arising from a series of trials in the same drug program.

5.7   EXAMPLE: META-ANALYSIS

This example considers meta-analysis data which are taken from Thompson and Pocock (1991). The data come from nine trials comparing a diuretic treatment with a control treatment in relation to the incidence of pre-eclampsia. The numbers of women with pre-eclampsia within each trial and treatment group are shown in Table 5.2.

5.7.1   Analyses

The four analysis models shown in Table 5.3 are considered. Models 1 and 2 were fitted using fixed effects models (GLMs). In these models the data are analysed in binomial form (i.e. using the trial-treatment frequencies). However, identical results would have been obtained had the data been analysed in Bernoulli form (i.e. using one observation per patient). Model 3 is a random effects model and the data are analysed in Bernoulli form using a GLMM. Using this form allows the 'shrunken' treatment effects at each trial to be estimated using the SAS macro

Table 5.2   Frequencies of pre-eclampsia/numbers randomised in trials included in meta-analysis.

Trial	Diuretic	Control	Odds ratio
Weseley	14/131	14/136	1.04
Flowers	21/385	17/134	0.40
Menzies	14/57	24/48	0.33
Fallis	6/38	18/40	0.23
Cuadros	12/1011	35/760	0.25
Landesman	138/1370	175/1336	0.74
Krans	15/506	20/524	0.77
Tervila	6/108	2/103	2.97
Campbell	65/153	40/102	1.14
Total	291/3759	345/3183	0.72

Table 5.3 Variance component and treatment odds ratio estimates.

Model	Method (data form)	Fixed effects	Random effects
1	GLM (binomial)	Treatment, trial, trial·treatment	—
2	GLM (binomial)	Treatment, trial	—
3	P-L (Bernoulli)	Treatment	Trial, trial·treatment
3(a)	P-L (binomial)	Treatment	Trial, trial·treatment
4	Bayes (Bernoulli)	Treatment	Trial, trial·treatment

Model	Variance components Trial	Trial·treatment	Dispersion parameter	Treatment odds ratio (95% CI)	p-value
1	—	—	1^F	0.63 (0.47−0.84)	0.002
2	—	—	1^F	0.66 (0.56−0.79)	<0.0001
3	1.42	0.16	0.98	0.60 (0.36−1.01)*	0.05
3(a)	1.43	0.16	1^F	0.60 (0.36−1.01)*	0.05
4	1.32	0.21	1^F	0.60 (0.35−1.01)	0.06

Notes: ^{F} = Parameter is fixed.
 * = Calculated using $t_{8,0.975}$ statistics.

GLIMMIX. However, very similar results can be obtained by analysing the data as binomial frequencies, since there are no baseline values and no categories are uniform. We illustrate this by fitting Model 3(a) which is identical to Model 3 except that the data are in binomial form and the dispersion parameter is fixed at one (allowing variance at the residual level to be modelled by the trial·treatment variance component). Model 4 is the same random effects model as Model 3 but is fitted using the Bayesian approach. This was done using the Gibbs sampler in the BUGS package (see Section 2.3.5) in the same way as described for the multi-centre analysis performed in Section 2.5 (Model 5).

Odds ratios and confidence intervals are calculated by exponentiating the treatment difference estimates and their confidence intervals on the logit scales (see Section 3.3.10). The treatment effect is tested using asymptotic Wald chi-square tests in Models 1 and 2, and an F test in Models 3 and 3(a) (see Section 3.3.9 for details on GLM and GLMM significance testing). In Model 4 twice the probability of the treatment difference being greater than zero is taken to provide a 'Bayesian' p-value (see Section 2.3.3).

5.7.2 Results

The results are shown in Table 5.3. Results from Models 1 and 2 do not take account of any additional variation in the treatment effect between trials and therefore should be formally related only to the trials included. The trial·treatment interaction was highly significant in Model 1 ($p = 0.0006$) and this would cast doubt on any formal extrapolation of the results from these models. In Model 1 the overall treatment effect is calculated as an unweighted average of the treatment effects at each trial (see Section 5.2). Such an estimate is clearly inappropriate, since the trial sizes differ widely. This problem does not arise in Model 2, where centre-treatment effects are omitted. Note that in the above models the confidence intervals are based on exponentiating the treatment estimate $\pm 1.96 \times$ SE, because of the asymptotic normality of the estimate.

Models 3, 3(a) and 4 take account of the extra variation in the treatment effect between trials by fitting trial and trial-treatment effects as random. In these models we are assuming that the random effects are normally distributed, and as the variance components are estimated with 8 DF, the confidence intervals are more appropriately estimated as treatment estimate $\pm t_{8,0.975} \times$ SE. Model 3(a) gives very similar results to Model 3, indicating that it makes little difference here whether the data are analysed in Bernoulli or binomial form. The trial variance component is fairly large in all of the models, indicating that the overall incidence of pre-eclampsia varies greatly between trials. Thus, it is likely that quite different inclusion criteria were used for the trials or that pre-eclampsia was defined differently by the different practitioners. The positive trial-treatment components indicates some variation in the treatment effect across trials. This is reflected in the size of the treatment confidence intervals, which are wider than those in Models 1 and 2. The results from these analyses can be generalised with some confidence to the full population of pre-eclampsia sufferers.

The Bayesian analysis (Model 4) gives almost identical treatment ORs and confidence intervals to the pseudo-likelihood analyses (Models 3 and 3(a)). The differences in variance component estimates between Models 3 and 4 are not unexpected, since we have taken them to be the medians of the marginal posterior distributions in Model 4, whereas in Model 3 the estimates are the values which maximise the pseudo-likelihood surface.

5.7.3   Treatment estimates in individual trials

Another advantage of using a random effects model is that shrunken estimates of the treatment effect at each trial can be obtained. Because shrinkage is greater when there are fewer observations per trial, any spurious outlying estimates caused by random variation will not occur. The shrunken estimates also allow us to check whether results from any particular trial are outlying. If they are, it may be an indication that the trial was not suitable for inclusion in the meta-analysis. The shrunken estimates obtained from Model 3 are given in Table 5.4 along with the (unshrunken) GLM estimates obtained from Model 1 (on the logit scale). These estimates are all shrunken towards the overall treatment estimate of -0.51 ($\exp(-0.51) = 0.60$). The standard errors are smaller for the shrunken estimates because they utilise information from the whole sample, not just that from the individual trials. Greatest shrinkage occurs for the smallest trials. For example, the unshrunken estimate from Tervila appears extreme, but

Table 5.4 Shrunken and unshrunken treatment effect estimates (standard errors) at each trial.

Trial	Shrunken (Model 3)	Unshrunken (Model 1)	Number of patients
Weseley	−0.14 (0.33)	0.04 (0.40)	267
Flowers	−0.82 (0.30)	−0.92 (0.34)	519
Menzies	−0.91 (0.34)	−1.12 (0.42)	105
Fallis	−1.04 (0.40)	−1.47 (0.55)	78
Cuadros	−1.16 (0.28)	−1.39 (0.34)	1771
Landesman	−0.31 (0.12)	−0.30 (0.12)	2766
Krans	−0.32 (0.30)	−0.26 (0.35)	1030
Tervila	0.07 (0.46)	1.09 (0.83)	211
Campbell	0.03 (0.24)	0.14 (0.26)	255

the shrunken estimate is much more reasonable and would not cause us to suspect the quality of the trial.

SAS code and output

Variables

treat = Treatment,

trial = Trial number,

eclam = Number of women with pre-eclampsia,

n   = Number of women in treatment group at trial.

Model 1

```
DATA a; INPUT trial treat eclam n;
CARDS;
1 1 14 131
1 2 14 136
2 1 21 385
2 2 17 134
3 1 14 57
3 2 24 48
4 1 6 38
4 2 18 40
5 1 12 1011
5 2 35 760
6 1 138 1370
6 2 175 1336
7 1 15 506
7 2 20 524
8 1 6 108
8 2 2 103
9 1 65 153
9 2 40 102
;
```

```
PROC GENMOD; CLASS treat trial;
MODEL eclam/n = treat trial trial*treat/ DIST=B TYPE3 WALD;
```

The output is not shown but has a similar form to that from Model 2 below.

Model 2

```
PROC GENMOD; CLASS treat trial;
MODEL eclam/n = treat trial/ DIST=B TYPE3 WALD;
```

Model Information

Description	Value
Data Set	WORK.A
Distribution	BINOMIAL
Link Function	LOGIT
Dependent Variable	ECLAM
Dependent Variable	N
Observations Used	18
Number Of Events	636
Number Of Trials	6942

Class Level Information

Class	Levels	Values
TREAT	2	1 2
TRIAL	9	1 2 3 4 5 6 7 8 9

Criteria For Assessing Goodness Of Fit

Criterion	DF	Value	Value/DF
Deviance	8	29.3761	3.6720
Scaled Deviance	8	29.3761	3.6720
Pearson Chi-Square	8	28.7996	3.6000
Scaled Pearson X2	8	28.7996	3.6000
Log Likelihood	.	-1877.3795	.

The deviance and Pearson chi-square are measures of model fit and have similar roles to the residual sum of squares in normal data models.

Analysis Of Parameter Estimates

Parameter		DF	Estimate	Std Err	ChiSquare	Pr>Chi
INTERCEPT		1	-0.1137	0.1379	0.6798	0.4096
TREAT	1	1	-0.4104	0.0885	21.4844	0.0000
TREAT	2	0	0.0000	0.0000	.	.
TRIAL	1	1	-1.8457	0.2380	60.1433	0.0000
TRIAL	2	1	-2.1344	0.2118	101.5184	0.0000
TRIAL	3	1	-0.2363	0.2409	0.9618	0.3267
TRIAL	4	1	-0.5054	0.2779	3.3073	0.0690
TRIAL	5	1	-3.2740	0.1958	279.6685	0.0000
TRIAL	6	1	-1.7287	0.1420	148.1486	0.0000

```
TRIAL       7    1    -3.0515      0.2150      201.3631    0.0000
TRIAL       8    1    -2.9293      0.3830       58.4970    0.0000
TRIAL       9    0     0.0000      0.0000           .          .
SCALE            0     1.0000      0.0000           .          .
```

NOTE: The scale parameter was held fixed.

```
                Wald Statistics For Type 3 Analysis

            Source       DF    ChiSquare   Pr>Chi
            TREAT         1      21.4844   0.0000
            TRIAL         8     444.0331   0.0000
```

Asymptotic Wald chi-square tests are performed for each fixed effect parameter. These tests should be interpreted cautiously in small datasets.

Model 3

Variables

outcome = Success of treatment (1/0),
one   = 1 for all observations,
trial   = Trial number,
treat   = Treatment group,
freq   = Number of women with this outcome.

```
%inc 'glimmix.sas';
%GLIMMIX(STMTS=
   %STR(CLASS trial treat;
   MODEL outcome/one=treat;
   RANDOM=trial treat*trial;),
   ERROR=binomial,
   FREQ=freq);
```

This code analyses the data in Bernoulli form so that there is a 0/1 observation corresponding to each patient in the trial. The FREQ option is used here to indicate the number of times each observation in our dataset is repeated — this saves setting up a large dataset containing 6942 observations, many of which would be the same.

```
                Goodness-of-Fit Statistics

         Description                        Value
         Deviance                         3729.1679
         Scaled Deviance                  3824.7592
         Pearson Chi-Square               6753.5024
         Scaled Pearson Chi-Square        6926.6178
         Extra-Dispersion Scale              0.9750
```

As in PROC GENMOD the deviance and Pearson chi-square are measures of model fit and the scaled deviance and scaled Pearson chi-square are equal to their unscaled values divided by the scale (dispersion) parameter (0.9750).

```
                 Covariance Parameter Estimates

                 Cov Parm            Estimate

                 TRIAL               1.42317699
                 TRIAL*TREAT         0.16213468
                 Residual            0.97500722

                    Parameter Estimates

                        Standard
Parameter   Estimate     Error      ChiSquare    Pr>ChiSq       Mu

INTERCEPT   -1.8137     0.4289        17.8842     0.0000     0.1402
TREAT 1     -0.5105     0.2266         5.0758     0.0243     0.3751
TREAT 2      0.0000        .              .          .       0.5000
```

The 'Covariance Parameter Estimates' are presented in the same format as in the
PROC MIXED table. The 'Parameter Estimates' tables can be interpreted in the
same way as in PROC GENMOD.

Model 3(a)

Variables

As in Models 1 and 2.

```
%GLIMMIX(STMTS=%STR(
   CLASS trial treat;
   MODEL eclam/n=treat;
   RANDOM trial trial*treat;
   PARMS (0.0) (0.0) (1.0)/ EQCONS=3;
   ESTIMATE 'A-B' treat 1 -1;
   ),
   ERROR=B);
```

The PARMS statement sets the initial values of the variance parameters and the
EQCONS option fixes the dispersion parameter (the third variance parameter)
at 1.0.

```
                 Class Level Information

           Class    Levels    Values

           TRIAL       9       1 2 3 4 5 6 7 8 9
           TREAT       2       1 2

             Covariance Parameter Estimates

             Cov Parm            Estimate

             TRIAL               1.42501872
             TRIAL*TREAT         0.15928102
```

```
            GLIMMIX Model Statistics

        Description                      Value

        Deviance                         3.9559
        Scaled Deviance                  3.9559
        Pearson Chi-Square               3.6973
        Scaled Pearson Chi-Square        3.6973
        Extra-Dispersion Scale           1.0000

                Tests of Fixed Effects

        Source    NDF    DDF   Type III F      Pr > F

        TREAT      1      8        5.11         0.0537

            ESTIMATE Statement Results

Parameter    Estimate   Std Error    DF      t    Pr > |t|      Mu

   A-B        -0.5105     0.2259      8     -2.26    0.0537    0.3751
```

Estimating treatment effect by trial from Models 1 and 3

Model 1

```
%GLIMMIX(DATA=a,
    STMTS=%STR(
    CLASS trial treat;
    MODEL outcome/one=treat trial trial*treat;
    ESTIMATE 'overall' treat 1 -1;
    ESTIMATE 'c1' treat 1 -1 trial*treat 1 -1;
    ESTIMATE 'c2' treat 1 -1 trial*treat 0 0 1 -1;
    ESTIMATE 'c3' treat 1 -1 trial*treat 0 0 0 0 1 -1;
    ESTIMATE 'c4' treat 1 -1 trial*treat 0 0 0 0 0 0 1 -1;
    ESTIMATE 'c5' treat 1 -1 trial*treat 0 0 0 0 0 0 0 0 1 -1;
    ESTIMATE 'c6' treat 1 -1 trial*treat 0 0 0 0 0 0 0 0 0 0 1 -1;
    ESTIMATE 'c7' treat 1 -1 trial*treat 0 0 0 0 0 0 0 0 0 0 0 0
     1 -1;
    ESTIMATE 'c8' treat 1 -1 trial*treat 0 0 0 0 0 0 0 0 0 0 0 0
     0 0 1 -1;
    ESTIMATE 'c9' treat 1 -1 trial*treat 0 0 0 0 0 0 0 0 0 0 0 0
     0 0 0 1 -1;
    ),
    FREQ=freq,
    ERROR=B
    );
```

Note that although this model analyses the data in Bernoulli form (i.e. there is one observation per patient), it will give the same results as the code given earlier for Model 1, which analyses the data in binomial form. Bernoulli form is used to allow the GLIMMIX macro to be used with ESTIMATE statements to give the treatment effect estimates at each centre.

Model 3

```
%GLIMMIX(DATA=a,
   STMTS=%STR(
   CLASS trial treat;
   MODEL outcome/one=treat;
   RANDOM trial trial*treat;
   ESTIMATE 'overall' treat 1 -1;
   ESTIMATE 'c1' treat 1 -1 | trial*treat 1 -1;
   ESTIMATE 'c2' treat 1 -1 | trial*treat 0 0 1 -1;
   ESTIMATE 'c3' treat 1 -1 | trial*treat 0 0 0 0 1 -1;
   ESTIMATE 'c4' treat 1 -1 | trial*treat 0 0 0 0 0 0 1 -1;
   ESTIMATE 'c5' treat 1 -1 | trial*treat 0 0 0 0 0 0 0 0 1 -1;
   ESTIMATE 'c6' treat 1 -1 | trial*treat 0 0 0 0 0 0 0 0 0 0 1 -1;
   ESTIMATE 'c7' treat 1 -1 | trial*treat 0 0 0 0 0 0 0 0 0 0 0 0
    1 -1;
   ESTIMATE 'c8' treat 1 -1 | trial*treat 0 0 0 0 0 0 0 0 0 0 0 0
    0 0 1 -1;
   ESTIMATE 'c9' treat 1 -1 | trial*treat 0 0 0 0 0 0 0 0 0 0 0 0
    0 0 0 0 1 -1;
   ),
   FREQ=freq,
   ERROR=B
   );
```

The estimates produced by these analyses are given in Table 5.4.

6

Repeated Measures Data

In this chapter we consider mixed model approaches to analysing repeated measures data. An introduction to repeated measures data and its analysis is given in Section 6.1. In Section 6.2 covariance pattern models are described and two worked examples follow in Sections 6.3 and 6.4. Random coefficients models are described in Section 6.5, which is followed by a worked example in Section 6.6. Methods for sample size estimation are introduced in Section 6.7. In this chapter we will just be considering simple designs where the repeated measures are defined on a single timescale. Occasionally, designs have a more complex pattern of repeated measurements; for example, repeated measurements may be taken within each of several visits. An example of such a design will be given in Section 8.1.

6.1 INTRODUCTION

Any dataset in which subjects are measured repeatedly over time can be described as repeated measures data. Repeated measurements can either be made at predetermined intervals (e.g. at fortnightly visits, or at specified times following a drug dose), or in an uncontrolled fashion so there are variable intervals between the repeated measurements. Whether or not the intervals are fixed and, if so, whether these intervals are constant, will influence the type of analysis model chosen.

6.1.1 Reasons for repeated measurements

There are many reasons for collecting repeated measures data. Some examples are:

- To ensure that a treatment is effective over a specified period of time. Often this will be done using a carefully planned trial with fixed timings for visits.
- To monitor safety aspects of the treatment over a specified period (repeated efficacy measurements are then incidental).

- To see how long a single dose of a drug takes to become effective by measuring drug concentration or a physiological response at fixed intervals (often over 24 hours). In this situation repeated measurements are taken at a single visit.

- Repeated observations are sometimes inherent in the measurement itself; for example, blood pressure monitors can take measurements as frequently as every 10 seconds.

- To monitor particular groups of patients over time. Often these are retrospective studies where repeated measurements have been recorded in an unplanned fashion. For example, repeated observations on patients with a particular disease may be available from a hospital clinic.

6.1.2   Analysis objectives

The objectives in analysing repeated measures data will differ depending on the purpose of the study. Ideally, they should be clarified at the design stage. Some examples of common objectives are as follows:

- To measure the average treatment effect over time.

- To assess treatment effects separately at each time point and to test whether treatment interacts with time.

- To assess specific features of the treatment response profile, e.g. area under curve (AUC), maximum or minimum value, time to the maximum value.

- To identify any covariance patterns in the repeated measurements.

- To determine a suitable model to describe the relationship of a measurement with time.

Before considering mixed models we first review some fixed effect approaches to analysing repeated measures data that are sometimes satisfactory.

6.1.3   Fixed effects approaches

It is important that any analysis of a parallel group study compares treatment effects against a background of between-patient variation. This is because treatment effects are contained within patient effects. Each of these fixed effect approaches below has the potential to compare treatments in this way.

Analysis of mean response over time

This method is satisfactory when the overall treatment effect is of interest, the times are fixed, and there are no missing data. However, it does not give any information on whether the treatment effect changes over time. When there are missing data, the analysis is only likely to be satisfactory if the response variable does not change with time.

Separate analyses at each time point

Here, separate analyses are carried out to compare treatments at each time point. Treatment standard errors are then correctly estimated at the between-patient level. One of many drawbacks to this approach is that repeated testing is taking place and therefore a significant treatment effect is more likely to occur at some time point by chance. Also, the tests will be correlated. There may also be problems of interpretation if a treatment effect is significant at some time points but not at others. Another, admittedly less important, disadvantage is that the treatment standard errors will be less accurate, since they are based only on the observations at one time point, rather than using data from all time points. This analysis strategy is often observed in medical journals, but it is a strategy which should be discouraged.

Analyses of response features

Features summarising each patient's response profile (e.g. area under the curve (AUC), minimum or maximum value, time to maximum value) can be analysed. This approach is satisfactory if these summary features are of particular interest and if there is not a great deal of missing data. When there are missing data it may not be possible to obtain a satisfactory estimate of some features (e.g. an AUC estimate would be biased if some of the observations were missing, although it is possible that an approach such as interpolation would help overcome this; the maximum value may be unrepresentative if observations around the true maximum were missing). If summary statistics are used, then restraint should be exercised in their selection to avoid problems of multiple testing.

Analysis of raw data fitting patient effects as fixed

In this model, patient, treatment, time and treatment·time effects can be fitted as fixed effects. However, treatment standard errors should not be obtained from the residual mean square, since this represents 'within'-patient variation. When there are no missing data, standard errors based on between-patient variation can be calculated manually using the patient sum of squares in the ANOVA table (packages such as SAS (PROC GLM) will, by default, calculate treatment standard errors from the residual mean square). When there are no missing data, this model will give identical results to an equivalent random effects model fitting patients as random. It also has the advantage of being able to assess the treatment effects over time. However, it is not appropriate when there are missing data unless special adjustments are made.

6.1.4   Mixed model approaches

Mixed models have the following advantages:

- A single model can be used to estimate overall treatment effects and to estimate treatment effects at each time point. Treatment effects are correctly compared against a background of between-patient variation. There is no need to calculate mean values across all time points (to obtain the overall treatment effects), or to analyse time points separately (to obtain treatment effects at each time point). Standard errors for treatment effects at individual time points are calculated using information from all time points and are therefore more robust than standard errors calculated from separate time points.

- The presence of missing data causes no problems provided they can be assumed missing at random.

- The covariance pattern of the repeated measurements can be determined and taken account of (e.g. the model can tell us whether the measurements across all time points have a constant correlation, or whether the pattern of correlations is more complex and varies with time).

There are several ways a mixed model can be used to analyse repeated measures data. The simplest approach is to use a random effects model with patient effects fitted as random. This will allow for a constant correlation between all observations on the same patient. However, often, the correlation between observations on the same patient is not constant. For example, correlation may decrease as visits become more widely separated in time. A covariance pattern model can be used to allow for this or, alternatively, for a more complex pattern of correlation. These models are considered in Section 6.2. When the relationship of the response variable with time is of interest, a random coefficients model is appropriate. Here, regression curves are fitted for each patient and the regression coefficients are allowed to vary randomly between the patients. These models are considered in Section 6.5.

6.2   COVARIANCE PATTERN MODELS

The basic structure of covariance pattern models has been described in Section 2.1.5. Here, we consider their use for analysing repeated measures data in more detail and describe some more complex types of covariance pattern.

As described in Section 2.1.5, in covariance pattern models the covariance structure is defined directly by specifying a covariance pattern rather than by using random effects. Observations within each category of a chosen *blocking effect* (e.g. patients) are assumed to have a specific pattern of covariance which is defined across a time effect such as period or visit. For example, in a repeated measures trial a pattern across periods could be specified for covariances between observations occurring on the same patients. The covariance pattern is defined within the residual matrix, **R**. This matrix is blocked by patients so that only observations on the same patient are correlated. **R** can be written

$$
\mathbf{R} = \begin{pmatrix}
\mathbf{R}_1 & \mathbf{0} & \mathbf{0} & \mathbf{0} & \mathbf{0} & \mathbf{0} & \mathbf{0} & \cdots \\
\mathbf{0} & \mathbf{R}_2 & \mathbf{0} & \mathbf{0} & \mathbf{0} & \mathbf{0} & \mathbf{0} & \cdots \\
\mathbf{0} & \mathbf{0} & \mathbf{R}_3 & \mathbf{0} & \mathbf{0} & \mathbf{0} & \mathbf{0} & \cdots \\
\mathbf{0} & \mathbf{0} & \mathbf{0} & \mathbf{R}_4 & \mathbf{0} & \mathbf{0} & \mathbf{0} & \cdots \\
\mathbf{0} & \mathbf{0} & \mathbf{0} & \mathbf{0} & \mathbf{R}_5 & \mathbf{0} & \mathbf{0} & \cdots \\
\mathbf{0} & \mathbf{0} & \mathbf{0} & \mathbf{0} & \mathbf{0} & \mathbf{R}_6 & \mathbf{0} & \cdots \\
\mathbf{0} & \mathbf{0} & \mathbf{0} & \mathbf{0} & \mathbf{0} & \mathbf{0} & \mathbf{R}_7 & \cdots \\
\cdot & \cdot & \cdot & \cdot & \cdot & \cdot & \cdot & \cdot \\
\cdot & \cdot & \cdot & \cdot & \cdot & \cdot & \cdot & \cdot \\
\cdot & \cdot & \cdot & \cdot & \cdot & \cdot & \cdot & \cdot
\end{pmatrix}.
$$

The $\mathbf{R}_i$ are submatrices of covariances corresponding to each patient and have dimension equal to the number of observations occurring on each patient. The **0**'s represent matrix blocks of 0's denoting zero correlation between observations on different patients. We will now consider different ways to define covariance patterns in the $\mathbf{R}_i$ matrix blocks.

6.2.1 Covariance patterns

A large selection of covariance patterns are available for use in mixed models. Most of the patterns are dependent on measurements being taken at fixed times and some are also easier to justify when the observations are evenly spaced. There are also patterns where covariances are based on the exact value of time (rather than, say, visit number), and these are most useful in situations where the time intervals are irregular. Some examples of covariance patterns are given below. Still further possible types of covariance patterns can be found in the SAS PROC MIXED documentation.

Simple covariance patterns

Some simple covariance patterns for the $\mathbf{R}_i$ matrices for a trial with four time points are shown below. In the general pattern (i), sometimes also referred to as 'unstructured', the variances of responses, σ_i^2, differ for each time period i, and the covariances, θ_{jk}, differ between each pair of periods j and k. For the first-order autoregressive model (ii) the variances are equal and the covariances decrease exponentially depending on their separation $|j - k|$, so $\theta_{jk} = \rho^{|j-k|}\sigma^2$. This is sometimes an appropriate model when time periods are evenly spaced. It can then be seen as a 'natural' model from a time series viewpoint. However, it may be justified empirically in circumstances where the observations are not evenly spaced. For example, in monitoring the acute effect of drugs it is common to take measurements at short intervals soon after drug administration, when the level of observations may be changing rapidly, with increasingly separated intervals later on as observations change more slowly. Under such circumstances, adjoining observations may well show similar covariances, despite unequal time periods, with exponentially decreasing covariances for increasingly separated observation numbers.

For the compound symmetry covariance model (iii), all covariances are equal. The Toeplitz model (iv) uses a separate covariance for each level of separation between the time points. This model is also known as the general autoregressive model.

(i)   General

$$\mathbf{R}_i = \begin{pmatrix} \sigma_1^2 & \theta_{12} & \theta_{13} & \theta_{14} \\ \theta_{12} & \sigma_2^2 & \theta_{23} & \theta_{24} \\ \theta_{13} & \theta_{23} & \sigma_3^2 & \theta_{34} \\ \theta_{14} & \theta_{24} & \theta_{34} & \sigma_4^2 \end{pmatrix}.$$

(ii) First-order autoregressive

$$\mathbf{R}_i = \sigma^2 \begin{pmatrix} 1 & \rho & \rho^2 & \rho^3 \\ \rho & 1 & \rho & \rho^2 \\ \rho^2 & \rho & 1 & \rho \\ \rho^3 & \rho^2 & \rho & 1 \end{pmatrix}.$$

(iii) Compound symmetry

$$\mathbf{R}_i = \begin{pmatrix} \sigma^2 & \theta & \theta & \theta \\ \theta & \sigma^2 & \theta & \theta \\ \theta & \theta & \sigma^2 & \theta \\ \theta & \theta & \theta & \sigma^2 \end{pmatrix}.$$

(iv) Toeplitz

$$\mathbf{R}_i = \begin{pmatrix} \sigma^2 & \theta_1 & \theta_2 & \theta_3 \\ \theta_1 & \sigma^2 & \theta_1 & \theta_2 \\ \theta_2 & \theta_1 & \sigma^2 & \theta_1 \\ \theta_3 & \theta_2 & \theta_1 & \sigma^2 \end{pmatrix}.$$

Before software for fitting covariance patterns was readily available, repeated measures analyses were often performed either using a random effects model or by fitting a multivariate normal distribution to the repeated measurements. A random effects model gives identical results to a compound symmetry pattern model (iii) (provided the patient variance component is not negative). Equality of the correlation terms in the compound symmetry structure was often assessed using a test of sphericity (e.g. Greenhouse and Geisser, 1959). However, if a lack of sphericity were found, there were limited alternative analyses available. If the data were complete, then fitting a multivariate normal distribution would give the same results as using a general pattern (i). This model could require a lot of covariance parameters. Also, if there were missing data, then fitting a multivariate normal distribution would not be satisfactory as all information is lost on patients with incomplete data in most packages. Covariance pattern models overcome the above limitations by providing a flexible choice of covariance patterns which can be fitted to either complete or incomplete data.

Different variances for each time point

Sometimes variability in a measurement will differ between the time points. This was allowed for in the general covariance pattern above (i). Some additional patterns allowing for differing variances are given below. In pattern (v) time points have different variances but observations on the same patient are uncorrelated. This should only be used if preliminary analyses with more parameterised patterns indicate a lack of correlation between the repeated observations. Patterns (vi)–(viii) have similar forms to the autoregressive, compound symmetry and Toeplitz patterns shown above, except that different variances for each time point are now used.

(v) Heterogeneous uncorrelated

$$\mathbf{R}_i = \begin{pmatrix} \sigma_1^2 & 0 & 0 & 0 \\ 0 & \sigma_2^2 & 0 & 0 \\ 0 & 0 & \sigma_3^2 & 0 \\ 0 & 0 & 0 & \sigma_4^2 \end{pmatrix}.$$

(vi) Heterogeneous compound symmetry

$$\mathbf{R}_i = \begin{pmatrix} \sigma_1^2 & \rho\sigma_1\sigma_2 & \rho\sigma_1\sigma_3 & \rho\sigma_1\sigma_4 \\ \rho\sigma_1\sigma_2 & \sigma_2^2 & \rho\sigma_2\sigma_3 & \rho\sigma_2\sigma_4 \\ \rho\sigma_1\sigma_3 & \rho\sigma_2\sigma_3 & \sigma_3^2 & \rho\sigma_3\sigma_4 \\ \rho\sigma_1\sigma_4 & \rho\sigma_2\sigma_4 & \rho\sigma_3\sigma_4 & \sigma_4^2 \end{pmatrix}.$$

(vii) Heterogeneous first-order autoregressive

$$\mathbf{R}_i = \begin{pmatrix} \sigma_1^2 & \rho\sigma_1\sigma_2 & \rho^2\sigma_1\sigma_3 & \rho^3\sigma_1\sigma_4 \\ \rho\sigma_1\sigma_2 & \sigma_2^2 & \rho\sigma_2\sigma_3 & \rho^2\sigma_2\sigma_4 \\ \rho^2\sigma_1\sigma_3 & \rho\sigma_2\sigma_3 & \sigma_3^2 & \rho\sigma_3\sigma_4 \\ \rho^3\sigma_1\sigma_4 & \rho^2\sigma_2\sigma_4 & \rho\sigma_3\sigma_4 & \sigma_4^2 \end{pmatrix}.$$

(viii) Heterogeneous Toeplitz

$$\mathbf{R}_i = \begin{pmatrix} \sigma_1^2 & \rho_1\sigma_1\sigma_2 & \rho_2\sigma_1\sigma_3 & \rho_3\sigma_1\sigma_4 \\ \rho_1\sigma_1\sigma_2 & \sigma_2^2 & \rho_1\sigma_2\sigma_3 & \rho_2\sigma_2\sigma_4 \\ \rho_2\sigma_1\sigma_3 & \rho_1\sigma_2\sigma_3 & \sigma_3^2 & \rho_1\sigma_3\sigma_4 \\ \rho_3\sigma_1\sigma_4 & \rho_2\sigma_2\sigma_4 & \rho_1\sigma_3\sigma_4 & \sigma_4^2 \end{pmatrix}.$$

Separate covariance patterns for each treatment group

Sometimes measurements on different treatments will have different variances and covariances. For example, it may be the case that measurements are more variable on an active treatment than on a placebo. This can be allowed for by using separate sets of covariance parameters for each treatment group. For example, if the first three patients in a trial received treatments A, B and A and were each measured at three time points, then the **R** matrix for these patients with separate *compound symmetry* structures for each treatment would be

$$
\mathbf{R} = \begin{pmatrix}
\sigma_A^2 & \theta_A & \theta_A & 0 & 0 & 0 & 0 & 0 & 0 \\
\theta_A & \sigma_A^2 & \theta_A & 0 & 0 & 0 & 0 & 0 & 0 \\
\theta_A & \theta_A & \sigma_A^2 & 0 & 0 & 0 & 0 & 0 & 0 \\
0 & 0 & 0 & \sigma_B^2 & \theta_B & \theta_B & 0 & 0 & 0 \\
0 & 0 & 0 & \theta_B & \sigma_B^2 & \theta_B & 0 & 0 & 0 \\
0 & 0 & 0 & \theta_B & \theta_B & \sigma_B^2 & 0 & 0 & 0 \\
0 & 0 & 0 & 0 & 0 & 0 & \sigma_A^2 & \theta_A & \theta_A \\
0 & 0 & 0 & 0 & 0 & 0 & \theta_A & \sigma_A^2 & \theta_A \\
0 & 0 & 0 & 0 & 0 & 0 & \theta_A & \theta_A & \sigma_A^2
\end{pmatrix}.
$$

Alternatively, if a *general structure* were used for each treatment group, then the **R** matrix would be

$$
\mathbf{R} = \begin{pmatrix}
\sigma_{A,1}^2 & \theta_{A,12} & \theta_{A,13} & 0 & 0 & 0 & 0 & 0 & 0 \\
\theta_{A,12} & \sigma_{A,2}^2 & \theta_{A,23} & 0 & 0 & 0 & 0 & 0 & 0 \\
\theta_{A,13} & \theta_{A,23} & \sigma_{A,3}^2 & 0 & 0 & 0 & 0 & 0 & 0 \\
0 & 0 & 0 & \sigma_{B,1}^2 & \theta_{B,12} & \theta_{B,13} & 0 & 0 & 0 \\
0 & 0 & 0 & \theta_{B,12} & \sigma_{B,2}^2 & \theta_{B,23} & 0 & 0 & 0 \\
0 & 0 & 0 & \theta_{B,13} & \theta_{B,23} & \sigma_{B,3}^2 & 0 & 0 & 0 \\
0 & 0 & 0 & 0 & 0 & 0 & \sigma_{A,1}^2 & \theta_{A,12} & \theta_{A,13} \\
0 & 0 & 0 & 0 & 0 & 0 & \theta_{A,12} & \sigma_{A,2}^2 & \theta_{A,23} \\
0 & 0 & 0 & 0 & 0 & 0 & \theta_{A,13} & \theta_{A,23} & \sigma_{A,3}^2
\end{pmatrix}.
$$

If most of the covariances were small or negative, then observations on the same patient could be made *uncorrelated*, while different variances were still allowed for each treatment.

$$
\mathbf{R} = \begin{pmatrix}
\sigma_A^2 & 0 & 0 & 0 & 0 & 0 & 0 & 0 & 0 \\
0 & \sigma_A^2 & 0 & 0 & 0 & 0 & 0 & 0 & 0 \\
0 & 0 & \sigma_A^2 & 0 & 0 & 0 & 0 & 0 & 0 \\
0 & 0 & 0 & \sigma_B^2 & 0 & 0 & 0 & 0 & 0 \\
0 & 0 & 0 & 0 & \sigma_B^2 & 0 & 0 & 0 & 0 \\
0 & 0 & 0 & 0 & 0 & \sigma_B^2 & 0 & 0 & 0 \\
0 & 0 & 0 & 0 & 0 & 0 & \sigma_A^2 & 0 & 0 \\
0 & 0 & 0 & 0 & 0 & 0 & 0 & \sigma_A^2 & 0 \\
0 & 0 & 0 & 0 & 0 & 0 & 0 & 0 & \sigma_A^2
\end{pmatrix}.
$$

Banded covariances

It will sometimes be apparent from the covariance parameter estimates that the correlation between widely separated observations is negligible. In this situation it may be appropriate to 'band' the $\mathbf{R}_i$ matrices by setting correlations between the observations that are widely separated in time to zero. For example, a general covariance pattern with band size 3 would be

$$
\mathbf{R}_i =
\begin{pmatrix}
\sigma_1^2 & \theta_{12} & \theta_{13} & 0 \\
\theta_{12} & \sigma_2^2 & \theta_{23} & \theta_{24} \\
\theta_{13} & \theta_{23} & \sigma_3^2 & \theta_{24} \\
0 & \theta_{24} & \theta_{34} & \sigma_4^2
\end{pmatrix}.
$$

Banding can be done for any covariance pattern and has the advantage of reducing the number of covariance parameters which need to be fitted. It is therefore of greatest use for trials with a large number of time points.

Covariance patterns defined using time as a continuous measure

A covariance pattern can be defined according to the exact separation of observations in time. This is the only type of pattern apart from compound symmetry which is appropriate when time points do not occur at pre-determined intervals. However, it can also be useful for time points that are fixed but unevenly separated, or when there is some flexibility in the timing of fixed interval time points (for example, an interval of $12-16$ days may be allowed for a nominal two-week interval). There are many ways to define covariances from the time interval. Two examples are

$$
r_{ijk} = \sigma^2 \rho^{d_{ijk}} \quad \text{power,}
$$

$$
r_{ijk} = \sigma^2 \exp(d_{ijk}^2/\rho^2) \quad \text{Gaussian,}
$$

where,

r_{ijk} = covariance for observations j and k on patient i,

d_{ijk} = distance (usually time) between observations j and k on patient i.

Both these structures cause the covariance to decrease exponentially with the time interval between pairs of observations on the same patient.

6.2.2   Choice of covariance pattern

There are many covariance patterns available and choosing the most appropriate one is not always easy. The ideal is to select the pattern that best fits the true covariance of data. As well as providing appropriate standard errors for fixed effect estimates, this can yield additional information about the action of treatments or the phenomenon being studied. However, as more covariance parameters are included in the pattern, the chances of overfitting increase (i.e. the covariance pattern matches the observed pattern but may not be the true pattern). The likelihood (or quasi-likelihood) statistic is a basic measure of model fit but it will increase as more covariance parameters are added, so therefore cannot be used directly for making comparisons. There are two alternative approaches to making a model choice. One is to compare models based on measures of fit that are adjusted for the number of covariance parameters. Another is to use likelihood ratio tests to find whether additional parameters cause a statistically significant improvement in the model. Our preference is for the second approach, in which covariance parameters are only included if they are proved to be necessary.

Measures of model fit

The likelihood statistic is expected to become larger as more parameters are included in the model. The two statistics below are based on the likelihood but make allowance for the number of covariance parameters fitted. They can be used to make direct comparisons between models which fit the same fixed effects. Akaike's information criterion (AIC) (Akaike, 1974) is given by

$$AIC = \log(L) - q,$$

where q is the number of covariance parameters. Schwarz's information criterion (SIC) (Schwarz, 1978), takes into account the number of fixed effects, p, the number of observations, N, and the number of covariance parameters, q. It is given by

$$SIC = \log(L) - (q \log(N - p))/2.$$

Models with larger values of AIC and SIC denote better fits. However, it is unclear to us which criterion is preferable. Both the criteria are calculated within PROC MIXED.

We anticipate that these criteria will also be approximately satisfactory when based on the quasi-likelihood statistics arising out of GLMMs and categorical mixed models.

Statistical comparisons between models

Models can be compared statistically using likelihood ratio tests provided that they fit the same fixed effects and their covariance parameters are nested. Nesting of covariance parameters occurs when the covariance parameters in the simpler model can be obtained by restricting some of the parameters in the more complex model (e.g. a compound symmetry pattern is nested within a Toeplitz pattern, but it is not nested within a first-order autoregressive pattern). The likelihood ratio test statistic is given by

$$2(\log(L_1) - \log(L_2)) \sim \chi^2_{DF}.$$

where $DF = $ difference in number of covariance parameters fitted.

If the covariance parameters in the models compared are not nested, statistical comparison using a likelihood ratio test is not valid. In this situation comparisons of each model with a simpler model which is nested within both models could be made and the model giving the most significant improvement selected. Alternatively, Akaike's or Schwarz' criteria defined above could be used.

Again, we anticipate that these tests will be approximately satisfactory when based on the quasi- or pseudo-likelihood statistics arising out of GLMMs and categorical mixed models, but are unaware of formal justification for this.

Which covariance patterns to consider?

It is not usually practical to test a large number of covariance patterns in a single application. A safe strategy would be to start with simple patterns such as the compound symmetry or first-order autoregressive. A general pattern could then be used to give some indication of whether any other more complex patterns are likely to be appropriate. These more complex patterns can be tested and should be accepted only if they lead to a significant improvements in the likelihood. Particularly complex patterns are only likely to be significant in larger datasets where more information on the true covariance pattern is available.

In many datasets, especially those with only a few repeated measurements, estimates of overall treatment effects will differ little between models using different covariance patterns. If obtaining a reliable treatment estimate and standard error is the only objective, a compound symmetry pattern is likely to be robust. A rough check can be made of this by comparing the results with those obtained using a general pattern. If the differences are small, then the compound symmetry pattern can be used with reasonable confidence.

6.2.3   Choice of fixed effects

Treatment, time, treatment·time and baseline effects (if recorded) can all be fitted as fixed. Estimates of the overall treatment effect will differ depending on whether treatment·time effects are included in the model. When they are, treatment effects are calculated as the average of the treatment estimates obtained from each time point. When they are omitted, a weighted average of estimates from each time point is obtained. The weights are related to the variances of estimates at each time point, which are in turn related to the number of observations at each time point. The decision as to which estimate to use should rest with the desired interpretation and whether the treatment·time effects are significant. The unweighted estimate may be more appealing if there are missing data at later time points, so that bias towards earlier time points is reduced. On the other hand, if a treatment effect appears relatively constant over time, then the weighted estimate has the advantage of being less influcnced by potentially inaccurate estimates at time points with fewer observations. If treatment·time effects are significant, it would be wise to present additional treatment estimates at each time point. Less importance should then be attached to the overall treatment estimate.

The above discussion has assumed that we are interested in comparing treatments, though of course the applications can be much wider. In studies where patients are grouped by a variable other than treatment (e.g. an epidemiological study to compare patients suffering from different disease types), the comparator variable should simply be substituted for the treatment effects above.

6.2.4   General points

Missing data

Missing data, in the form of gaps in the series of observations or caused by patient dropout, may occur frequently in repeated measures data. They are a less serious problem, however, when a mixed model is used, unless there are very substantial between-treatment group differences with respect to the pattern of dropout. The reason missing values are less problematical is that observations at each time point influence estimates of treatment effects at every other time point, due to the specification of a covariance pattern. Thus, patients whose observations are limited to early time points because of dropout, will nevertheless be taken into account when estimates are made of treatment effects at later time points. Clearly, such individuals will not influence these estimates as greatly as individuals whose data is complete, so the pattern of missing data in different treatment groups cannot be completely ignored. There will also be potential biases if patients show patterns of rapid deterioration prior to dropout. In such cases, their early observations may be 'good' leading to a corresponding 'good' influence on the unobserved time points after dropout, when this is clearly inappropriate. *Ad hoc* approaches to inputting missing values, or analyses including and excluding dropouts may then need to be employed. Alternatively, a method that allows for non-random missing data in repeated measures analysis could be considered (e.g. Diggle and Kenward, 1994). Thus, it is too simplistic to say that missing values do not matter in the mixed model analysis of repeated measures data, but the method is quite robust, even when the data may not be entirely missing at random.

Significance testing

Fixed effects can be tested using F tests as described in Section 2.4.4. Again, Satterthwaite's DF should be used wherever possible. If suitable software is not available for calculating an appropriate DF we suggest that treatment effects should be compared using the patient DF, which is likely to produce a conservative test. Comparisons of treatment at individual time points can also be compared using the patient DF.

Fixed effect standard error bias

Downward bias of fixed effect standard errors will occur as described in Section 2.4.3, because the covariance parameters are estimated and not known. A more robust variance estimate for fixed effects, $\hat{\boldsymbol{\alpha}}$, known as the 'empirical' variance estimator, was suggested for covariance pattern models in Section 2.4.3. This variance takes into account the observed covariance in the data and may help alleviate some of the small sample bias. It does, however, cause the fixed effects variances to reflect the observed covariance in the data rather than that specified by the covariance pattern. For example, it will allow variances to differ for each treatment group. If a covariance pattern has been selected statistically by using likelihood ratio tests to determine when more complex patterns are justified, it would seem appropriate to use the variances produced by the model (i.e. $(\mathbf{X}\mathbf{V}^{-1}\mathbf{X})^{-1}$), which are based directly on the covariance pattern specified. However, the empirical variance may be preferable when limited time is available to determine the most appropriate covariance pattern. This may be the best approach for those working in the pharmaceutical industry where statistical methods need to be specified in advance in the protocol. Even when a simple covariance pattern such as compound symmetry is used in estimating the fixed effects, the empirical variance will still reflect the observed covariance pattern of the data.

The empirical variance is calculated in SAS by using the EMPIRICAL option in the PROC MIXED statement. For non-normal data it is calculated by default when the REPEATED statement in PROC GENMOD is used. (See Sections 9.2 and 9.3 for more detail on PROC MIXED and PROC GENMOD.)

Model checking

In covariance pattern models it is assumed that the residuals have a multivariate normal distribution with zero means and covariance matrix **R**. Although this assumption is difficult to check formally, plots of the residuals should be sufficient to identify any important outliers or deviations from normality. To check the covariances more fully, plots of pairs of time points can also be used. However, picking up a non-normal covariance from these may not be easy and any outliers are likely to already have shown up in the plots of all residuals. If separate variances are used for treatments or time points, then plots of the residuals by treatment or time will help determine whether these are appropriate. If there is evidence of non-normality or outliers, then the suggestions described in Section 2.4.6 on model checking should be used.

6.3   EXAMPLE: COVARIANCE PATTERN MODELS FOR NORMAL DATA

The multi-centre hypertension trial analysed in Sections 1.3 and 4.3 is now considered as repeated measures data. DBP recorded at each of the fortnightly post-treatment visits will be analysed and the effect of centres will be ignored. The primary objective is to obtain an overall estimate of the treatment difference. The numbers of patients attending at each visit is summarised by treatment below. Visits 3–6 are the four post-treatment visits and visit 2 values are used as the baseline covariate.

	Treatment			
Visit	A	B	C	Total
1	106	101	104	311
2	106	100	102	308
3	100	96	94	290
4	95	91	94	280
5	87	88	93	268
6	83	84	91	258

6.3.1   Analysis models

Treatment, time, treatment·time and baseline effects are fitted as fixed effects in all the models considered.

The patterns listed below are fitted and compared statistically using likelihood ratio tests. The covariance pattern used by Model 6 was suggested by the results from Models 1–5.

Model	Covariance pattern
1	Compound symmetry
2	First-order autoregressive
3	Toeplitz
4	General
5	Separate compound symmetry for each treatment group
6	Separate Toeplitz pattern for each treatment group

6.3.2   Selection of covariance pattern

The covariance patterns and measures of model fit resulting from each analysis are shown in Table 6.1. Correlations between visits are positive in all models, indicating that it is important to take account of the correlations between the repeated measurements.

Models 1 and 2 are the simplest covariance patterns. Since they each use two covariance parameters, we choose Model 1, which has the highest likelihood. It seems unlikely that the correlation between periods decays exponentially as they become more widely separated as modelled in Model 2.

Table 6.1   Results from using different covariance patterns.

Model	Covariance parameters (variances and correlation matrix)		$-2\log(L)$ (no. parameters)	Akaike's information criterion (AIC)
1	76	$\begin{pmatrix} 1 & & & \\ 0.53 & 1 & & \\ 0.54 & 0.53 & 1 & \\ 0.53 & 0.53 & 0.53 & 1 \end{pmatrix}$	7463.35(2)	-3733.7
2	76	$\begin{pmatrix} 1 & & & \\ 0.57 & 1 & & \\ 0.57^2 & 0.57 & 1 & \\ 0.57^3 & 0.57^2 & 0.57 & 1 \end{pmatrix}$	7485.26(2)	-3744.6
3	76	$\begin{pmatrix} 1 & & & \\ 0.57 & 1 & & \\ 0.48 & 0.57 & 1 & \\ 0.46 & 0.48 & 0.57 & 1 \end{pmatrix}$	7450.57(4)	-3729.3
4	67 71 86 73	$\begin{pmatrix} 1 & & & \\ 0.51 & 1 & & \\ 0.48 & 0.61 & 1 & \\ 0.46 & 0.50 & 0.61 & 1 \end{pmatrix}$	7442.32(10)	-3731.2
5	A 85	$\begin{pmatrix} 1 & & & \\ 0.54 & 1 & & \\ 0.54 & 0.54 & 1 & \\ 0.54 & 0.54 & 0.54 & 1 \end{pmatrix}$	7447.55(6)	-3729.8
	B 68	$\begin{pmatrix} 1 & & & \\ 0.39 & 1 & & \\ 0.39 & 0.39 & 1 & \\ 0.39 & 0.39 & 0.39 & 1 \end{pmatrix}$		
	C 76	$\begin{pmatrix} 1 & & & \\ 0.63 & 1 & & \\ 0.63 & 0.63 & 1 & \\ 0.63 & 0.63 & 0.63 & 1 \end{pmatrix}$		
6	A 85	$\begin{pmatrix} 1 & & & \\ 0.58 & 1 & & \\ 0.48 & 0.58 & 1 & \\ 0.50 & 0.48 & 0.58 & 1 \end{pmatrix}$	7423.98(12)	-3724.0
	B 68	$\begin{pmatrix} 1 & & & \\ 0.42 & 1 & & \\ 0.33 & 0.42 & 1 & \\ 0.42 & 0.33 & 0.42 & 1 \end{pmatrix}$		
	C 76	$\begin{pmatrix} 1 & & & \\ 0.69 & 1 & & \\ 0.61 & 0.69 & 1 & \\ 0.46 & 0.61 & 0.69 & 1 \end{pmatrix}$		

Model 3 with a Toeplitz pattern indicates that the correlation between visits may be less when they are not adjacent. The compound symmetry pattern is nested within the Toeplitz pattern used by Model 1 and the models can therefore be compared statistically using a likelihood ratio test. This test gave $\chi_2^2 = 12.78$, indicating that the Toeplitz structure is a significant improvement ($p = 0.002$).

Model 4 with a general pattern also indicates that the correlation between visits is less when they are not adjacent. We determine whether the extra parameters used lead to a significant improvement over Model 3. The likelihood ratio test gives $\chi_6^2 = 8.25(p = 0.22)$, which shows that the use of the extra six parameters in the general pattern is not necessary.

Model 5 has separate compound symmetry patterns for each treatment and indicates that covariances may differ between treatments. The likelihood ratio test shows that this model is a significant improvement over Model 1, $\chi_4^2 = 15.80(p = 0.003)$. However, it cannot be compared statistically to Model 3 (Toeplitz), since the two models are not nested.

On the basis of the fact that Model 5 indicates that separate covariances for each treatment group may be necessary and that Model 3 suggests a Toeplitz pattern, Model 6 incorporating both these features was tested. Models 3 and 5 are nested within Model 6 and Model 6 shows significant improvements over them both, $\chi_8^2 = 26.59(p = 0.0008)$ and $\chi_6^2 = 23.57(p = 0.0006)$.

Thus, we have statistically justified the use of a fairly complex covariance pattern. This is likely to be partly because the trial is relatively large, so the covariance parameters are estimated with a reasonable accuracy. Model 6 has given us statistical evidence that the treatment groups have different variances. Also, the Toeplitz structures indicate that correlations between repeated measurements are highest for treatment C and lowest for treatment B. These differences are likely to be, in some way, due to the different actions of the treatments. In smaller trials, however, it is often not possible statistically to justify any pattern more complex than the compound symmetry or first-order autoregressive. This is not necessarily because a more complex pattern does not exist, but because there is insufficient information to determine it.

6.3.3   Assessing fixed effects

The overall treatment effect estimates obtained from Models 1 and 6 are summarised in Table 6.2. The 'empirical' standard errors are given in addition to the usual 'model-based' standard errors (calculated by $(\mathbf{XV}^{-1}\mathbf{X})^{-1}$). In Model 6, where an appropriate covariance pattern has been selected statistically, the empirical standard errors are very close to the model-based standard errors. However, in Model 1, which uses a simple compound symmetry covariance pattern which does not fully reflect the true covariance of the repeated measurements, the standard error estimates are noticeably different. The empirical estimates reflect the different covariances between treatment groups even though this is not modelled by the compound symmetry covariance pattern. Note,

Table 6.2   Comparing treatment effects between Models 1 and 6.

	Treatment difference	Mean difference	Model-based SE	Empirical SE
Model 1	A – B	1.22	1.03	0.98
(compound	A – C	3.01	1.02	1.07
symmetry)	B – C	1.79	1.03	0.97
Model 6	A – B	1.25	0.99	0.99
(separate	A – C	3.04	1.08	1.08
Toeplitz pattern	B – C	1.79	1.00	0.98
for each				
treatment)				
Model 6 omitting	A – B	1.23	0.99	0.99
treatment·visit	A – C	3.02	1.07	1.06
effects	B – C	1.79	0.99	0.98

therefore, that the empirical estimates are similar, whatever model was fitted. The slight differences arise from the small changes in the fixed effects estimates themselves with different models. Thus, if the only objective of the trial were to obtain estimates of treatment effects and their standard errors, then it would not appear necessary to go to the trouble of obtaining the most appropriate covariance pattern. However, in some situations, knowledge of the covariance pattern may itself be of some interpretational value.

In the last row of Table 6.2, results from fitting Model 6 omitting the treatment.time effects are given. In this model the treatment effects are calculated as weighted averages of the effects at each time point. This contrasts with the model fitting the treatment·time interaction, which causes treatment effects to

Table 6.3   Treatment effects (empirical standard errors) at each visit, Model 6.

Visit	Treatment difference	Treatment effects (SE)
3	A – B	1.36 (1.23)
	A – C	3.42 (1.25)
	B – C	2.06 (1.26)
4	A – B	0.56 (1.22)
	A – C	1.89 (1.24)
	B – C	1.34 (1.20)
5	A – B	3.00 (1.38)
	A – C	4.77 (1.41)
	B – C	1.77 (1.29)
6	A – B	0.09 (1.30)
	A – C	2.10 (1.33)
	B – C	2.01 (1.21)

be unweighted averages of the effects at each time point. There is little difference between the two sets of results, because there is only a relatively small amount of missing data. It is a matter of personal choice as to which approach is preferable.

The treatment·time interaction was almost significant in Model 6 ($p = 0.07$), so it may be helpful to present the treatment effect estimates at each visit, in addition to the overall treatment estimates. These are given for Model 6 in Table 6.3. There are noticeable differences in the treatment effect estimates between the visits, but an absence of any coherent pattern.

6.3.4   Model checking

Residual plots are used to detect any outliers or a general lack of normality.

Residual plots by treatments

Plots of residuals against their predicted values are given for all observations, and for each treatment group separately (Figure 6.1).

The residual of over 40 in the plot for treatment A is a potential outlier. We check its influence by reanalysing the data with the corresponding patient's

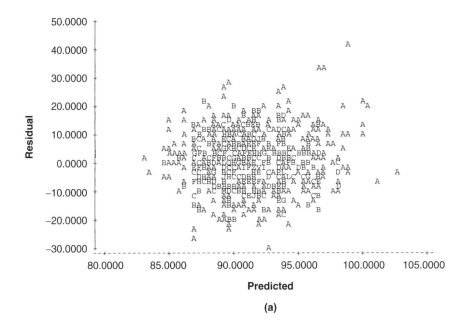

(a)

Figure 6.1   Plots of residuals against their predicted values: (a) All residuals; (b) treatment A; (c) treatment B; (d) treatment C. A = 1 obs, B = 2 obs, etc.

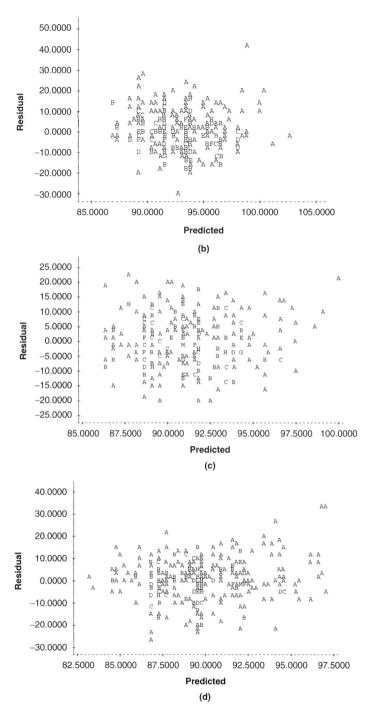

Figure 6.1   *(continued).*

Table 6.4   Comparing results from Model 6 with and without outlier.

	Without outlier	With outlier
Treatment differences (SE)		
A – B	1.15 (0.97)	1.25 (0.99)
A – C	2.93 (1.06)	3.04 (1.08)
B – C	1.77 (1.00)	1.79 (1.98)

Variance and correlation matrices

Treatment A

$$
77 \begin{pmatrix} 1 & & & \\ 0.55 & 1 & & \\ 0.44 & 0.55 & 1 & \\ 0.45 & 0.44 & 0.55 & 1 \end{pmatrix} \quad 85 \begin{pmatrix} 1 & & & \\ 0.58 & 1 & & \\ 0.48 & 0.58 & 1 & \\ 0.50 & 0.48 & 0.58 & 1 \end{pmatrix}
$$

B

$$
68 \begin{pmatrix} 1 & & & \\ 0.43 & 1 & & \\ 0.33 & 0.43 & 1 & \\ 0.42 & 0.33 & 0.43 & 1 \end{pmatrix} \quad 68 \begin{pmatrix} 1 & & & \\ 0.42 & 1 & & \\ 0.33 & 0.42 & 1 & \\ 0.42 & 0.33 & 0.42 & 1 \end{pmatrix}
$$

C

$$
76 \begin{pmatrix} 1 & & & \\ 0.70 & 1 & & \\ 0.61 & 0.70 & 1 & \\ 0.47 & 0.61 & 0.70 & 1 \end{pmatrix} \quad 76 \begin{pmatrix} 1 & & & \\ 0.69 & 1 & & \\ 0.61 & 0.69 & 1 & \\ 0.46 & 0.61 & 0.69 & 1 \end{pmatrix}
$$

data removed (patient 314). A reanalysis using Model 6 gave the results shown in Table 6.4. The treatment effect estimates have altered but not by enough to change the overall conclusions about treatment. It may therefore be reasonably safe to report the original analysis including the outlying observation. The covariance parameters have altered slightly for treatment A. If the interpretation of covariance patterns is of particular interest, then further likelihood ratio tests on analyses with patient 314 deleted should be carried out to check whether use of the Toeplitz covariance pattern grouped by treatments is still justified.

Bivariate plots of visit pairs

Bivariate plots of residuals at each pair of time points are additionally used to help assess the assumption of multivariate normality for the repeated measurements (Figure 6.2).

None of these plots shows a marked deviation from the elliptical shape expected of bivariate normal distributions. Note that the observation with a residual of over 40 considered earlier (patient 314, visit 5) does not now appear quite as outlying (appears in plots of visits 3 and 4 against visit 5).

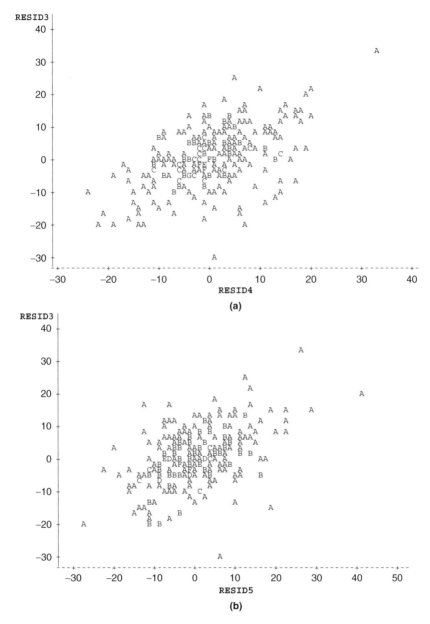

Figure 6.2 Plots of residuals against their predicted values, separately for each visit pair: (a) Visit 3 vs. visit 4 (b) Visit 3 vs. visit 5 (c) Visit 3 vs. visit 6 (d) Visit 4 vs. visit 5 (e) Visit 4 vs. visit 6; (f) Visit 5 vs. visit 6 A = 1 obs, B = 2 obs, etc.

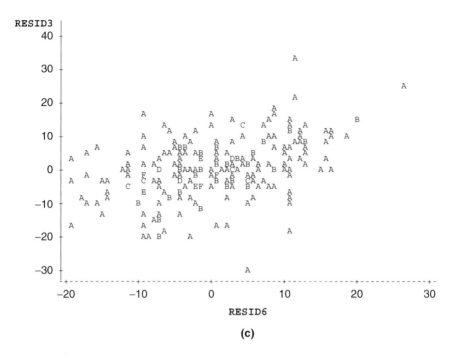

(c)

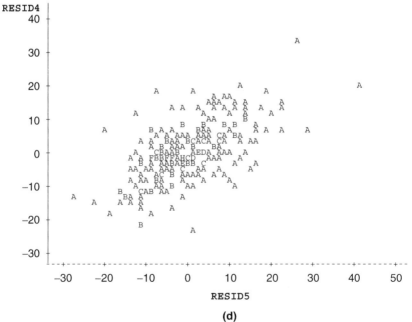

(d)

Figure 6.2   (*continued*).

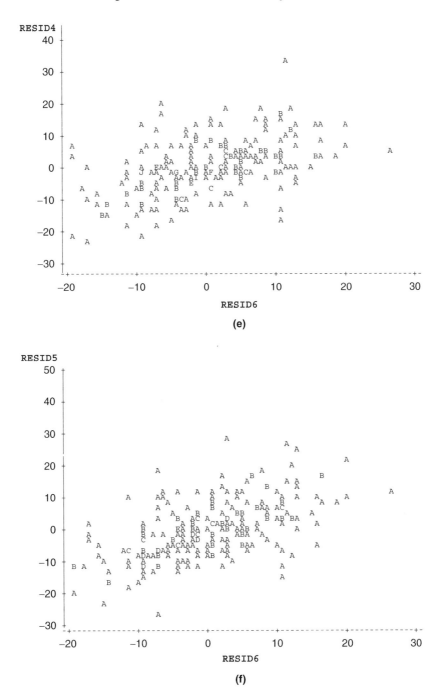

Figure 6.2 (*continued*).

SAS code and output

Variables

treat = treatment (A,B,C),
pat   = patient number,
dbp   = diastolic blood pressure,
dbp1 = baseline diastolic blood pressure,
visit = visit number.

SAS code is given below for Model 6. Code for the other models is identical except that different REPEATED statements are used. These statements are given after the Model 6 output.

```
PROC MIXED EMPIRICAL NOCLPRINT; CLASS pat treat visit;
MODEL dbp = dbp1 treat visit treat*visit/ DDFM=SATTERTH;
REPEATED visit/ SUBJECT=pat TYPE=TOEP GROUP=treat R=1,3,4 RCORR=1,3,4;
LSMEANS treat/ DIFF PDIFF CL;
ESTIMATE 'a-b,v3' treat 1 -1  0 treat*visit 1 0 0 0 -1  0  0  0  0  0  0  0;
ESTIMATE 'a-c,v3' treat 1  0 -1 treat*visit 1 0 0 0  0  0  0  0 -1  0  0  0;
ESTIMATE 'b-c,v3' treat 0  1 -1 treat*visit 0 0 0 0  1  0  0  0 -1  0  0  0;
ESTIMATE 'a-b,v4' treat 1 -1  0 treat*visit 0 1 0 0  0 -1  0  0  0  0  0  0;
ESTIMATE 'a-c,v4' treat 1  0 -1 treat*visit 0 1 0 0  0  0  0  0  0 -1  0  0;
ESTIMATE 'b-c,v4' treat 0  1 -1 treat*visit 0 0 0 0  0  1  0  0  0 -1  0  0;
ESTIMATE 'a-b,v5' treat 1 -1  0 treat*visit 0 0 1 0  0  0 -1  0  0  0  0  0;
ESTIMATE 'a-c,v5' treat 1  0 -1 treat*visit 0 0 1 0  0  0  0  0  0  0 -1  0;
ESTIMATE 'b-c,v5' treat 0  1 -1 treat*visit 0 0 0 0  0  0  1  0  0  0 -1  0;
ESTIMATE 'a-b,v6' treat 1 -1  0 treat*visit 0 0 0 1  0  0  0 -1  0  0  0  0;
ESTIMATE 'a-c,v6' treat 1  0 -1 treat*visit 0 0 0 1  0  0  0  0  0  0  0 -1;
ESTIMATE 'b-c,v6' treat 0  1 -1 treat*visit 0 0 0 0  0  0  0  1  0  0  0 -1;
```

The NOCLPRINT option suppresses the lengthy printing of patient categories. The EMPIRICAL option causes the empirical estimates of fixed effects standard errors to be given. This option should be omitted to obtain standard errors directly from the covariance parameters fitted to the model. Use of the R and RCORR options displays the covariance parameters in a more meaningful way as matrices. If no subject numbers are specified, matrices for the first patient only will be printed. Here, we have requested matrices for patients 1, 3 and 4 so that a covariance and correlation matrix is printed for a patient on each treatment.

```
                REML Estimation Iteration History
   Iteration  Evaluations     Objective    Criterion
          0            1    5811.6217215
          1            2    5441.1837877   0.00009243
          2            1    5440.9144957   0.00000131
          3            1    5440.9108920   0.00000000

                    Convergence criteria met.

                     R Matrix for PAT 1
     Row         COL1           COL2           COL3           COL4
       1    76.11692434   52.76243212   46.49245995   35.37454800
       2    52.76243212   76.11692434   52.76243212   46.49245995
```

3	46.49245995	52.76243212	76.11692434	52.76243212
4	35.37454800	46.49245995	52.76243212	76.11692434

R Correlation Matrix for PAT 1

Row	COL1	COL2	COL3	COL4
1	1.00000000	0.69317609	0.61080319	0.46473959
2	0.69317609	1.00000000	0.69317609	0.61080319
3	0.61080319	0.69317609	1.00000000	0.69317609
4	0.46473959	0.61080319	0.69317609	1.00000000

R Matrix for PAT 3

Row	COL1	COL2	COL3	COL4
1	68.21004269	28.93225936	22.47599007	28.69210421
2	28.93225936	68.21004269	28.93225936	22.47599007
3	22.47599007	28.93225936	68.21004269	28.93225936
4	28.69210421	22.47599007	28.93225936	68.21004269

R Correlation Matrix for PAT 3

Row	COL1	COL2	COL3	COL4
1	1.00000000	0.42416422	0.32951145	0.42064340
2	0.42416422	1.00000000	0.42416422	0.32951145
3	0.32951145	0.42416422	1.00000000	0.42416422
4	0.42064340	0.32951145	0.42416422	1.00000000

R Matrix for PAT 4

Row	COL1	COL2	COL3	COL4
1	84.98088475	49.51862752	41.07367987	42.21374108
2	49.51862752	84.98088475	49.51862752	41.07367987
3	41.07367987	49.51862752	84.98088475	49.51862752
4	42.21374108	41.07367987	49.51862752	84.98088475

R Correlation Matrix for PAT 4

Row	COL1	COL2	COL3	COL4
1	1.00000000	0.58270313	0.48332846	0.49674396
2	0.58270313	1.00000000	0.58270313	0.48332846
3	0.48332846	0.58270313	1.00000000	0.58270313
4	0.49674396	0.48332846	0.58270313	1.00000000

Covariance Parameter Estimates (REML)

Cov Parm			Estimate	Std Error	Z	Pr>|Z|
VISIT Diagonal	TREAT A		84.98088475	8.58805298	9.90	0.0001
TOEP(2)	TREAT A		49.51862752	8.47505428	5.84	0.0001
TOEP(3)	TREAT A		41.07367987	8.66220101	4.74	0.0001
TOEP(4)	TREAT A		42.21374108	9.75303141	4.33	0.0001
VISIT Diagonal	TREAT B		68.21004269	6.17855341	11.04	0.0001
TOEP(2)	TREAT B		28.93225936	5.84057953	4.95	0.0001
TOEP(3)	TREAT B		22.47599007	6.22420340	3.61	0.0003
TOEP(4)	TREAT B		28.69210421	7.44132452	3.86	0.0001
VISIT Diagonal	TREAT C		76.11692434	8.32316933	9.15	0.0001
TOEP(2)	TREAT C		52.76243212	8.25160977	6.39	0.0001

```
         TOEP(3)    TREAT C    46.49245995    8.25472777    5.63    0.0001
         TOEP(4)    TREAT C    35.37454800    8.93336142    3.96    0.0001
Residual                        1.00001307          .          .        .
```

Model Fitting Information for DBP

Description	Value
Observations	1092.000
Variance Estimate	1.0000
Standard Deviation Estimate	1.0000
REML Log Likelihood	-3711.99
Akaike's Information Criterion	-3723.99
Schwarz's Bayesian Criterion	-3753.89
-2 REML Log Likelihood	7423.980
Null Model LRT Chi-Square	370.7108
Null Model LRT DF	11.0000
Null Model LRT P-Value	0.0000

Tests of Fixed Effects

Source	NDF	DDF	Type III F	Pr>F
DBP1	1	598	20.45	0.0001
TREAT	2	180	4.05	0.0191
VISIT	3	429	12.83	0.0001
TREAT*VISIT	6	327	1.95	0.0727

ESTIMATE Statement Results

Parameter	Estimate	Std Error	DDF	T	Pr>|T|
a-b,v3	1.35784362	1.23328138	392	1.10	0.2716
a-c,v3	3.42119997	1.25093356	328	2.73	0.0066
b-c,v3	2.06335635	1.25624193	407	1.64	0.1013
a-b,v4	0.55566888	1.22199669	366	0.45	0.6496
a-c,v4	1.89169936	1.23576575	307	1.53	0.1268
b-c,v4	1.33603048	1.20115585	337	1.11	0.2668
a-b,v5	3.00259108	1.37711910	561	2.18	0.0296
a-c,v5	4.76939204	1.41493254	512	3.37	0.0008
b-c,v5	1.76680096	1.28909619	440	1.37	0.1712
a-b,v6	0.08612705	1.29899896	428	0.07	0.9472
a-c,v6	2.09571878	1.33077408	390	1.57	0.1161
b-c,v6	2.00959174	1.20620217	330	1.67	0.0967

Least Squares Means

Level	LSMEAN	Std Error	DDF	T	Pr>|T|	Alpha	Lower
TREAT A	92.74370315	0.76265048	98	121.61	0.0001	0.05	91.2302
TREAT B	91.49314550	0.63584666	90.9	143.89	0.0001	0.05	90.2301
TREAT C	89.69920062	0.75352104	88.8	119.04	0.0001	0.05	88.2019

Least Squares Means

Upper
94.2572
92.7562
91.1965

Differences of Least Squares Means

Level 1	Level 2	Difference	Std Error	DDF	T	Pr>|T|	Alpha	Lower	Upper
TREAT A	TREAT B	1.25055766	0.98867210	182	1.26	0.2075	0.05	-0.7002	3.2013
TREAT A	TREAT C	3.04450254	1.07877156	193	2.82	0.0053	0.05	0.9168	5.1722
TREAT B	TREAT C	1.79394488	0.98057323	169	1.83	0.0691	0.05	-0.1418	3.7297

REPEATED statements for Models 1–5

```
1. REPEATED visit/ SUBJECT=pat TYPE=CS R RCORR;
2. REPEATED visit/ SUBJECT=pat TYPE=AR(1) R RCORR;
3. REPEATED visit/ SUBJECT=pat TYPE=TOEP R RCORR;
4. REPEATED visit/ SUBJECT=pat TYPE=UN R RCORR;
5. REPEATED visit/ SUBJECT=pat TYPE=CS GROUP=treat R=1,3,4
   RCORR=1,3,4;
```

Checking model assumptions in Model 6

```
PROC MIXED NOCLPRINT; CLASS treat visit pat;
MODEL dbp = dbp1 treat visit treat*visit/ PREDICTED;
REPEATED visit/ SUBJECT=pat TYPE=TOEP GROUP=treat;
MAKE 'PREDICTED' OUT=pred NOPRINT;
ID visit treat pat dbp dbp1;

DATA pred; SET pred;
PROC SORT; BY TREAT visit pat;
PROC PRINT NOOBS; VAR pat treat visit resid pred dbp dbp1;
TITLE 'RESIDUALS AND PREDICTED VALUES';
PROC PLOT; PLOT resid*pred;
TITLE 'RESIDUALS VS PREDICTED VALUES';
PROC PLOT; BY treat; PLOT resid*pred;
TITLE 'RESIDUALS VS PREDICTED VALUES BY TREATMENT GROUP';

DATA v3(KEEP=pat treat resid3 pred3)
 v4(KEEP=pat treat resid4 pred4)
 v5(KEEP=pat treat resid5 pred5)
 v6(KEEP=pat treat resid6 pred6);
SET pred;
IF visit=3 THEN DO; resid3=resid; pred3=pred; OUTPUT v3; END;
IF visit=4 THEN DO; resid4=resid; pred4=pred; OUTPUT v4; END;
IF visit=5 THEN DO; resid5=resid; pred5=pred; OUTPUT v5; END;
IF visit=6 THEN DO; resid6=resid; pred6=pred; OUTPUT v6; END;

DATA v3v4; MERGE v3 v4; BY treat pat;
TITLE 'VISIT 3 RESIDUALS VS VISIT 4 RESIDUALS';
PROC PLOT; PLOT resid3*resid4;

DATA v3v5; MERGE v3 v5; BY treat pat;
TITLE 'VISIT 3 RESIDUALS VS VISIT 5 RESIDUALS';
PROC PLOT; PLOT resid3*resid5;

DATA v3v6; MERGE v3 v6; BY treat pat;
TITLE 'VISIT 3 RESIDUALS VS VISIT 6 RESIDUALS';
PROC PLOT; PLOT resid3*resid6;
```

```
DATA v4v5; MERGE v4 v5; BY treat pat;
TITLE 'VISIT 4 RESIDUALS VS VISIT 5 RESIDUALS';
PROC PLOT; PLOT resid4*resid5;

DATA v4v6; MERGE v4 v6; BY treat pat;
TITLE 'VISIT 4 RESIDUALS VS VISIT 6 RESIDUALS';
PROC PLOT; PLOT resid4*resid6;

DATA v5v6; MERGE v5 v6; BY treat pat;
TITLE 'VISIT 5 RESIDUALS VS VISIT 6 RESIDUALS';
PROC PLOT; PLOT resid5*resid6;
```

The first part of the code fits the mixed model and outputs the residuals and predicted values to a dataset pred to be plotted. Residuals are then split into four datasets corresponding to each visit. Datasets for each pair of visits are then merged and residuals from each visit are plotted against those from every other visit.

The PROC MIXED output is identical to that given above. The following output is from the PRINT procedure.

```
              RESIDUALS AND PREDICTED VALUES
     PAT    TREAT   VISIT    RESID        PRED      DBP     DBP1

       4     A       3     -5.2789     93.2789      88      100
       5     A       3     -6.6491     95.6491      89      105
       7     A       3     13.0844     99.9156     113      114
      10     A       3      5.7730     94.2270     100      102
      14     A       3    -13.2789     93.2789      80      100
      18     A       3    -18.2789     93.2789      75      100
      22     A       3     -9.3827     91.3827      82       96
      23     A       3      0.7211     93.2789      94      100
      27     A       3    -13.7010     94.7010      81      103
      29     A       3      4.1952     92.8048      97       99
      33     A       3     21.7730     94.2270     116      102
      35     A       3     -2.3308     92.3308      90       98
      44     A       3      6.3509     95.6491     102      105
      45     A       3      9.7211     93.2789     103      100
      54     A       3     10.0844     99.9156     110      114
     ETC
```

The residual plots are given in the main text.

6.4 EXAMPLE: COVARIANCE PATTERN MODELS FOR COUNT DATA

This was a placebo controlled trial of an anti-convulsant treatment for epilepsy involving 59 patients. The data are taken from Thall and Vail (1990). The number of epileptic seizures was counted over an eight-week period prior to treatment and then over four two-week periods following treatment. None of the patients dropped out of the study. A histogram of the number of epileptic episodes is

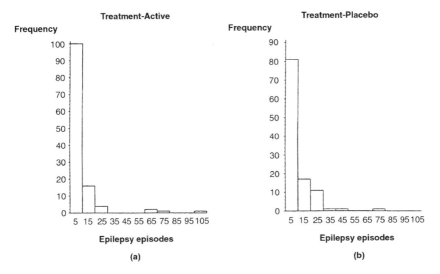

Figure 6.3   Histograms of the number of epilepsy episodes by treatment group.

shown in Figure 6.3 by treatment group. These show that many patients have few seizures, while a few have a large number. This distribution indicates that a Poisson error and a log link function may be appropriate. However, it is possible that the small number of very large frequencies will produce outlying residuals. Note that here an offset variable (see Section 3.1.2) is not needed because the trial periods are strictly two weeks long (it does not matter that the baseline period is longer — this is taken into account by the baseline effect estimate).

6.4.1   GLMM analysis models

Models were fitted using PROC GENMOD with a REPEATED statement. Initially, a model fitting baseline, treatment, visit and treatment·visit effects as fixed was fitted with a compound symmetry structure. The treatment·visit interaction was not significant ($p = 0.56$) and was removed from the model, causing the overall treatment effect to be a weighted average of the effects at each visit. However, because the data are complete, this estimate is expected to differ little from the unweighted estimate (obtained if treatment·visit effects are retained, see Section 6.1). The four models below were fitted with baseline, treatment and visit effects taken as fixed in each.

Model	Covariance pattern
1	Compound symmetry (CS)
2	First-order autoregressive (AR1)
3	Toeplitz
4	General

Table 6.5 Results from all models.

Model	Covariance parameters	Treatment difference	Model-based SE	Empirical SE
1 CS	1 0.42 1 0.42 0.42 1 0.42 0.42 0.42 1	0.099	0.147	0.196
2 AR(1)	1 0.50 1 0.50^2 0.50 1 0.50^3 0.50^2 0.50 1	0.119	0.139	0.195
3 Toeplitz	1 0.50 1 0.36 0.50 1 0.28 0.36 0.50 1	0.111	0.147	0.195
4 General	1 0.46 1 0.41 0.58 1 0.28 0.31 0.44 1	0.112	0.146	0.190

Results

Results from the four models are shown in Table 6.5. The GENMOD procedure did not output a quasi-likelihood value so it is not possible for us to compare the models statistically. The covariance terms in all the models are positive, indicating that observations on the same patient are correlated. The correlation terms vary for each visit pair in the general structure (Model 4) and this indicates that a more complex pattern than compound symmetry may be required. The empirical standard errors (see Section 2.4.3) are much higher than the model-based standard errors in all models. This is an indicator that none of the models may adequately account for the observed covariance in the data. This may be because the covariance parameters vary between the treatment groups (we were unable to allow for this using PROC GENMOD), or because the Poisson assumption may not be appropriate. The model assumptions will therefore need to be checked carefully. Although the appropriateness of the model is under doubt, we will proceed with examining the treatment effect estimates.

No evidence of a significant treatment effect was apparent. Relative rates and 95% confidence intervals can be calculated from the mean treatment difference and its standard error. As the empirical standard errors are larger, we will take the conservative approach and use them. In Model 1 the confidence interval for the treatment effect on the linear scale is

$$95\% \text{ CI} = 0.099 \pm t_{58,0.975} \times 0.196.$$

The DF is taken as 58, because this would be the DF for the patient error stratum (60 minus one for the intercept and one for the treatment effect). Using

$t_{58,0.975} = 2.00$ we obtain:

$$95\% \text{ CI} = 0.099 \pm 2.00 \times 0.196 = (-0.293, 0.491).$$

A comparison of the treatments in terms of a relative rate is obtained by exponentiating the effect estimate.

$$RR = \frac{\text{seizure rate on placebo}}{\text{seizure rate on active}} = \exp(0.099) = 1.10.$$

Confidence intervals for the relative rate are calculated by exponentiating the confidence intervals calculated on the linear scale, $\exp(-0.293, 0.491) = (0.74, 1.63)$.

Plots of the Pearson residuals (see Section 3.3.11) against their predicted values (on the linear scale, i.e. given by $\mathbf{X}\hat{\alpha}$) are used to provide a rough check of model assumptions and to look for outliers. This is done for Model 1 (Figure 6.4). This plot

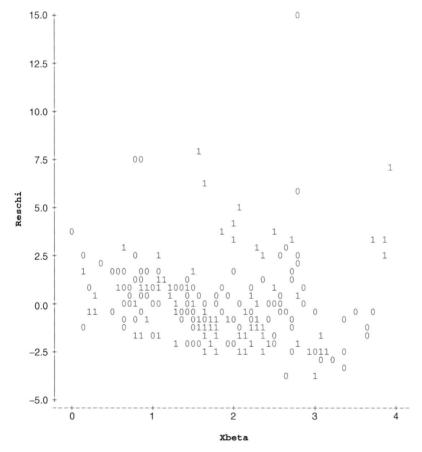

Figure 6.4 Plots of the Pearson residuals against their predicted values ($1 = $ active treatment, $0 = $ placebo).

shows that there are many large positive residuals and therefore that the Poisson assumption may not be appropriate. The potential influence of these observations on the results was assessed in a very simple way by refitting Model 1 with the largest residual removed (patient 25, visit 3). This model gave a treatment effect estimate of 0.046 with an empirical standard error of 0.180. This is noticeably different from the treatment effect of 0.099 (0.196) estimated with the outlying observation included and further confirms our suspicion that the assumption of a Poisson distribution is not appropriate. A transformation of the data is unlikely to overcome this problem since there are a large number of zero observations. In the absence of non-parametric mixed models, one possibility would be to categorise the number of epilepsy attacks and to use a categorical mixed model. We will now consider using this approach.

6.4.2   Analysis using a categorical mixed model

The number of post-treatment epilepsy attacks were categorised into four groups as shown below. The groupings are chosen so that each category is of a reasonable size.

Group	Attacks
1	0 (10%)
2	1–3 (29%)
3	4–10 (39%)
4	11+ (22%)

The baseline attack rate was not categorised and was again fitted as log(attacks). Attempts were made to fit compound symmetry, Toeplitz and general covariance patterns to the data using the SAS macro written by Lipsitz *et al.* (1994) (see Section 9.1). However, convergence was achieved only for the compound symmetry model. The treatment·time interaction was not significant and this effect was removed from the model.

Covariance parameter estimates

Matrices of correlations between each pair of partitions are given below. Recall from Section 4.2 that instead of a single correlation parameter to model the compound symmetry pattern, there is now a matrix of rank $(c - 1) \times (c - 1)$ giving a correlation parameter for each pair of partitions. The correlation values in our model were

Partition	1	2	3
1	0.67		
2	0.03	0.17	
3	0.00	0.16	0.28

Thus, there appears to be some correlation between observations on the same patients.

Fixed effects estimates

The fixed effect estimates were as follows:

Effect	Estimate	Empirical SE	Model-based SE
Intercept 1	−0.32	0.56	0.52
2	1.95	0.50	0.42
3	4.67	0.65	0.61
Baseline	−0.078	0.01	0.01
Treatment	−0.81	0.35	0.35
Visit 1	−0.34	0.31	0.26
Visit 2	−0.64	0.37	0.26
Visit 3	0.29	0.36	0.25

The three intercept terms (arising from the three possible partitions of the four categories), and the visit terms are of little interest. The large size of the baseline term relative to its standard error indicates that the model has benefited from its inclusion. The overall treatment effect is significant ($p = 0.02$). This differs from the GLMM analysis and appears to indicate that the analysis of the categorised attack rate is more sensitive. This is likely to be because there are three extremely large values (>60 attacks) in the active treatment group compared with only one in the placebo group. In the GLMM analysis these will have a large effect on the variance of the treatment effects, whereas this does not occur in the categorical analysis since they are grouped with other values of >10. Similarly, they will have a reduced influence on the estimated magnitude of the treatment effect.

The coefficient for the treatment effect is difficult to interpret directly, but by exponentiation we can calculate an odds ratio in an analogous way to Section 3.1.5 where GLMMs were considered. In this case, $\exp(-0.81) = 0.44$ and this is the estimate of the odds ratio for the probability of a 'favourable' outcome on placebo compared with active treatment, whether 'favourable' is defined as 0 attacks, $<=3$ attacks, or $<=10$ attacks. Note that the odds ratio from this model is defined in terms of a 'favourable' outcome, whereas in the GLMM it is in terms of the rate of epilepsy attacks. Note also that it is an inherent assumption of this model that the same odds ratio applies to every partition between the categories. 95% confidence intervals can be calculated as before from $\exp(-0.81 \pm t_{58,0.975} \times 0.35) = \exp(-0.81 \pm 2.00 \times 0.35) = (0.22, 0.90)$. (Note that these use a t distribution with the patient DF of 58 as used in the GLMM.)

SAS code and output

Variables

pat   = patient,
time = time (1,2,3,4),
treat = treatment (1=anti-convulsant drug, 0=placebo),
epis = number of epilepsy attacks,
lbase = log(baseline epilepsy attacks).

SAS code is given below for Model 1. The SAS code for the other models is identical except that different REPEATED statements are used. These statements are printed after the Model 1 output.

Model 1

```
PROC GENMOD; CLASS pat time treat;
MODEL epis=lbase treat time/ DIST=P;
REPEATED SUBJECT=pat/ WITHIN=time MODELSE TYPE=CS CORRW;
```

The GENMOD Procedure

Model Information

Description	Value
Data Set	WORK.A
Distribution	POISSON
Link Function	LOG
Dependent Variable	EPIS
Observations Used	236

Class Level Information

Class	Levels	Values
PAT	59	1 2 3 4 5 6 7 8 9 10 11 12 13
		14 15 16 17 18 19 20 21 22 23
		24 25 26 27 28 29 30 31 32 33
		34 35 36 37 38 39 40 41 42 43
		44 45 46 47 48 49 50 51 52 53
		54 55 56 57 58 59
TIME	4	1 2 3 4
TREAT	2	0 1

Parameter Information

Parameter	Effect	TIME	TREAT
PRM1	INTERCEPT		
PRM2	LBASE		
PRM3	TREAT		0
PRM4	TREAT		1
PRM5	TIME	1	
PRM6	TIME	2	
PRM7	TIME	3	
PRM8	TIME	4	

```
         Criteria For Assessing Goodness Of Fit
    Criterion              DF        Value      Value/DF
    Deviance              230     972.0308       4.2262
    Scaled Deviance       230     972.0308       4.2262
    Pearson Chi-Square    230    1105.8872       4.8082
    Scaled Pearson X2     230    1105.8872       4.8082
    Log Likelihood          .    2936.5976          .

    .       Analysis Of Initial Parameter Estimates
    Parameter      DF  Estimate  Std Err  ChiSquare  Pr>Chi
    INTERCEPT       1   -2.1687   0.1297   279.7233  0.0001
    LBASE           1    1.1765   0.0309  1453.1961  0.0001
    TREAT    0      1    0.1032   0.0453     5.1897  0.0227
    TREAT    1      0    0.0000   0.0000        .       .
    TIME     1      1    0.2030   0.0649     9.7777  0.0018
    TIME     2      1    0.1344   0.0659     4.1539  0.0415
    TIME     3      1    0.1405   0.0659     4.5504  0.0329
    TIME     4      0    0.0000   0.0000        .       .
    SCALE           0    1.0000   0.0000        .       .
    NOTE:  The scale parameter was held fixed.
```

Note that these parameter estimates are those obtained from PROC GENMOD without using a REPEATED statement and therefore can be ignored.

```
                GEE Model Information
        Description                     Value
        Correlation Structure           Exchangeable
        Within-Subject Effect           TIME (4 levels)
        Subject Effect                  PAT (59 levels)
        Number of Clusters              59
        Correlation Matrix Dimension    4
        Maximum Cluster Size            4
        Minimum Cluster Size            4

             Working Correlation Matrix
                  COL1      COL2      COL3      COL4
        ROW1    1.0000    0.4158    0.4158    0.4158
        ROW2    0.4158    1.0000    0.4158    0.4158
        ROW3    0.4158    0.4158    1.0000    0.4158
        ROW4    0.4158    0.4158    0.4158    1.0000
```

This is the within-patient correlation matrix, **P**, estimated by the GLMM with a single parameter to give the compound symmetry pattern.

```
            Analysis Of GEE Parameter Estimates
            Empirical Standard Error Estimates
                Empirical  95% Confidence Limits
    Parameter Estimate Std Err  Lower    Upper      Z  Pr>|Z|
    INTERCEPT  -2.1823 0.5027 -3.1675  -1.1971  -4.342  0.0000
```

```
LBASE          1.1805 0.1501  0.8863  1.4747  7.8654 0.0000
TREAT     0    0.0994 0.1955 -0.2837  0.4825  0.5086 0.6110
TREAT     1    0.0000 0.0000  0.0000  0.0000  0.0000 0.0000
TIME      1    0.2030 0.0987  0.0096  0.3964  2.0571 0.0397
TIME      2    0.1344 0.0762 -0.0149  0.2837  1.7646 0.0776
TIME      3    0.1405 0.1228 -0.1003  0.3812  1.1436 0.2528
TIME      4    0.0000 0.0000  0.0000  0.0000  0.0000 0.0000
Scale          2.1663      .       .       .       .       .
```

NOTE: The scale parameter for GEE estimation was
computed as the square root of the normalized
Pearson's chi-square.

Analysis Of GEE Parameter Estimates

Model-Based Standard Error Estimates

Model 95% Confidence Limits

```
Parameter Estimate Std Err   Lower    Upper      Z  Pr>|Z|

INTERCEPT  -2.1823  0.4043 -2.9746 -1.3899 -5.398 0.0000
LBASE       1.1805  0.1001  0.9844  1.3766 11.797 0.0000
TREAT     0 0.0994  0.1469 -0.1885  0.3874  0.6768 0.4985
TREAT     1 0.0000  0.0000  0.0000  0.0000  0.0000 0.0000
TIME      1 0.2030  0.1077 -0.0081  0.4140  1.8850 0.0594
TIME      2 0.1344  0.1093 -0.0798  0.3486  1.2299 0.2187
TIME      3 0.1405  0.1091 -0.0734  0.3544  1.2871 0.1981
TIME      4 0.0000  0.0000  0.0000  0.0000  0.0000 0.0000
Scale       2.1663      .       .       .       .       .
```

Estimates are listed both with model-based standard errors and empirical standard errors with corresponding z statistics and p-values. However, we would suggest that t tests are preferable for testing fixed effects (see Section 3.3.9).

REPEATED statements for other models

```
REPEATED SUBJECT=pat/ WITHIN=time MODELSE TYPE=AR(1) CORRW;
REPEATED SUBJECT=pat/ WITHIN=time MODELSE TYPE=TOEP CORRW;
REPEATED SUBJECT=pat/ WITHIN=time MODELSE TYPE=UN CORRW;
```

Residual plot using Model 1

```
PROC GENMOD; CLASS pat time treat;
MODEL epis=lbase treat time/ DIST=P;
REPEATED SUBJECT=pat/ WITHIN=time MODELSE TYPE=CS CORRW OBSTATS;
MAKE 'GEEOBSTATS' OUT=resid NOPRINT;

PROC PLOT DATA=resid; PLOT reschi*xbeta=treat;
```

This plot is given in main text.

Categorical analysis

The SAS macro written by Lipsitz *et al.* (1994) was used to fit these models (see Section 9.1). This macro and the code used can be obtained from Web page [www.med.ed.ac.uk/phs/mixed].

6.5   RANDOM COEFFICIENTS MODELS

6.5.1   Introduction

A random coefficients model is an alternative approach to modelling repeated measures data. Here, a model is devised to describe arithmetically the relationship of a measurement with time. The statistical properties of random coefficients models have already been introduced in Sections 1.5 and 2.1.4. Here, we will consider in more depth the practical details of fitting these models and the situations in which they are most appropriate.

The most common applications are those in which a linear relationship is assumed between the outcome variable of interest and time. The main question of interest is then likely to be whether the rate of change in this outcome variable differs between the 'treatment' groups. Such an example was reported by Smyth *et al.* (1997). They carried out a randomised controlled trial of glutathione versus placebo in patients with ovarian cancer who were being treated with cis-platinum. This drug has proven efficacy in the treatment of ovarian cancer, but has a number of adverse effects as well. Amongst these is a toxic effect on the kidneys. This effect can be monitored by the creatinine levels in the patients' blood. One of the hoped-for secondary effects of glutathione was to reduce the rate of decline of renal (kidney) function. This was assessed using a random coefficients model, but analysis showed no statistically significant difference between the rates of decline in the two treatment arms. Such an analysis may find widespread application in the analysis of 'safety' variables in clinical trials, because it is important to establish what effect new drugs may have on a range of biochemical and haematological variables. If these variables are measured serially, analysis is likely to be more efficient if based on all observations, using a method which will be sensitive to a pattern of rise or decline in the 'safety' variables. A further example in which the rate of decline of CD4 counts is compared in two groups of HIV-infected haemophiliacs will be presented in detail in Section 6.6.1.

In fitting linear random coefficients models, as described above, we will wish to fit fixed effects to represent the average rate of change of our outcome variables over time (i.e. a time effect) and we will assess the extent to which treatments differ in the average rate of change by fitting a treatment·time interaction. We will also require fixed effects to represent the average intercepts for each treatment (i.e. a treatment effect). In addition to the fixed effects representing average slopes and intercepts, the random coefficients model allows the slopes and intercepts to vary randomly between patients and cause a separate regression line to be fitted for each patient. This is achieved by fitting patient effects as random (to allow intercepts to vary) and patient·time as random to allow slopes to vary. These effects are used in the calculation of the standard errors of the time and treatment·time effects, which are our main focus of interest. Our basic model is therefore

Fixed effects : time, treatment, treatment·time,
Random effects : patient, patient·time.

The effects described above represent a minimum set of effects which will be considered in the model. Other patient characteristics, such as age and sex and their interactions with time, can readily be incorporated into the model, and we will see later that polynomial relationships and the effect of baseline levels can also be incorporated.

When the repeated measures data are obtained at fixed points in time, there will be a choice between the use of covariance pattern models and random coefficients models. This choice may be influenced by how well the dependency of the observations on time can be modelled, and whether interest is centred on the changing levels of the outcome variable over time, or on its absolute levels. In many instances, the random coefficients model will be the 'natural' choice, as in the examples presented. If the times of observation are not standardised, or if there are substantial discrepancies between the scheduled times and actual time of observation, then random coefficients models are more likely to be the models of choice.

Modelling non-linear relationships with time

The models considered above assume a linear relationship with time. In many applications the linear model described is sufficient for assessing whether there is a time trend, or whether the trend is varying across treatment groups. However, it is also possible to model non-linear relationships with time; for example, by using polynomial or exponential functions. Here, we will only consider models that can be fitted using polynomial functions of time.

We suggest that a model based on polynomials of time is built up by adding polynomials of increasingly higher order one at a time, both as fixed effects and random coefficients. If a variance component for a random coefficient is negative, the last random coefficient added to the model should be removed and no further random coefficients should be added. However, higher order polynomials can still be considered as fixed effects if appropriate. This model-building process is illustrated in the worked example in Section 6.6.2 and readers may find it helpful to refer to this example and the example in Section 6.6.1 before considering the material in the remainder of this section.

6.5.2   General points

Negative variance components

The usual action when a negative variance component estimate is obtained for a random coefficient would be to refit the model with the random coefficient removed. However, the user should be warned that not all packages will produce a negative variance component estimate. For example, in PROC MIXED we have found that non-convergence or a message stating that the **G** matrix is not positive semi-definite are usually indications of a negative variance component. (A matrix, **A**, is positive semi-definite if $\mathbf{x'Ax}$ is a non-negative number for all vectors, $\mathbf{x}$.) The recommended action is then to remove the random coefficients one by one in decreasing order of complexity until all variance components become positive.

Use of baseline measurements

If there is a pre-treatment, baseline observation, then there are two distinct ways in which it can be used. In one approach it can be specified as a fixed effect, so that it is considered as a covariate in the analysis. However, it will often be more natural to think of such an observation as the first repeated measurement at time zero, with time measured from the start of treatment. Such an example occurs in Section 6.6.1 when we analyse the change in CD4 counts over time.

Non-comparative datasets

Repeated measures data is not always collected to compare specific groups of patients (e.g. treatment groups); it may be collected simply to monitor a group of patients over time. In this situation the linear model (i) introduced earlier would simplify to

> Fixed effects : time,
> Random effects : patient, patient·time,

and interest would lie primarily with the time effect estimate. Its standard error would reflect the variation in time slope occurring between patients.

Shrunken random coefficient estimates

Estimates of the random coefficients (i.e. intercept and time effects for each patient) are not usually of interest. However, it is possible to estimate them and they will be shrunken towards the overall intercept and time effects. This avoids the potential problem of unrealistic slope estimates which may occur when there are only a few observations per patient.

Significance testing

The points relating to significance testing given in Section 2.4.4 apply also to random coefficients models. Time and treatment·time effects can be tested using F tests with Satterthwaite's approximate DF used for the denominator. If software is not available for calculating Satterthwaite's DF, then the patient DF can be used as a conservative estimate of the denominator DF (since time and treatment·time effects are contained within the random patient·time coefficients whose DF are equal to the patient DF).

If required, the significance of the variance components can also be tested using likelihood ratio tests to compare models including and not including the corresponding random coefficients. However, even if the variance component is non-significant, it is usually desirable to retain a random coefficient in the model provided its variance component is positive.

Model checking

The residuals can be checked for normality by plotting them against their predicted values and by using normal plots. Additionally, plots of the residuals against time will help check whether their variance is constant over time. The random coefficients are assumed to have a multivariate normal distribution with zero means and covariance matrix **G**. This assumption is difficult to check formally. However, normal plots of the random coefficients and plots against their predicted values should be sufficient to detect marked non-normality and to show up any outliers. To check the multivariate normal assumption more fully, bivariate plots of the random coefficients can be used (e.g. patient residuals against patient·time residuals). If there is evidence of non-normality or of outliers, the suggestions described in Section 2.4.6 should be followed.

6.5.3   Comparisons with fixed effects approaches

Here, we consider two fixed effects approaches that have been used for modelling linear relationships with time for repeated measures data, and show how they differ from the random coefficients model.

The first approach is an extremely statistically naive one, but one which appears from time to time in the medical literature. Here, treatment, time and treatment·time are fitted as fixed effects and the effects of patient and patient·time are totally ignored. Thus, all observations are treated as independent and the standard errors of intercept and slope·effects will be erroneously small because they take no account of between patient variability.

The second fixed effects approach is more robust. A two-stage model is used: first, time slopes are calculated for each patient; then an analysis is performed on the slope estimates (e.g. Rowell and Walters, 1976). This has the advantage over the first approach in that random variation in slope effects is allowed between patients. However, a drawback is that slopes estimated for patients with only a few observations can be unrealistic, and the slopes from all patients are given equal weight regardless of their accuracy. A suitably weighted analysis could, of course, be considered, but it will usually be simpler and more efficient to apply a random coefficients model.

When a random coefficients model is used the problems encountered with the fixed effects approaches above do not occur. The standard errors of intercept and slope effects take into account between-patient variability; shrunken random coefficient estimates avoid the problem of unrealistic slope estimates which may occur when there are only a few observations per patient; slopes from individual patients are appropriately weighted.

6.6   EXAMPLES OF RANDOM COEFFICIENTS MODELS

In this section we will present two examples. In the first of these we will fit (after a transformation) a linear model. This corresponds to the most common type of random coefficients model which is likely to be encountered in practice. In the second example we examine the use of polynomial models. Only in this example do we go through in detail the model checking procedures which we recommend be undertaken in all analyses.

6.6.1   A linear random coefficients model

One of the measures of disease severity in patients with HIV infections is the CD4 count. This has been found to decline with the time since infection with HIV. Many patients with haemophilia have become HIV positive, often because of receiving infected blood products, but it has been reported that treatment of their haemophilia with high-purity monoclonally immunopurified factor VIII concentrates can slow or halt their rate of decline in CD4 count. In Britain, such patients have been treated for some years with high-purity factor VIII concentrate, but some centres have been supplied with concentrate which is monoclonally immunopurified, while others have used concentrates which are ion-exchange purified. This permitted a 'natural experiment' which was monitored prospectively from the time of a patient's transfer from intermediate-purity factor VIII concentrate to high-purity concentrate, for a period of three years (Hay *et al.*, 1998).

Various transformations of the CD4 count have been proposed in the literature. In this case, in agreement with most authors, a square root transformation was found to give approximately normal distributions, and to produce linear rates of change over time. CD4 counts were taken at entry to the study (i.e. at the time of change to one of the forms of high-purity factor VIII concentrate) and at approximately six-monthly intervals thereafter for a period of three years. The exact times when the samples were taken were used in the following analysis.

Of the 116 patients with severe haemophilia A who entered the study, 79 received the monoclonally immunopurified product, while 37 received ion-exchange-purified factor VIII. By the end of the study 25 patients had died, 15 in the monoclonal group and 10 in the ion-exchange group, and three patients were lost to follow-up. One patient died before any post-treatment CD4 counts were obtained, but the data from all other patients were used in the analysis.

The median CD4 counts at entry to the study were $0.30 \times 10^9/l$ in the monoclonal group and $0.16 \times 10^9/l$ in the ion-exchange group. From a design viewpoint, this difference was unhelpful, but was a consequence of the absence of randomisation. One of the centres using the ion-exchange product had patients who had been infected earlier, with consequently lower CD4 counts. The final CD4 counts had median values of 0.16 vs. $0.08 \times 10^9/l$. The median reductions were $0.08 \times 10^9/l$ in the monoclonal group and $0.03 \times 10^9/l$ in the ion-exchange group.

In the analysis presented below using a random coefficients model, no covariates are fitted. The terms fitted are therefore

Fixed effects : time, treatment, treatment·time,
Random coefficients : patient, patient·time.

The results of fitting the model are shown in Table 6.6. The principal interest will be in the magnitude of the treatment·time interaction and its level of significance, because this indicates whether the two treatments differ in their effect on the rate of decline of CD4 counts. The estimate is similar in magnitude to its standard error, indicating an absence of statistical significance ($p = 0.24$). The treatment estimate is just over twice its standard error, and this is statistically significant at conventional levels ($p = 0.03$). This test is of little interest to us, however, as this simply confirms that the CD4 counts in the monoclonal group are higher than those in the ion-exchange group, as was noted in the CD4 levels prior to the commencement of either form of high-purity factor VIII concentrate.

The terms in the **G** matrix are positive along the diagonal, but there is a negative covariance term. The diagonal terms show that there is between-patient variation in the slopes and intercepts, as we would expect. The negative covariance term is not surprising because of the well-known negative correlation between estimates of slopes and intercepts in regression analysis. However, it is perfectly possible in random coefficients models for covariance terms to be positive on occasions.

Table 6.6   Results from analysis of change in CD4 counts.

	Fixed effects (SE)	Covariance parameters in G matrix		Residual
Intercept	0.523 (0.026)	0.0509	−0.0003	0.0073
Treatment	−0.102 (0.047)	−0.0003	0.0027	
Time	−0.050 (0.008)			
Treatment·time	0.016 (0.014)			

In reporting a random coefficients analysis, it is helpful to give the point estimates for average change in each treatment group, together with their standard errors or confidence intervals. If the analysis is being undertaken using SAS, these are not immediately available, but can soon be obtained. The SAS output, given in more detail at the end of this chapter, has the following section:

```
                  Solution for Fixed Effects
     Effect      TREAT    Estimate   Std Error   DF     t  Pr>|t|

     INTERCEPT          0.52283369 0.02632760 112 19.86 0.0001
     TREAT       1     -0.10248803 0.04708074 112 -2.18 0.0316
     TREAT       2      0.00000000        .     .    .     .
     TIME              -0.04961546 0.00758255 103 -6.54 0.0001
     TIME*TREAT  1      0.01582718 0.01353944 107  1.17 0.2450
     TIME*TREAT  2      0.00000000        .     .    .     .
```

The line labelled TIME gives the mean change and standard error in the 'reference' category for treatment, which in this case is treatment group 2 (monoclonal). Thus, the mean rate of decline is $0.050(\text{CD4 count})^{1/2}$ per year with a standard error of 0.008. To obtain the corresponding figures for the ion-exchange group the program needs to be rerun with the labelling of the groups reversed to give

```
                  Solution for Fixed Effects
     Effect      TREAT    Estimate   Std Error   DF     t  Pr>|t|

     INTERCEPT          0.42034566 0.03903145 113 10.77 0.0001
     TREAT       1      0.10248803 0.04708074 112  2.18 0.0316
     TREAT       2      0.00000000        .     .    .     .
     TIME              -0.03378828 0.01121701 108 -3.01 0.0032
     TIME*TREAT  1     -0.01582718 0.01353944 107 -1.17 0.2450
     TIME*TREAT  2      0.00000000        .     .    .     .
```

It will be seen that the interaction term remains unchanged, apart from its sign, but the row for TIME now gives the rate of decline and standard error for the ion-exchange group. This is $0.034(\text{CD4 count})^{1/2}$ per year with a standard error of 0.011.

Examinations of residuals are not shown here, but they indicated that the model produced an acceptable fit.

6.6.2   A polynomial random coefficients model

In this example, antibody levels to a herpes virus were measured in 45 children suffering from one of two types of cancer: solid lump tumour (18) or leukaemia (27). The measurements were taken during hospital visits for courses of chemotherapy treatment. The duration of treatment ranged from one month to three years (median 12 months) and the intervals between treatments differed between the children. The aim of the study was to establish whether virus antibody levels were affected by chemotherapy treatment and whether this change was related to cancer type. It is known that the herpes virus antibody is present in nearly all children (all children in this study had it) and its average level decreases as children become older. Virus antibody levels for individual patients are plotted below. These indicate that levels of the antibody fluctuate widely in some children (see Figure 6.5). The relationship of antibody level with time can be assessed by using a random coefficients model. This will help to determine whether chemotherapy is having an adverse effect on virus levels.

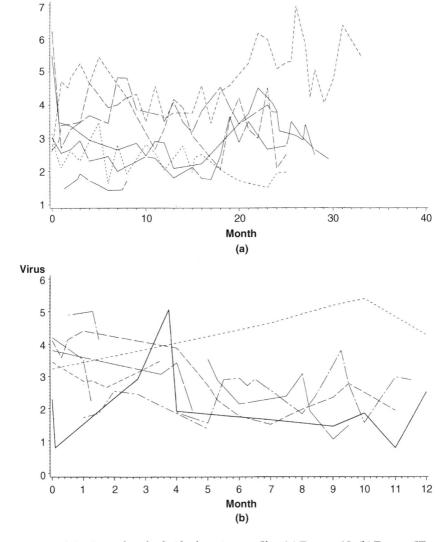

Figure 6.5   Examples of individuals patient profiles: (a) Type = AL. (b) Type = ST.

Building a polynomial model

As the relationship of virus level with time may be non-linear, a model using polynomials of time will be considered. However, note that in many applications the linear model described in Section 6.6.1 is deemed sufficient to assess whether there is a time trend, or whether the trend is varying across treatment groups.

The model is built up by adding polynomials of increasing order one by one into the model, both as fixed effects and as random coefficients. Random coefficients will be retained provided their variance components are positive. However, fixed effects polynomials are added until they are non-significant in order to obtain the best model over time. The age distribution will change at different treatment durations depending on which patients have observations present. To help overcome this variation, age at the start of treatment is included in all models. The following linear model (Model 1) is fitted initially.

Fixed effects : type (solid lump or leukaemia), age, time,
Random coefficients : patient (intercepts), patient·time (slopes).

Results from this analysis are shown in Table 6.7. The variance components corresponding to the two random coefficients are both positive. This indicates that there is more variation between the regression lines for each patient than expected by chance, i.e. patients vary in their rates of virus decay. Next, quadratic time effects are added into the model (Model 2):

Fixed effects : type, age, time, time^2,
Random coefficients : patient, patient·time, patient·time^2.

The three variance components obtained from this model were all positive, indicating more variation between the quadratic curves for each patient than expected by chance.

Table 6.7 Results from Models 1–3.

Model	Fixed effects		G matrix and residual
1 (linear)	Intercept	3.65 (0.23)	0.44
	Type	−0.23 (0.25)	0.013 0.0042
	Age	−0.046 (0.037)	
	Time	−0.032 (0.013)	0.56
2 (quadratic)	Intercept	3.70 (0.24)	0.59
	Type	−0.08 (0.25)	−0.043 0.025
	Age	−0.051 (0.037)	0.0016 −0.0007 0.00002
	Time	−0.081 (0.030)	
	Time^2	0.0025 (0.0010)	0.53
3 (cubic)	Intercept	3.74 (0.24)	0.60
	Type	−0.060 (0.25)	−0.045 0.024
	Age	−0.049 (0.037)	0.0017 −0.0007 0.00002
	Time	−0.118 (0.035)	
	Time^2	0.0065 (0.0023)	0.53
	Time^3	−0.00011 (0.00006)	

When cubic time effects were added as random coefficients the model did not converge (using PROC MIXED). Results provided at the last iteration indicated that the patient·time^3 variance component estimate was becoming negative. Thus, random coefficients up to the quadratic term only are considered. However, a fixed cubic effect can still be included (Model 3):

Fixed effects : type, age, time, time^2, time^3,
Random coefficients : patient (intercepts), patient·time, patient·time^2.

The fixed cubic effect was almost significant ($p = 0.06$) and is retained in the model. A further model with a fixed quartic time effect was tested, but this was non-significant so no further models were considered.

Tests of interactions with type and age

Interactions were tested between time effects (time, time^2 and time^3) and cancer type, and found to be non-significant. Therefore, we can be reasonably confident in assuming that changes in the virus antibody level are similar for the two types of cancer.

The age effect was not significant in any model. However, since it is known that average antibody levels decrease with age, it is important to retain it. Interactions with age were found to be non-significant so we assume that the pattern of antibody change over time is unrelated to age.

Checking model assumptions

The random coefficients model assumes that the residuals have a normal distribution, and that the random coefficients have a multivariate normal distribution. We check these assumptions for Model 3.

Residuals   Residuals are checked by plotting them against their predicted values, and by using a normal plot (Figure 6.6). A plot against time is also used to determine whether the residual variance varies over time.

These plots show one potential outlying observation with a residual of over three (patient 67, month 0). On closer examination it appeared possible that this was caused by a recording error (the recorded value of 7.22 was way above all other values for this patient and it seemed likely that the true value was 1.22). On removal of this observation and refitting Model 3, the fixed effects results changed fairly noticeably (Table 6.8). Since a recording error appeared likely, we will base our conclusions on the analysis with the outlier removed.

Random coefficients

The three sets of random coefficients are plotted against each other to help check that their joint distribution is multivariate normal, and to check for any outlying patients (Figure 6.7). These plots indicate no marked deviation from bivariate normal distributions. The patient·time and patient·time^2 residuals are

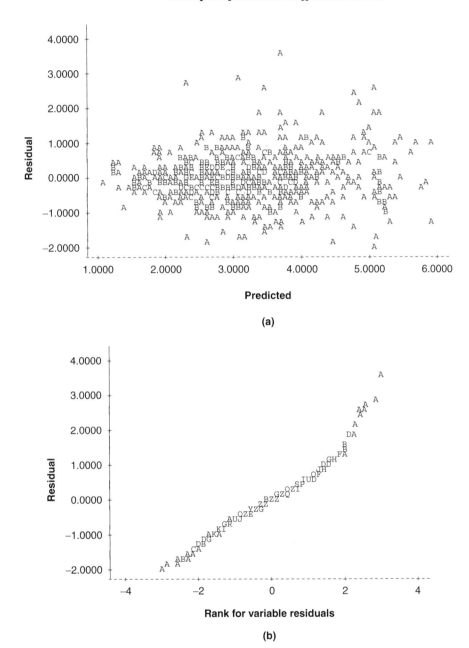

Figure 6.6 Plots of (a) Residuals against their predicted values; (b) Normal plot; (c) Residuals against time (months); A = 1 obs, B = 2 obs, etc.

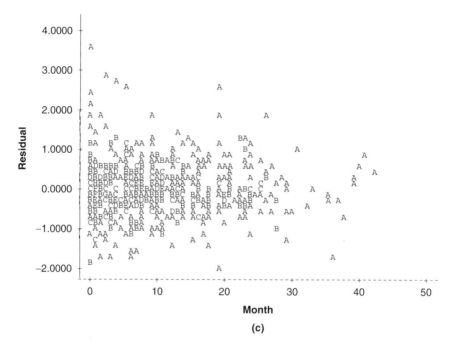

Figure 6.6 (*continued*).

Table 6.8 Results from Model 3 with and without the outlier.

Model	Fixed effects		G matrix and residual		
3 — (without outlier)	Intercept	3.75 (0.25)	0.60		
	Type	−0.11 (0.26)	−0.025	0.016	
	Age	−0.059 (0.038)	0.0011	−0.0005	0.00001
	Time	−0.097 (0.032)			
	Time^2	0.0025 (0.0022)	0.50		
	Time^3	−0.00009 (0.00005)			
3	Intercept	3.74 (0.24)	0.60		
	Type	−0.060 (0.25)	−0.045	0.024	
	Age	−0.049 (0.037)	0.0017	−0.0007	0.00002
	Time	−0.118 (0.035)			
	Time^2	0.0065 (0.0023)	0.53		
	Time^3	−0.00011 (0.00006)			

highly correlated, as indicated by their correlation of −0.98 (calculated from the covariance parameters by $-0.000709/(0.0236 \times 0.0000224)^{1/2}$).

Plot of predicted virus antibody level

The results become more meaningful when the predicted antibody level is plotted against time. This is done for Models 2 and 3 with the outlying observation

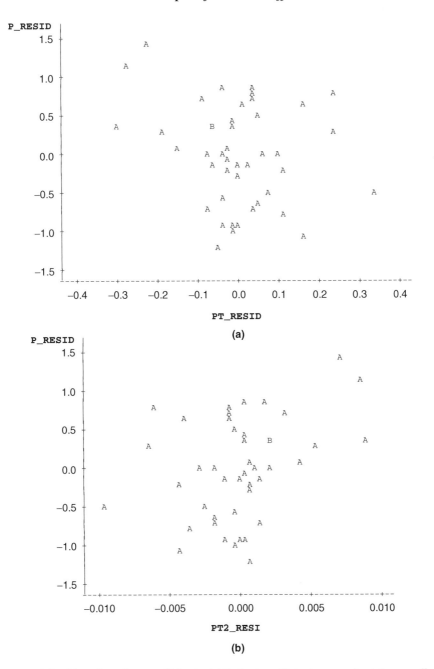

Figure 6.7 Plot of random coefficients: (a) Patient coefficients vs. patient·time coefficients; (b) Patient coefficients vs. patient·time^2 coefficients; (c) patient·time coefficients vs. patient·time^2 coefficients: A = 1 obs, B = 2 obs, etc.

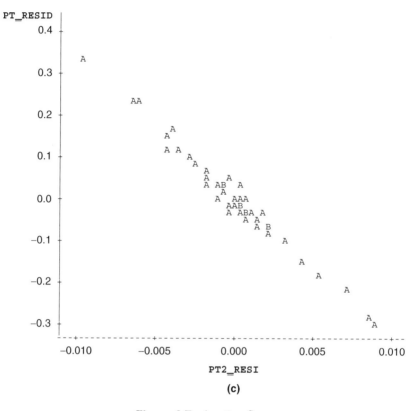

(c)

Figure 6.7 (*continued*).

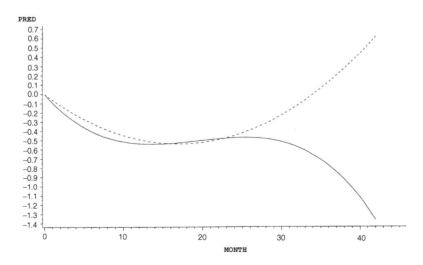

Figure 6.8 Predicted virus antibody level vs. time (Models 2 and 3). Curve ———— Cubic ----- Quadratic.

Table 6.9   Results from Model 1 with time centred about its mean of 10 months.

Model	Fixed effects		G matrix and residual
1	Intercept	3.65 (0.23)	$\begin{pmatrix} 0.44 & \\ 0.013 & 0.0042 \end{pmatrix}$
	Type	−0.23 (0.25)	
	Age	−0.046 (0.037)	
	Time	−0.032 (0.013)	0.56
1 — with altered time origin	Intercept	3.34 (0.26)	$\begin{pmatrix} 1.13 & \\ 0.056 & 0.0042 \end{pmatrix}$
	Type	−0.23 (0.25)	
	Age	−0.046 (0.037)	
	Time	−0.032 (0.013)	0.56

removed (Figure 6.8). The quadratic model (Model 2) curve is also plotted to make the conclusions more robust since the cubic coefficient was only of borderline significance.

The curves for the two models differ markedly for higher values of time. However, only a small proportion of observations were made after 24 months (5%) and the cubic coefficient will be based largely on these. The models are therefore only really plausible up to 24 months, for which they are quite similar. The virus antibody levels decrease most rapidly initially then flatten. The divergent curves illustrate how over-interpretation might occur when part of them is only based on a small amount of data.

Illustration of invariance to time origin

We pointed out in Section 2.1.4 that the fixed effects and **V** matrix estimates were invariant to the origin used for time. This is, of course, important if the methods are to be robust. We illustrate this property by fitting Model 1 with time centred about its mean of 10 months (i.e. using time = time − 10). Results from this model are shown in Table 6.9.

Thus, the variance parameters connected with the patient random coefficients (intercepts) and the fixed intercept effect, have altered to adjust for the change in time origin. However, the fixed effects and covariance parameters are unaltered. The terms in $\mathbf{V} = \text{var}(\mathbf{y})$ will also be found to be identical between the models when calculated for specific values of time.

SAS code and output

Linear random coefficients model (6.6.1)
Variables

```
patient  = patient number,
treat    = type of Factor VIII,
time     = time in years from start of treatment,
cd4_sqrt = square root of CD4 count.
```

```
PROC MIXED NOCLPRINT DATA=cd4;
CLASS patient treat;
MODEL cd4_sqrt = treat time treat*time / SOLUTION DDFM=SATTERTH;
RANDOM INT time / SUBJECT=patient TYPE=UN SOLUTION;
TITLE 'SQUARE ROOT OF CD4 COUNTS OVER TIME';
TITLE3 'RANDOM COEFFICIENTS MODEL';
```

The use of the RANDOM statement to fit patient and patient·time effects as random coefficients is not immediately obvious. Specification of patient as a SUBJECT effect (SUBJECT = patient) blocks the **G** matrix by patients and causes interactions between the effects specified (INT and time) and patient to be fitted as random coefficients (hence patient and patient·time are fitted). The TYPE = UN option causes the random coefficients specified to have a multivariate normal distribution (i.e. a general covariance structure). Here, the distribution will be bivariate normal as only two random coefficients are specified.

```
              SQUARE ROOT OF CD4 COUNTS OVER TIME
                  RANDOM COEFFICIENTS MODEL
              REML Estimation Iteration History
      Iteration   Evaluations      Objective      Criterion
             0           1        -1199.047616
             1           3        -2058.150274     0.00283032
             2           1        -2061.519531     0.00024121
             3           1        -2061.783459     0.00000248
             4           1        -2061.786033     0.00000000

                    Convergence criteria met.

           Covariance Parameter Estimates (REML)
           Cov Parm     Subject      Estimate
             UN(1,1)     PATIENT     0.05090228
             UN(2,1)     PATIENT    -0.00026112
             UN(2,2)     PATIENT     0.00273965
             Residual                0.00734653
```

UN(1,1) and UN(2,2) are the variance component estimates for the patient and patient·time random coefficients. UN(2,1) is the covariance between the random coefficients. Note that the relative sizes of the patient·time variance component cannot be compared directly with the residual because it involves time. In this analysis all the variance components are positive. However, in the situation where a variance component is negative, SAS would not converge and the variance component estimates output from the final iteration would usually show that one variance component estimate was becoming very close to zero.

```
              Model Fitting Information for CD4_SQRT
              Description                       Value
              Observations                    673.0000
              Res Log Likelihood              416.1231
```

```
Akaike's Information Criterion   412.1231
Schwarz's Bayesian Criterion     403.1116
-2 Res Log Likelihood            -832.246
Null Model LRT Chi-Square         862.7384
Null Model LRT DF                  3.0000
Null Model LRT P-Value             0.0000
```

Solution for Fixed Effects

Effect	TREAT	Estimate	Std Error	DF	t	Pr>|t|
INTERCEPT		0.52283369	0.02632760	112	19.86	0.0001
TREAT	1	-0.10248803	0.04708074	112	-2.18	0.0316
TREAT	2	0.00000000	.	.	.	.
TIME		-0.04961546	0.00758255	103	-6.54	0.0001
TIME*TREAT	1	0.01582718	0.01353944	107	1.17	0.2450
TIME*TREAT	2	0.00000000	.	.	.	.

Solution for Random Effects

Effect	PATIENT	Estimate	SE Pred	DF	t	Pr>|t|
INTERCEPT	101	0.19830632	0.06420411	435	3.09	0.0021
TIME	101	0.00876935	0.02996521	270	0.29	0.7700
INTERCEPT	102	0.29878360	0.05750004	506	5.20	0.0001
TIME	102	0.03688603	0.02749137	336	1.34	0.1806
INTERCEPT	103	0.05649755	0.05888391	453	0.96	0.3378
TIME	103	0.02603969	0.03425349	202	0.76	0.4480
INTERCEPT	104	0.23851724	0.05816984	500	4.10	0.0001
TIME	104	-0.00047932	0.02765206	331	-0.02	0.9862

etc.

Note that the output immediately above has been generated by the use of the SOLUTION option in the RANDOM statement. The INTERCEPT and TIME terms do not give the intercepts and slopes directly. To achieve this, these terms would need to be added to the relevant fixed effects estimates.

Tests of Fixed Effects

Source	NDF	DDF	Type III F	Pr>F
TREAT	1	112	4.74	0.0316
TIME	1	107	37.95	0.0001
TIME*TREAT	1	107	1.37	0.2450

Polynomial random coefficients model (6.6.2)

Variables

pat   = patient number,
virus   = herpes antibody level,
type   = illness type: A = acute leukaemia; S = solid lump tumour,
month   = time, months since start of treatment,
month2 = $month^2$,
month3 = $month^3$,
month4 = $month^4$.

SAS code for cubic model selected

```
PROC MIXED; CLASS type pat;
MODEL virus=age type month month2 month3/ S DDFM=SATTERTH;
RANDOM INT month month2/ SUB=pat TYPE=UN;
```

```
                    Class Level Information
              Class     Levels  Values
              TYPE          2   AL ST
              PAT          44   3 5 13 61 63 65 67 69 71 73
                               77 79 81 85 87 91 93 95 97
                               101 103 105 107 109 113 115
                               117 119 123 127 129 141 142
                               143 144 145 146 147 148 149
                               150 151 152 155
```

```
                 REML Estimation Iteration History
        Iteration   Evaluations     Objective      Criterion
            0            1       924.62491860
            1            2       463.71096754     0.00904136
            2            1       461.03583088     0.00345899
            3            1       460.06043790     0.00073103
            4            1       459.86896964     0.00004682
            5            1       459.85770126     0.00000027
            6            1       459.85763855     0.00000000
```

```
                     Convergence criteria met.
              Covariance Parameter Estimates (REML)
              Cov Parm    Subject       Estimate
              UN(1,1)     PAT          0.59595502
              UN(2,1)     PAT         -0.04469787
              UN(2,2)     PAT          0.02356320
              UN(3,1)     PAT          0.00170506
              UN(3,2)     PAT         -0.00070932
              UN(3,3)     PAT          0.00002239
              Residual                 0.52577415
```

UN(1,1), UN(2,2) and UN(3,3) are the variance component estimates for the patient, patient·time and patient·time^2 random coefficients. UN(2,1), UN(3,1) and UN(3,2) are the covariances between the random coefficients. Note that the relative sizes of the patient·time and patient·time^2 variance components cannot be compared directly with other variance components because they involve time.

```
               Model Fitting Information for VIRUS
          Description                        Value
          Observations                     625.0000
          Res Log Likelihood               -798.752
          Akaike's Information Criterion    -805.752
          Schwarz's Bayesian Criterion      -821.250
          -2 Res Log Likelihood            1597.504
          Null Model LRT Chi-Square          464.7673
```

```
            Null Model LRT DF                  6.0000
            Null Model LRT P-Value            0.0000

                     Solution for Fixed Effects
     Effect      TYPE     Estimate  Std Error    DF     t Pr>|t|
     INTERCEPT             3.73850744 0.24266009   50 15.41 0.0001
     AGE                  -0.04852342 0.03682608 43.3 -1.32 0.1946
     TYPE        AL       -0.06019123 0.24854561   43 -0.24 0.8098
     TYPE        ST        0.00000000    .    .    .    .
     MONTH                -0.11821206 0.03532625 39.6 -3.35 0.0018
     MONTH2                0.00654249 0.00234497  123  2.79 0.0061
     MONTH3               -0.00010720 0.00005547 79.5 -1.93 0.0569

                     Tests of Fixed Effects
         Source      NDF    DDF   Type III F    Pr>F
         AGE          1    43.3         1.74   0.1946
         TYPE         1     43          0.06   0.8098
         MONTH        1    39.6        11.20   0.0018
         MONTH2       1    123          7.78   0.0061
         MONTH3       1    79.5         3.73   0.0569
```

This table shows that the linear and quadratic time effects are significant, and that the cubic effect is nearly significant.

SAS *Code for other models tested (without output)*

```
PROC MIXED; CLASS type pat;
MODEL virus=age type month/ S DDFM=SATTERTH;
RANDOM int month/ SUB=pat TYPE=UN;

PROC MIXED; CLASS type pat;
MODEL virus=age type month month2/ S DDFM=SATTERTH;
RANDOM int month month2/ SUB=pat TYPE=UN;

PROC MIXED; CLASS type pat;
MODEL virus=age type month month2 month3 type*month type*month2
type*month3/ S DDFM=SATTERTH;
RANDOM int month month2/ SUB=pat TYPE=UN;

PROC MIXED; CLASS type pat;
MODEL virus=age type month month2 month3 age*month age*month2
age*month3/ S;
RANDOM int month month2/ SUB=pat TYPE=UN;
```

Model checking using Model 3

```
PROC MIXED; CLASS type pat;
MODEL virus=age type month month2 month3/ S DDFM=SATTERTH
PREDICTED PM;
RANDOM int month month2/ SUB=pat TYPE=UN SOLUTION;
ID pat month;
MAKE 'PREDICTED' OUT=resid NOPRINT;
MAKE 'PREDMEANS' OUT=predm NOPRINT;
MAKE 'SOLUTIONR' OUT=solut NOPRINT;
```

```
PROC PRINT NOOBS DATA=resid; VAR pat month_resid_ _pred_ virus;
TITLE 'RESIDUALS AND PREDICTED VALUES';
PROC PLOT DATA=resid; PLOT_resid_*_pred_;
TITLE 'RESIDUALS AGAINST THEIR PREDICTED VALUES';
PROC PLOT DATA=resid; PLOT _resid_*month;
TITLE 'RESIDUALS AGAINST TIME (MONTHS)';
PROC RANK DATA=resid OUT=norm NORMAL=TUKEY; VAR _resid_; RANKS
s_resid;
PROC PLOT DATA=norm; PLOT _resid_*s_resid;
TITLE 'RESIDUALS - NORMAL PLOT';

DATA solut; SET solut;
patx=pat*1; * obtain numeric patient variable;
DROP pat;

DATA p_resid(KEEP=pat p_resid) pt_resid(KEEP=pat pt_resid)
  pt2_resi(KEEP=pat pt2_resi); SET solut;
pat=patx;
IF _effect_='INTERCEPT' THEN DO;
  p_resid=_est_;
  OUTPUT p_resid;
END;
ELSE IF _effect_='MONTH2' THEN DO;
  pt2_resi=_est_;
  OUTPUT pt2_resi;
END;
ELSE DO;
  pt_resid=_est_;
  OUTPUT pt_resid;
END;

PROC SORT DATA=predm; BY pat;
PROC MEANS NOPRINT DATA=predm; BY pat;
VAR _pred_; OUTPUT OUT=predm MEAN=p_pred N=freq;

DATA a; MERGE p_resid pt_resid pt2_resi predm; BY pat;
PROC PRINT NOOBS; VAR pat p_resid pt_resid pt2_resi p_pred freq;
TITLE 'RANDOM COEFFICIENTS AND PREDICTED VALUES FOR EACH
PATIENT';
PROC PLOT; PLOT p_resid*pt_resid;
TITLE 'PATIENT COEFFICIENTS VS PATIENT.TIME COEFFICIENTS';
PROC PLOT; PLOT p_resid*pt2_resi;
TITLE 'PATIENT COEFFICIENTS VS PATIENT.TIME2 COEFFICIENTS';
PROC PLOT; PLOT pT_resid*pt2_resi;
TITLE 'PATIENT.TIME COEFFICIENTS VS PATIENT.TIME2 COEFFICIENTS';
```

This code may not at first sight be straightforward to understand. The steps used
are summarised below.

- Fit Model 3.
- Use MAKE statements to: output the residuals (and predicted values given by
  $\mathbf{X}\hat{\boldsymbol{\alpha}} + \mathbf{Z}\hat{\boldsymbol{\beta}}$ which are not required here) to dataset resid; output the predicted
  values given by $\mathbf{X}\hat{\boldsymbol{\alpha}}$ to dataset predm; output the random effects estimates to

dataset `solut`. The ID statement causes the patient, centre and treat variables to be included in the datasets `resid` and `predm`.

- Produce a print and plots of residuals.
- Create datasets `p_resid`, `pt_resid` and `pt2_resid` containing random coefficient estimates for the patient and patient·time and patient·time^2 effects, respectively.
- Obtain predicted means for each patient (based on the predicted values $\mathbf{X}\hat{\alpha}$) and merge these with the datasets of random coefficients.
- Produce print and plots of the random coefficients.

```
                     RESIDUALS AND PREDICTED VALUES

      PAT    MONTH      _RESID_        _PRED_       VIRUS
        3     0.00       1.5577        3.9323       5.4900
        3     0.75      -0.3390        3.8041       3.4651
        3     2.00      -0.2553        3.6118       3.3566
        3     4.00      -0.4266        3.3569       2.9302
        3     7.00      -0.4503        3.0860       2.6357
        3     9.00      -0.1350        2.9722       2.8372
        3    10.00      -0.4761        2.9335       2.4574
        3    11.00      -0.0145        2.9060       2.8915
        3    12.00      -0.0441        2.8890       2.8450
        3    13.00      -0.8045        2.8821       2.0775
        3    16.00      -0.6973        2.9144       2.2171
        3    21.00       0.6139        3.1070       3.7209
        3    22.00       1.3394        3.1606       4.5000
        3    23.00       0.9924        3.2176       4.2100
        3    23.50       0.8027        3.2473       4.0500
        3    24.00       0.5224        3.2776       3.8000
        3    24.25      -0.0929        3.2929       3.2000
        3    26.00      -0.3435        3.4035       3.0600
        3    26.75      -0.5320        3.4520       2.9200
        3    27.00      -0.0782        3.4682       3.3900
        3    28.00      -1.0532        3.5332       2.4800
        5     0.00       0.3092        2.8225       3.1318
        5     0.50      -0.2923        2.7807       2.4884
      ETC

  RANDOM COEFFICIENTS AND PREDICTED VALUES FOR EACH PATIENT

      PAT    P_RESID      PT_RESID    PT2_RESI    P_PRED  FREQ
        3    0.35105      -0.05926    0.0022881   3.07590   21
        5   -0.71021       0.03226   -.0018661    3.05075   22
       13    0.87930       0.04089    0.0002709   3.06776   32
       61   -0.21677       0.11841   -.0043331    3.07713   22
       63    0.72495       0.03671   -.0007843    2.82919   20
       65    0.70395      -0.09316    0.0032769   2.82021   10
       67    0.38618      -0.30506    0.0090391   2.93704    9
       69   -0.27759      -0.00126    0.0006366   2.72964   26
       71    0.64374       0.16498   -.0040015    2.90710   25
       73   -0.13393      -0.05631    0.0015240   3.39389    7
```

77	1.17545	-0.27557	0.0087241	3.46299	2
79	0.49003	0.04483	-.0004417	3.04049	24
ETC					

All the residual plots appear in the main text.

6.7   SAMPLE SIZE ESTIMATION

Sample sizes for repeated measures studies are often calculated as if a simple between-patient trial with no repeated measurements was planned. However, it is possible to take into account the correlation that occurs between the repeated observations in the sample size estimate. This will lead to a smaller sample size than that calculated for a simple between-patient study. It therefore seems desirable ethically and on the grounds of cost that correlations within patients are taken into account. Obviously, the covariance pattern of the data will not be known in advance, but the assumption of a constant correlation between patients (compound symmetry pattern) is likely to be adequate. When no previous estimate of the within-patient correlation is available, a conservative prediction of the correlation could be used (i.e. a higher correlation than anticipated).

6.7.1   Normal data

To obtain a formula for sample size estimation we require the variance of the mean of measurements on individual patients. The variance of the sum of the observations on any single patient, i, is

$$\text{var}\left(\sum_j y_{ij}\right) = m\,\text{var}(y_{ij}) + m(m-1)\,\text{cov}(y_{ij}, y_{ik})$$

$$= m\sigma^2 + m(m-1)\rho\sigma^2$$

$$= m\sigma^2\{1 + (m-1)\rho\}.$$

This gives the variance of each patient mean as

$$\text{var}(\bar{y}_i) = \sigma^2\{1 + (m-1)\rho\}/m,$$

where

$m = $ number of repeated measures,

$\sigma^2 = $ between-patient variation (sum of variance parameters when compound symmetry pattern fitted using PROC MIXED),

$\rho = $ correlation between observations on same patient (compound symmetry variance parameter divided by sum of variance parameters in PROC MIXED).

Using the usual sample size estimation equation,

$$\Delta = (z_{1-\alpha/2} + z_\beta) \times SE\,(t_i - t_j),$$

we obtain the number of patients required per group as

$$n = 2(z_{1-\alpha/2} + z_\beta)^2 \sigma^2 \{1 + (m - 1)\rho\}/m\Delta^2,$$

where

    α = significance level,
    β = power,
    Δ = difference to be detected,
    t_i = ith treatment effect.

If a small trial is planned, for example with less than about 10 patients per group, a more accurate sample size could be obtained by substituting t statistics for the z statistics above (with DF equal to the number of patients minus the number of treatment groups).

Example

The analysis of the repeated DBP measurements in the hypertension trial using a compound symmetry covariance pattern model gave a residual variance of $\sigma^2 = 76$ and the repeated measures had correlation $\rho = 0.53$ (see Section 6.3). We calculate the sample size required for a future study involving four post-treatment visits required to detect a difference in DBP of 5 mmHg at the 5% significance level with 80% power. The number of patients required per treatment group is

$$n = 2(1.96 + 0.84)^2 \times 76 \times (1 + 3 \times 0.53)/(4 \times 25)$$
$$= 31.$$

Had no account been taken of the repeated measurements, then n would have been

$$n = 2(z_{1-\alpha/2} + z_\beta)^2 \sigma^2 / \Delta^2$$
$$= 2(1.96 + 0.84)^2 \times 76/25$$
$$= 48.$$

If there is flexibility in the number of repeated measurements, then it might be worth calculating sample sizes for varying numbers of repeated measurements. For example, if 10 repeated measures were used, then

$$n = 2(1.96 + 0.84)^2 \times 76 \times \{1 + 9 \times 0.53\}/(10 \times 25)$$
$$= 28.$$

However, this reduction in the number of patients required would be unlikely to justify the use of six additional repeated measures.

6.7.2   Non-normal data

When the variable of primary interest is non-normal the sample size formula given for normal data can be used with $\phi a/b$ substituted for σ^2, and Δ taken as the required treatment difference defined on the 'linked' scale (e.g. using logits for binary data or logs for count data). The number of patients per treatment group is

$$n = 2(z_{1-\alpha/2} + z_\beta)^2 v\{1 + (m - 1)\rho\}/m\Delta^2,$$

where

$v = \phi a/b,$

$a =$ denominator term for binomial data or offset term for count data
   (see Section 3.1),

$b =$ expected variance, e.g. $\mu(1 - \mu)$ for binary data or μ for count data
   (see Section 3.1),

$\phi =$ dispersion parameter.

a will be one in most situations and an average value can be used for μ when setting b (for example, if the expected proportions for two treatments were 0.4 and 0.2, μ could be taken as 0.3 and b as $0.3 \times 0.7 = 0.21$). If a dispersion parameter is not available, then ϕ can be taken as one.

Example: binary data

We calculate the sample size required for a future study involving four post-treatment visits required to detect a doubling in the odds ratio at the 5% significance level with 80% power. The difference in logit (log(odds ratio)) required is then $\log(2) - \log(1) = 0.693$. A correlation of $\rho = 0.50$ between the repeated measurements and a dispersion parameter of one will be assumed. We will assume that the expected rate of 'positives' is 0.4 and hence $b = 0.4 \times 0.6 = 0.24$. The number of patients required per treatment group is

$$n = 2(1.96 + 0.84)^2 \times (1.00/0.24) \times (1 + 3 \times 0.50)/(4 \times 0.693^2)$$

$$= 85.$$

Example: count data

In the epilepsy example (Section 6.4) the compound symmetry covariance pattern model gave $\phi = 2.17$ and $\rho = 0.42$. We calculate the sample size required for a future study involving four post-treatment visits required to detect a reduction of 50% in the epilepsy attack rate at the 5% significance level with

80% power. To obtain a 50% reduction we need log(epilepsy rate) to change by $\log(0.5) = -0.693$. We will assume the average number of epilepsy attacks per period is four and therefore b is taken as four. The number of patients required per treatment group is

$$n = 2(1.96 + 0.84)^2 \times (2.17/4) \times \{1 + 3 \times 0.42\}/(4 \times 0.693^2)$$
$$= 10.0$$

6.7.3 Categorical data

Sample size estimation is always difficult when the variable of interest is categorical. If there are greater than about five categories, then the formula for continuous data is likely to provide a reasonable approximation. In other situations the best approach might be to partition the categories and use the formula for binary data.

7

Cross-Over Trials

7.1 INTRODUCTION

In earlier chapters we have considered parallel group designs, where each subject is randomised to receive one of a number of alternative treatments. By contrast, in cross-over trials, subjects are randomised to receive different sequences of treatments, with the outcome being assessed for each treatment period. As before, we have a choice in analysis between fixed effects models and random effects models. In this context, we describe the treatment effect as being *crossed* with a random effect (subjects).

The vast majority of cross-over trials which are carried out in practice have the same basic design. Every subject receives each of the treatments being evaluated, for a standard period of time, with the outcome variables being assessed in the same way in each period of treatment. The simplest and most commonly encountered such design employs just two treatments, and is often referred to as a 2×2 cross-over trial, or as an AB/BA design. The use of this design with normally distributed data will be covered in some depth in Section 7.3. The use of more than two treatments with patients receiving every treatment is known as a higher order complete block design and is covered in Section 7.4. More complicated designs are considered in Sections 7.5 and 7.6. In Section 7.7 we will show how covariance pattern models can be employed in the analysis of cross-over trials. The following two sections (7.8 and 7.9) will give examples of the analysis of binary data and categorical data in the setting of cross-over trials. Data following Poisson distributions are not directly covered, but follow the same generalised linear mixed model approach used for binary data. Section 7.10 will consider the use of information from random effects models in the planning of future studies. The chapter finishes with a discussion of some general points in relation to the analysis of cross-over trials (Section 7.11).

7.2 ADVANTAGES OF MIXED MODELS IN CROSS-OVER TRIALS

Random effects models can be expected to give more precise estimates of treatment effects in situations where it is possible to recover extra information on treatments from the between-patients error stratum. The most common situation in which this occurs is where there are missing data, irrespective of the particular cross-over design used. It also occurs for the unbalanced designs considered in Sections 7.5 and 7.6. In the balanced situations which we meet with complete block designs, the results of a fixed effects analysis and a random effects analysis will generally be identical in the absence of missing data.

A very different application of mixed models to cross-over trials arises from the covariance pattern approach. By regarding the results in successive treatment periods as a form of repeated measures data we can examine various ways to model the covariance between repeated observations on the same patients. This can lead to greater flexibility in the interpretation of the data than with conventional analyses, and we examine examples using this approach in Section 7.7.

7.3   THE AB/BA CROSS-OVER TRIAL

This design employs two treatments (A and B) and two treatment periods. Patients are randomised to receive either the AB sequence of treatments or the BA sequence. We met a simplified hypothetical example of such a trial in Section 1.2. At that time, for simplicity of presentation, we assumed that there was no effect of the period in which treatments were received. However, such an effect is always possible and we recommend that such an 'order' effect should be included in analysis. Our initial example was also restricted to single observations in each treatment period. In practice, the randomisation to the AB or BA sequence is often preceded by a run-in period. This approach has the advantage that patients showing poor compliance can be removed prior to randomisation, and the stability of the patient's condition can be assessed. With or without this run-in period, baseline levels for the outcome variables are usually recorded prior to randomisation. Following the first treatment period there is often a 'washout' period prior to the commencement of the second treatment, and a second 'baseline' observation may be made. Details of design considerations, and analytical methods for a fixed effects analysis are given in Senn (1993).

If all patients complete the trial without any missing values being generated for the outcome variables, the results of the fixed effects analysis and an analysis in which patient effects are regarded as random will usually be identical. (An exception occurs when the estimate of the patient variance component is negative, and is set to zero. The standard errors of the treatment differences will then be lower with the random effects model.) This arises because of balance over random effects in the design, as discussed in Section 1.6. It is common, however, for missing values to occur, usually because of premature patient withdrawal from the trial. In the fixed effects analysis of such a trial, observations from subjects with a missing value are not used, because all of the information in the single remaining observation would be needed to estimate the patient effect. When missing values do occur, and a random effects analysis is performed, the data from subjects with a single period of observation are utilised in the analysis in conjunction with the complete observations to improve the efficiency of treatment comparisons relative to the fixed effects analysis. This benefit was illustrated in Section 1.2 using the earlier example but with two observations deleted.

More generally, we will now consider a cross-over trial to compare treatments A and B, with N patients divided equally between the AB and the BA sequence, following Brown and Kempton (1994). We will also assume that a proportion, p, of patients only provide data for the first treatment period. On the assumption that these dropouts are also equally divided between the two treatment sequences, we will investigate the effect of p and the variance components on the relative efficiency of the random effects and fixed effects models in estimating treatment differences. To do this, we will look first at the variance of the estimate of treatment differences, $\text{var}_W (A - B)$, obtained from within-patient comparisons. This will give the variance appropriate to the fixed effects analysis. We will then look at the corresponding term from the between-patient comparison, using those patients who only have an observation in the first treatment period. We will then pool these two estimates and obtain the variance of the pooled estimate. In this situation, this will correspond to the results of fitting a random effects model, and we will compare the variances of the fixed effect and random effects model treatment estimates.

Within-patient comparisons   From our definitions there will be $N(1-p)$ patients with complete data. If the residual variance is σ_r^2, then

$$\text{var}_W (A - B) = \frac{2\sigma_r^2}{N(1 - p)}.$$

Between-patient comparisons   Each treatment sequence will yield $Np/2$ patients with observations in the first period only. The variance of individual observations will be the sum of the residual variance σ_r^2 and the between-patient component σ_p^2. Hence

$$\text{var}_B (A - B) = \frac{2(\sigma_r^2 + \sigma_p^2)}{Np/2} ,$$
$$= 4\sigma_r^2(1 + \gamma)/Np$$

where

$$\gamma = \sigma_p^2/\sigma_r^2.$$

Pooled comparisons   If we obtain a weighted average of the treatment effect from the within-patient and between-patient estimates, using weights inversely proportional to the variances, then we have the standard result that

$$\text{var}_p (A - B) = 1/(1/\text{var}_W + 1/\text{var}_B).$$

Thus,

$$1/\text{var}_p\,(A - B) = \frac{N(1 - p)}{2\sigma_r^2} + \frac{Np}{4\sigma_r^2(1 + \gamma)}$$

and

$$\text{var}_p\,(A - B) = \frac{4\sigma_r^2(1 + \gamma)}{Np + 2N(1 - p)(1 + \gamma)}.$$

Relative efficiency   The ratio of the variance of the treatment estimate using a fixed effects (within-patient) approach, to that using a random effects model (pooled), is plotted against p, the proportion of missing observations in the second period, for a range of values of γ, in Figure 7.1. From this figure we can see that the recovery of between-patient information is most beneficial when γ is small; that is, when the between-patient variance component is small. If the proportion of missing values is small, the benefit from analysis with a mixed model will be correspondingly small, although we can expect some reduction in the variance of the treatment estimate.

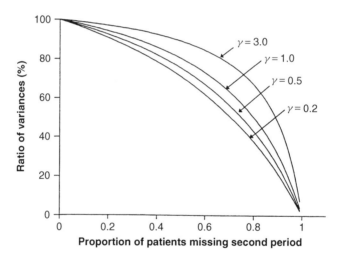

Figure 7.1   Ratio of variances of treatment differences, with and without recovery of between-patient information.

7.3.1 Example: AB/BA cross-over design

We illustrate the AB/BA cross-over design with results from an unpublished study comparing two diuretics in the treatment of mild to moderate heart failure. After initial screening for suitability, there was a period of not less than one day, and not more than seven days, where diuretic treatment was withheld. Immediately prior to randomisation to either the AB sequence of treatment, or the BA sequence, baseline observations were taken. Each treatment period lasted for five days, with an immediate transfer to the second treatment after the first treatment period was completed. As a washout was not employed between treatments, observations made in the first two days of each treatment period were not utilised in the analysis of the trial. The primary outcome measures were the frequency of micturition and the subjective assessment of urgency. As neither of these are suitable for illustrating the analysis of normally distributed data, we will use instead a secondary effectiveness variable; namely, oedema status, together with DBP. Oedema status is formed by the sum of the left and right ankle diameters. The DBP was calculated from the mean of three readings. Both of these variables are measured prior to randomisation, and at the end of each treatment period.

In total, 101 patients were recruited for the study, but seven withdrew prior to randomisation. Of the remaining 94 patients, only two failed to complete both treatment periods. Therefore, in order to illustrate the alternative methods of analysis, we have systematically removed approximately one in five of the observations from the second period. The structure of the data as analysed is shown in Table 7.1.

For each of our outcome variables, four analyses have been performed. In all of them, a treatment effect and a period effect were included as fixed. In two of the models the baseline level was also included in the model as a covariate. Whether or not the baseline is included in the model, separate models are considered with the patient effect being fitted either as random or as fixed. The results of the models are summarised in Table 7.2.

Examination of the variance component terms shows that for all models the patient term is larger than the residual term. This indicates that there may have been substantial benefits from employing a cross-over design rather than a parallel group design. We note that this is particularly striking for oedema status. Note also the effect of including the baseline as a covariate in the analysis. This has the

Table 7.1   Data structure for a cross-over trial comparing two diuretics in patients with heart failure.

			Baseline values		Post-treatment values	
Patient	Treatment	Period	Oedema	DBP	Oedema	DBP
1	B	1	45	60	45	55
1	A	2	45	60	45	60
2	A	1	51	50	48	60
2	B	2	51	50	48	65
3	A	1	53	70	50	70
3	B	2	53	70	52	80
4	B	1	49	68	47	60
4	A	2	49	68	47	60
5	A	1	46	65	45	60
6	A	1	61	95	60	95
6	B	2	61	95	59	97

Table 7.2   Analysis of cross-over trial of diuretics in heart failure.

Model	Fixed effects	Random effects
1	Treatment, period, patient	—
2	Treatment, period	Patient
3	Treatment, period, patient, baseline	—
4	Treatment, period, baseline	Patient

	Treatment effect: A − B (SE)	
Model	Oedema	DBP
1	0.304 (0.120)	0.812 (0.775)
2	0.301 (0.120)	0.926 (0.764)
3	0.304 (0.120)	0.812 (0.775)
4	0.309 (0.118)	1.013 (0.747)

	Variance components (SE)			
Model	Patient	Residual	Patient	Residual
1	—	0.528	—	22.19
2	66.825	0.530	76.77	22.25
3	—	0.528	—	22.19
4	3.763	0.526	25.60	21.91

effect of reducing the size of the patient variance component term in Model 4. The implications of this are that the benefits of the cross-over are somewhat reduced when a (highly correlated) baseline covariate is available and, conversely, that the use of a mixed model is likely to be most helpful in these circumstances if there are missing values.

We see this in the estimates of the treatment standard errors. Comparison of Models 1 and 2 for the oedema status show that the standard errors are identical (to the number of digits reported), indicating that the between-subject variation is so large that recovery of between-subject information is ineffective. With inclusion of the baseline level as a covariate we see that Model 3 gives the same result as Model 1. This result is well known, showing that a single baseline has no effect on a fixed effects analysis. It does, however, produce a small reduction in the treatment standard error when a mixed model is fitted, showing that some between-subject information has been utilised.

The results for DBP show the recovery of between-subject information more clearly because of the relatively smaller between-subject variation. Here, we see a detectable reduction in the treatment standard error, even when baselines are not used, and with the inclusion of baselines, a reduction of around 4% in the standard error is seen with the mixed model approach. This gain is modest but worthwhile.

The greatest advantage of the mixed model approach will unfortunately be gained in situations where a cross-over trial shows little benefit over a parallel-group study, i.e. where the between-subject variance component is small relative to the residual variance component.

Such a situation occurs in a trial reported by Jones and Kenward (1989). In this two-period, cross-over trial an oral mouthwash was compared with a placebo mouthwash. There were two six-week treatment periods with a three-week washout period separating them. The outcome variable reported was the average plaque score per tooth, with each tooth being assessed on an integer scale from zero to three. Results were presented for the 34 patients with data from both treatment periods. Interestingly, these data arose from a trial in which 41 patients were randomised, and 38 completed the trial. For the purposes of this illustration we have deleted the second observation from five randomly selected patients from the 34 with complete data.

Two models were fitted to the data using PROC MIXED. In both, a treatment effect and a period effect were included as fixed. In one, the patient effect was fitted as random and in the other it was fitted as fixed. The results are shown in Table 7.3.

Examination of the variance component terms shows that the patient term is appreciably smaller than the residual term. This indicates that the benefit of employing a cross-over design rather than a parallel-group design may be small. The estimate of the period effect (not shown) is small, but in accord with our recommendation in the previous section we retain it in the model. The main interest, of course, lies in the estimates of the treatment difference and the associated standard errors. Both analyses demonstrate a clear advantage to using the active mouthwash. For our purposes in comparing the results of the two analytical strategies it is the standard errors which interest us, because it is purely a matter of chance which method gives the larger point estimate of the treatment effect. We see that the use of the random effects model has reduced the standard error of the estimate of the treatment difference by around 6%.

Table 7.3 Analysis of oral mouthwash trial.

	Fixed patients	**Random patients**
Variance components		
Patients	—	0.029 (0.018)
Residual	0.069	0.066 (0.017)
Treatment difference (SE)	0.25 (0.069)	0.24 (0.065)

SAS code and output

The SAS code to generate analyses for oedema status (Table 7.2) is shown below for Models 3 and 4. Models 1 and 2 differ only in the exclusion of the baseline value from the model.

```
PROC MIXED NOCLPRINT; CLASS treat period patient;
TITLE 'FIXED EFFECT ANALYSIS WITH BASELINE';
MODEL oed=treat period patient oedbase;
LSMEANS treat /DIFF PDIFF;

PROC MIXED NOCLPRINT; CLASS treat period patient;
TITLE 'RANDOM EFFECTS ANALYSIS WITH BASELINE';
MODEL oed=treat period oedbase / DDFM=SATTERTH;
RANDOM patient;
LSMEANS treat /DIFF PDIFF;
```

The output reproduced below is from Model 4:

```
              RANDOM EFFECTS ANALYSIS WITH BASELINE
                 REML Estimation Iteration History

        Iteration  Evaluations     Objective     Criterion

            0            1      430.89194120
            1            3      319.06475822    0.00104614
            2            2      318.91514140    0.00001998
            3            1      318.91191725    0.00000001
                   Convergence criteria met.

              Covariance Parameter Estimates (REML)

             Cov Parm       Estimate
             PATIENT       3.76315687
             Residual      0.52603404

                Model Fitting Information for OED

          Description                           Value

          Observations                        168.0000
          Res Log Likelihood                  -310.162
          Akaike's Information Criterion      -312.162
          Schwarz's Bayesian Criterion        -315.262
          -2 Res Log Likelihood                620.3238

                     Tests of Fixed Effects
             Source      NDF    DDF   Type III F    Pr>F

             TREAT        1    75.3         6.80   0.0110
             PERIOD       1    74.9         4.02   0.0485
             OEDBASE      1    94.1      1433.73   0.0001

                     Least Squares Means
      Effect  TREAT    LSMEAN    Std Error   DF      t      Pr>|t|

      TREAT    A    55.36213650  0.21700531  108  255.12    0.0001
      TREAT    B    55.05360676  0.21681209  107  253.92    0.0001

                 Differences of Least Squares Means
     Effect TREAT _TREAT Difference    Std Error     DF    t   Pr>|t|
     TREAT   A      B    0.30852974   0.11831019   75.3  2.61  0.0110
```

7.4   HIGHER ORDER COMPLETE BLOCK DESIGNS

In these designs there are as many treatment periods as there are treatments to be compared, and each patient receives every treatment. If there are no missing data, then a conventional least squares analysis fitting treatment, period and patient effects is fully efficient. Whenever there are missing data, some of the within-patient treatment comparisons are unavailable for every patient. Therefore, additional between-patient information can be utilised.

7.4.1   Inclusion of carry-over effects

In any cross-over trial there is the possibility of carry-over effects. That is, the results in second or subsequent treatment periods may be influenced by treatment administered in earlier periods. In the simple two-period, cross-over trial considered previously there is no possibility of estimating carry-over. In all of the remaining designs considered, carry-over effects can be estimated, and in our examples we will consider results from models which include carry-over. However, we do this for completeness rather than in the belief that this is good practice and we return to this point in Section 7.11.

7.4.2   Example: Four-period, four-treatment cross-over trial

We consider the four-period, four-treatment, cross-over trial described by Jones and Kenward (1989). Three drugs, A, C and D, and a placebo B were compared to assess their effect on cardiac output, measured by the left ventricular ejection time. Each treatment was given for one week, with a one-week washout period between treatments. Observations were made at the end of each treatment period. Fourteen patients were used in the trial, yielding 56 observations. To demonstrate the use of the mixed model approach, we have arbitrarily set 13 of the 56 observations to be missing. The results of four analyses are presented in Table 7.4, from the combinations of inclusion or exclusion of carry-over effects, and handling the patient effects as fixed or random.

All pairwise comparisons of the carry-over effects were non-significant and will not be considered further, or the details presented. We see that between-patient variation is moderate, being around 50% higher than the residual variance component. The random effects analysis without carry-over effects produces an average 2% reduction in the standard errors of the paired treatment comparisons compared with the fixed effects model, a modest but worthwhile gain. Comparing these estimates with the analyses in which carry-over was also fitted, two points are clear. Firstly, the standard errors of the mean treatment differences are larger

Table 7.4 Estimates of variance components and treatment effects. Standard errors of estimates appear in brackets.

	Fixed patients	**Random patients**
Ignoring carry-over		
Variance components		
Patients	—	2721
Residual	1667	1657
Treatment differences		
A − B	77.4 (20.1)	72.5 (19.7)
A − C	36.8 (17.3)	32.8 (17.0)
A − D	77.3 (19.5)	74.6 (19.3)
B − C	−40.6 (19.9)	−39.7 (19.4)
B − D	−0.1 (21.2)	2.1 (20.9)
C − D	40.4 (18.7)	41.8 (18.4)
Including carry-over		
Variance components		
Patients	—	2750 (1394)
Residual	1840	1831 (577)
Treatment differences		
A − B	76.0 (22.6)	69.1 (21.9)
A − C	40.5 (20.4)	32.1 (19.8)
A − D	84.9 (22.5)	79.0 (22.1)
B − C	−35.4 (23.0)	−37.0 (22.2)
B − D	8.9 (25.5)	9.9 (24.7)
C − D	44.3 (23.0)	46.9 (22.2)

when carry-over terms are present, irrespective of whether a fixed effects model is fitted, or whether the patient term is regarded as random. Secondly, the reduction in the standard error of the mean treatment difference by fitting patient effects as random is larger when carry-over terms are also fitted. This latter result is general, and we will see more dramatic differences in later examples.

SAS code and output

```
/*create dummy variables for carryover effects*/
DATA new; carry=treat; SET mydata; RUN;

DATA new;SET new;
IF period eq 1 THEN carry=4;
* setting a carryover value for the first period.
  it is arbitrary which treatment is selected.
  the choice will only influence the absolute values
  of the fixed effect estimates and not the difference
  between them;

PROC MIXED; CLASS treat period patient;
TITLE 'Fixed Effect Analysis Without Carryover';
```

```
MODEL lvet=treat patient period;
LSMEANS treat/DIFF PDIFF;

PROC MIXED; CLASS treat period patient;
TITLE 'Random Effects Analysis Without Carryover';
MODEL lvet=treat period/DDFM=SATTERTH;
RANDOM patient;
LSMEANS treat/DIFF PDIFF;

PROC MIXED; CLASS treat period patient carry;
TITLE 'Fixed Effect Analysis Including Carryover';
MODEL lvet=treat patient period carry;
LSMEANS treat carry/DIFF PDIFF;

PROC MIXED; CLASS treat period patient carry;
TITLE 'Random Effects Analysis Including Carryover';
MODEL lvet=treat period carry/DDFM=SATTERTH;
RANDOM patient;
LSMEANS treat carry/DIFF PDIFF;
```

The following output is that generated by the last PROC MIXED procedure.

```
              Random Effects Analysis Including Carryover
                        The MIXED Procedure
                     Class Level Information
          Class      Levels  Values
          TREAT        4     1 2 3 4
          PERIOD       4     1 2 3 4
          PATIENT     14     1 2 3 4 5 6 7 8 9 10 11 12 13
                            14
          CARRY        4     1 2 3 4

                  REML Estimation Iteration History
       Iteration  Evaluations     Objective      Criterion
           0           1       330.25723527
           1           2       320.68175864     0.00102427
           2           1       320.49256587     0.00008011
           3           1       320.47902500     0.00000062
           4           1       320.47892487     0.00000000
                    Convergence criteria met.

              Covariance Parameter Estimates (REML)
              Cov Parm        Estimate
              PATIENT       2749.7104264
              Residual      1831.4737963

               Model Fitting Information for LVET
              Description                      Value
              Observations                    43.0000
              Res Log Likelihood             -190.564
              Akaike's Information Criterion -192.564
```

```
          Schwarz's Bayesian Criterion    -194.061
          -2 Res Log Likelihood            381.1289

                 Tests of Fixed Effects
          Source      NDF   DDF  Type III F   Pr>F
          REAT          3  22.3       5.70  0.0047
          PERIOD        3  21.8       3.99  0.0209
          CARRY         3  23.4       0.15  0.9292

          Random Effects Analysis Including Carryover
                    Least Squares Means
   Effect TREAT CARRY      LSMEAN    Std Error    DF     t  Pr>|t|

   TREAT    1          400.10420160 20.69102445 27.5 19.34  0.0001
   TREAT    2          331.01658291 22.81038053 30.4 14.51  0.0001
   TREAT    3          367.98542485 20.38822065 26.8 18.05  0.0001
   TREAT    4          321.12492718 21.27016357   29 15.10  0.0001
   CARRY          1    356.93813378 23.03087652 31.3 15.50  0.0001
   CARRY          2    364.73982746 24.02121398 31.6 15.18  0.0001
   CARRY          3    350.06932839 27.14112516   33 12.90  0.0001
   CARRY          4    348.48384693 21.23493940 27.5 16.41  0.0001

             Differences of Least Squares Means
 Effect TREAT CARRY _TREAT _CARRY   Difference   Std Error   DF    t  Pr>|t|
  TREAT   1            2        69.08761869 21.91860479 22.5  3.15 0.0045
  TREAT   1            3        32.11877675 19.76295910 22.3  1.63 0.1182
  TREAT   1            4        78.97927442 22.09840830 21.5  3.57 0.0017
  TREAT   2            3       -36.96884194 22.16674858 22.6 -1.67 0.1091
  TREAT   2            4         9.89165573 24.73986644 22.3  0.40 0.6931
  TREAT   3            4        46.86049767 22.16775810 22.8  2.11 0.0457
  CARRY         1         2    -7.80169368 25.32217794 23.6 -0.31 0.7607
  CARRY         1         3     6.86880539 29.79953801 24.4  0.23 0.8196
  CARRY         1         4     8.45428685 25.56929050 21.7  0.33 0.7441
  CARRY         2         3    14.67049907 28.46490027 23.2  0.52 0.6112
  CARRY         2         4    16.25598053 28.57668563 23.8  0.57 0.5748
  CARRY         3         4     1.58548146 32.52954441 25.4  0.05 0.9615
```

7.5   INCOMPLETE BLOCK DESIGNS

The previous example demonstrated a situation where we have a design which is intended to be balanced, but becomes unbalanced due to missing observations. In contrast, we now look at incomplete block designs, where the design itself is unbalanced. They are used in situations where, for practical reasons, the maximum possible number of treatment periods in a cross-over trial is less than the number of treatments to be evaluated, so complete balance is impossible. The principal reason for this constraint on the number of treatment periods will usually be the length of time for which any patient is in the trial. Some treatments require to be assessed over a period of several weeks, in order for there to be sufficient time for a 'steady-state' response to be reached, so the length of individual treatment periods in the trial can be considerable. In these circumstances it is not feasible to have multiple treatment periods because:

- The chance that a patient will withdraw before completing the trial protocol increases with the required time in the trial.
- In the programme of testing of a new treatment, excessively long studies will delay drug registration.
- Ethical considerations require that trial participation should not place an excessive burden on the patient.

It is readily seen that fitting a model with fixed patient effects to such a design will be inefficient, because it does not allow us to use between-patient information in our treatment comparisons. To demonstrate this, consider a two-period, cross-over trial with three treatments, A, B and C. Direct information on the comparison of treatments A and B is given from the within-patient differences in patients receiving both of these treatments. However, taking the random effects approach to modelling patient effects, we can see additionally that the distribution of the sum of the responses in patients receiving treatments A and C, and in those receiving B and C, also yields comparative information about treatments A and B. Thus, the random effects approach allows recovery of between-subject information. Although these designs cannot be completely balanced, they can be partially balanced by ensuring that all possible treatment sequences are used with equal frequency. As long ago as 1940 Yates described a method for recovering between-block information (the patient is the block in cross-over designs) for balanced incomplete block designs, but the benefits of applying this in cross-over studies has not been widely recognised until recently.

7.5.1   The three-treatment, two-period design (Koch's design)

Koch's design is the 'obvious' design to use with three treatments and two treatment periods. There are six possible treatment sequences — AB, AC, BA, BC, CA and CB — and in this design all six possibilities are used with equal frequency. The variances of treatment differences from within-patient comparisons (fixed effects model), between-patient comparisons and a combination of these (random effects model) have been investigated by Brown and Kempton (1994). They are summarised in Table 7.5 for models both with and without carry-over effects.

In the absence of carry-over, the within-patient (fixed effects) analysis is reasonably efficient in comparison with the fully efficient combined (random effects) analysis. The variance of the treatment estimates is never more than $4/3$ of the variance for the combined analysis. This upper limit occurs when γ, the ratio of the patient and residual variance components, is zero. For a moderate value of γ, such as 2, the ratio of the variances is 1.07:1, and for a high value such as 10, the ratio is 1.02:1. Of course, even such small gains in efficiency should be accepted, because the analysis is no longer difficult or time-consuming to perform. The

Table 7.5 Variances of estimates of differences of treatment and carry-over effects for Koch's design in multiples of σ^2/r. r is the number of replicates of the six treatment sequences (AB, BA, AC, CA, BC, CB). γ is the ratio of patient and residual variance components.

	Within-patient analysis (fixed effects)	Between-patient analysis	Combined analysis (random effects)
Treatment effects			
Omitting carry-over	$\dfrac{1}{3}$	$1 + 2\gamma$	$\dfrac{1 + 2\gamma}{2(2 + 3\gamma)}$
Including carry-over	$\dfrac{4}{3}$	$\dfrac{4}{3}(1 + 2\gamma)$	$\dfrac{2(1 + 2\gamma)(1 + \gamma)}{7 + 14\gamma + 3\gamma^2}$
Carry-over effects	4	$\dfrac{4}{3}(1 + 2\gamma)$	$\dfrac{2(1 + 2\gamma)(2 + 3\gamma)}{7 + 14\gamma + 3\gamma^2}$

major benefit from recovery of between-patient information, however, occurs if carry-over terms are fitted. This leads to a fourfold increase in treatment variance from the within-patient analysis, compared with the no carry-over situation, whereas the corresponding increase from the combined analysis is by a factor of only $(8 + 4\gamma)/7$ for small γ. The estimates of the carry-over effects themselves are even more dramatically affected by recovery of between-patient information. The ratio of the variance estimates from the within-patient analysis to the combined analysis has a maximum value of 7:1 when $\gamma = 0$, a value of 2.35:1 when $\gamma = 2$, and even when $\gamma = 10$ the ratio is 1.33:1.

7.5.2 Example: Two-period cross-over trial

Mead (1988) gives results of a two-period cross-over trial to compare three analgesic drugs labelled A, B, C. The trial involved 43 patients in total and the numbers receiving each treatment combination were as follows: AB 7; BA 5; AC 7; CA 8; BC 8; CB 8. The effectiveness of each treatment was assessed by the numbers of hours of pain relief provided. The design did not include a washout period between treatments, so there was a strong possibility of carry-over effects. The model fitted was

response = overall mean + patient effect + period effect + treatment effect

+ carry-over effect + random error.

Period, treatment and carry-over effects were taken as fixed. Analyses were carried out with patient effects specified first as fixed in a conventional least squares analysis, then as random in a mixed model analysis.

The estimated variance components and treatments effects are shown in Table 7.6, first omitting then including carry-over effects. Comparison of the two analyses omitting carry-over effects shows that the average standard error

Table 7.6 Estimates of variance components and treatment effects for a cross-over trial comparing three analgesic drugs. Standard errors of estimates appear in brackets.

	Fixed patients	**Random patients**
Ignoring carry-over		
Variance components		
Patients	—	1.3 (1.89)
Units within patients	10.8 (2.42)	10.7 (2.39)
Treatment effects		
A − B	3.4 (1.05)	3.5 (0.91)
A − C	2.0 (0.99)	1.9 (0.88)
B − C	−1.4 (0.98)	−1.6 (0.87)
Including carry-over		
Variance components		
Patients	—	1.3 (1.96)
Units within patients	11.3 (2.60)	10.9 (2.44)
Treatment effects		
A − B	3.7 (2.01)	3.8 (0.99)
A − C	2.6 (2.14)	1.9 (0.99)
B − C	−1.1 (2.08)	−1.9 (0.99)
Carry-over effects		
A − B	0.4 (3.43)	1.0 (1.45)
A − C	1.1 (3.66)	0.2 (1.41)
B − C	0.7 (3.72)	−0.8 (1.46)

for treatment differences from the combined analysis with recovery of between-patient information is on average 12% less than for the within-patient analysis. The comparison A − B which has the largest standard error shows the greatest increase in precision, so that the combined analysis also leads to treatment estimates with a smaller range of standard errors. For this trial the between-patient component of variance is relatively small and recovery of between-patient information is clearly worthwhile.

Although the estimated carry-over effects are generally small relative to their standard errors, the results from this model could be preferred on the basis of the trial design and the absolute magnitude of the estimated carry-over effects. With this model we see that the magnitude of the standard errors of the treatment effects using a mixed model are only half of those obtained from the fixed effects model. The standard errors of the carry-over effects show an even greater separation. Using the mixed model approach, the penalty in fitting carry-over terms is an increase of just over 10% in the standard errors of the treatment differences.

For this trial the between-patients variance component is small compared with the within-patients component ($\gamma = 0.12$), suggesting that there is little advantage in using a cross-over trial for testing these analgesics, even if carry-over effects can be avoided. Indeed, the predicted average standard error for treatment comparisons for a parallel group trial with the same number of patient sessions per treatment is 0.92, compared with 0.89 and 0.99 for the cross-over analysis ignoring and including carry-over effects, respectively.

SAS code and output

This is similar to that of the earlier examples and is omitted.

7.6   OPTIMAL DESIGNS

There has been substantial research into cross-over designs in which estimates of treatment effects and carry-over effects are both of interest. As indicated earlier in this chapter, we would question whether such an approach is likely to be desirable. However, for various combinations of numbers of treatments, and numbers of treatment periods, so-called optimal designs have been derived. They satisfy the property of giving uniformly most powerful unbiased estimates of treatment and carry-over effects.

One particular optimal design has been used in practice, and arguably has a stronger justification for its use than the other optimal designs. This is Balaam's design for the situation of two treatments and two treatment periods. Of course, the most common design for this situation is not Balaam's design but the simple AB/BA design. However, its critics would argue that a weakness is its inability to estimate carry-over effects and simultaneously use data from the second period in estimating treatment effects. Balaam's design resolves the problem by employing all four possible treatment sequences — AA, BB, AB and BA.

7.6.1   Example: Balaam's design

This is a well-known example initially described by Hunter *et al.* (1970). The aim was to determine the effect of Amantadine (treatment A) on subjects suffering from Parkinsonism. The trial was placebo controlled (treatment B). After a run-in period of one week during which baseline information was recorded, there were two four-weekly treatment periods, without a washout period. Weekly scores (0–4) were recorded for each of 11 physical signs and the data presented in Table 7.7 give the weekly average total scores in each treatment period. Seventeen patients were randomised, and the data have no missing values.

Table 7.8 presents the results of analyses with and without inclusion of a carry-over term, and with patient effects fitted as fixed or random.

An immediate point to note from the two mixed models is the very high patient variance component compared with the residual variance component (Table 7.8). This immediately suggests that little gain in efficiency will accrue from between-patient information. This is confirmed by comparison of the treatment standard errors, where even in the model where carry-over is fitted the reduction is only 4%. In most situations, where the between-patient variation is less extreme, the

Table 7.7 Average scores for amantadine trial.

Group	Subject	Baseline	Period 1	Period 2
1AA	1	14	12.50	14.00
	2	27	24.25	22.50
	3	19	17.25	16.25
	4	30	28.25	29.75
	Mean	**22.50**	**20.56**	**20.63**
2BB	1	21	20.00	19.51
	2	11	10.50	10.00
	3	20	19.50	20.75
	4	25	22.50	23.50
	Mean	**19.25**	**18.13**	**18.44**
3AB	1	9	8.75	8.75
	2	12	10.50	9.75
	3	17	15.00	18.50
	4	21	21.00	21.50
	Mean	**14.75**	**13.81**	**14.63**
4BA	1	23	22.00	18.00
	2	15	15.00	13.00
	3	13	14.00	13.75
	4	24	22.75	21.50
	5	18	17.75	16.75
	Mean	**18.60**	**18.30**	**16.60**

Table 7.8 Estimates of variance components and treatment effects. Standard errors of estimates appear in brackets.

	Fixed patients	**Random patients**
Ignoring carry-over		
Variance components		
Patients	—	30.3 (10.91)
Residual (within patients)	1.05	1.1 (0.38)
Treatment difference	1.29 (0.49)	1.24 (0.48)
Including carry-over		
Variance components		
Patients	—	30.5 (11.0)
Residual (within patients)	1.12	1.1 (0.42)
Treatment difference	1.42 (0.73)	1.28 (0.70)
Carry-over difference	0.25 (1.06)	0.10 (0.92)

existence of the AA and BB treatment groups would lead us to expect greater benefits from the mixed model approach.

In this study there is no evidence of any carry-over effect, and most statisticians would choose to report the model which excludes carry-over. However, having chosen a design for its optimal properties in estimating both treatment and carry-over effects, there is a strong case for reporting the fuller model.

The presentation here has been restricted to the analysis of the results in the two treatment periods, and the fact that baseline observations were also recorded has been ignored. Jones and Kenward (1989) present additional analyses utilising this baseline data, and in particular use interactions with the baseline to test for an effect of the baseline level on the treatment effect, period effect and carry-over effect. Although none of these terms was statistically significant at the 10% level of significance, they found indications that the treatment differences were higher with greater baseline levels. These authors also handled carry-over in a more involved way than we have employed in our analyses. They allowed for the possibility that carry-over would be different in those on the AA or BB sequence from those on the AB or BA sequence, but this term in the analysis of variance was clearly non-significant.

The conclusions from the trial will, in this instance, be qualitatively similar whichever of the previously described analytical methods is used, as long as carry-over is ignored in estimating treatment effects. Amantadine produces a reduction in the physical signs of Parkinson's disease which is statistically significant at the 5% level. Note, however, that inclusion of carry-over terms in the model produces a substantial increase in the standard error of the treatment effect, leading to non-significance of the treatment effect.

SAS code and output

The SAS code and the structure of the output is almost identical to that presented at the end of Section 7.4, except that two treatments are used instead of four. The key results have already been tabulated in Table 7.8.

7.7 COVARIANCE PATTERN MODELS

In the examples considered so far in this chapter, the mixed model approach has fitted the patient effects as random. As we have seen earlier, this implies that the observations within one patient are all assumed to have the same correlation and variance. However, it could be argued that in trials with three or more periods the correlation may vary with different pairs of periods. In particular, periods which were closer together might be expected to show higher correlations. In this section we explore the situation where the covariance patterns used are 'structured'.

7.7.1 Structured by period

This is perhaps the most obvious way to structure the residual covariance matrix. We have already mentioned the possibility that periods close together in time might have a higher correlation than periods far apart in time. Additionally, it is possible that the residual variance may itself change over successive periods. For example, in early periods of the trial, while the protocol is unfamiliar to patients, the observations may be more variable than in later periods. We have already seen that SAS offers a wide choice of covariance patterns, so a strategy for investigating alternatives is preferable to a blunderbuss approach of examining the full range available. A comparison of the compound symmetry structure (equivalent to simply fitting patient effects as random) and the general covariance matrix will usually be helpful in determining whether use of a more complicated covariance structure is likely to be useful. We will explore this further in the forthcoming example.

7.7.2   Structured by treatment

Although structuring by period is the most obvious way of introducing structure, this can also be applied to treatments. In parallel group trials we have already met situations where we might wish to fit separate variances for each of the treatment groups. There is an exact analogy in the cross-over situation, where the variances may differ for some of the treatments. Additionally, there may be good reason to suspect that the results from certain pairs of treatments may be more highly correlated than others if they have a similar mode of action. Thus, a more complicated structure for treatments than the simple compound symmetry may be highly plausible.

7.7.3   Example: Four-way cross-over trial

We will consider the four-way, cross-over trial analysed in Section 7.4 which compared the effects of three drugs A, C and D and a placebo B on blood flow. This time, no values are set to missing and therefore the study is more balanced (across random effects). Each treatment period lasted a week and was followed by a washout period also lasting a week. An analysis fitting a random effects model (i.e. a compound symmetry pattern) is compared with analyses fitting general covariance matrices. The covariance matrix is first structured according to periods and then according to treatments. Treatment and period effects are fitted as fixed effects in each analysis. In the first instance, we also include a carry-over effect as an additional fixed effect in the analysis.

   Examination of the second model in Table 7.9 shows that data are more variable in the first period than in other periods. This is a common occurrence in clinical trials and is one reason why a run-in period is often used. The correlations between periods indicate a tendency for more widely separated periods to have lower correlations than those close together. The third model shows a higher variation for the placebo treatment (B) than for the three active treatments. Interestingly, the correlations involving treatment A are substantially lower than the correlation between treatments B, C and D.

Table 7.9 Comparison of covariance pattern models for the four-way, cross-over trial, with inclusion of carry-over effects.

Covariance pattern	Period/ treatment	Variances (×10²)	Correlations	−2 Log likelihood no of (parameters)	Estimated treatment differences (model-based SE) (empirical SE)	Estimated carry-over differences (model-based SE) (empirical SE)
Compound symmetry	1–4	36	$\begin{pmatrix} 1 \\ 0.47 & 1 \\ 0.47 & 0.47 & 1 \\ 0.47 & 0.47 & 0.47 & 1 \end{pmatrix}$	519.2 (2)	A − B 38.0 (17.4) (22.6) A − C 13.4 (17.9) (16.6) A − D 68.7 (17.5) (16.1) B − C −24.6 (17.9) (13.8) B − D 30.7 (18.2) (17.6) C − D 55.3 (18.7) (12.7)	−27.5 (20.5) (15.9) −23.4 (21.1) (15.9) −8.9 (21.0) (14.9) 4.1 (21.5) (19.6) 18.6 (22.4) (19.2) 14.5 (22.4) (20.6)
General (across periods)	1 2 3 4	69 19 39 12	$\begin{pmatrix} 1 \\ 0.67 & 1 \\ 0.36 & 0.68 & 1 \\ 0.53 & 0.48 & 0.52 & 1 \end{pmatrix}$	502.2 (10)	A − B 37.1 (13.2) (15.3) A − C 23.6 (14.4) (13.1) A − D 64.1 (12.6) (11.5) B − C −13.5 (13.1) (10.1) B − D 27.0 (14.0) (9.8) C − D 40.5 (13.5) (9.5)	−20.0 (13.9) (11.9) −35.1 (14.7) (11.1) −5.6 (13.8) (12.5) −15.0 (14.2) (11.3) 14.4 (15.6) (15.4) 29.5 (15.4) (16.4)
General (across treatments)	A B C D	17 64 34 25	$\begin{pmatrix} 1 \\ 0.11 & 1 \\ 0.16 & 0.83 & 1 \\ -0.10 & 0.64 & 0.82 & 1 \end{pmatrix}$	498.0 (10)	A − B 43.8 (23.2) (22.2) A − C 8.3 (18.4) (19.5) A − D 66.4 (18.5) (17.5) B − C −35.5 (13.4) (11.6) B − D 22.5 (17.2) (15.6) C − D 58.1 (11.1) (10.7)	−23.5 (12.2) (12.2) −54.0 (16.5) (12.1) −4.3 (11.2) (7.7) −30.5 (15.6) (6.6) 19.3 (14.1) (14.2) 49.7 (18.0) (13.5)

Both general covariance models show significant improvements over the compound symmetry model when tested by likelihood ratio tests ($\chi^2_8 = 17.0$, $p = 0.03$ and $\chi^2_8 = 21.2, p = 0.07$) and might therefore be preferred.

A comparison of the three models with respect to the estimates of treatment differences shows that the estimates vary substantially when treatment C is involved. The same comment applies when the carry-over effects are compared.

The model-based standard errors are generally higher with the compound symmetry structure than with either of the alternative models. The use of empirical standard errors is usually expected to differ less between different models, because they use the observed covariances. We see, however, that they also differ disturbingly across the three models. This can be attributed to the variation in the estimates for treatment C across the three models.

Tests of the null hypothesis of equal treatment effects are all highly significant whichever model is used, whether or not they are based on the empirical covariance matrix and whether or not Satterthwaite's approximation to the denominator DF is used. The precise significance levels obtained though are much smaller with the empirical variances unless they are used with Satterthwaite's DF. For example, with the compound symmetry model, the *p*-values are 0.004 with the model-based covariance matrix and 0.0004 with the empirical covariance matrix when the Satterthwaite approximation is not used. The corresponding values when the Satterthwaite DF are used are 0.003 and 0.004.

The tests for equality of carry-over effects are even more dependent on the choice of model and the decision to use the empirical covariance matrix even when the Satterthwaite DF are used. With the model-based covariance, the three models yield *p*-values of 0.54, 0.14 and 0.03. With the empirical covariance, the corresponding values are 0.30, 0.10 and 0.0007.

What interpretation can we place then on such conflicting findings? The first point to make is that they are not due to outliers in the data. Examination of residuals produced no suspicious observations. The second point is to recognise that this is a trial with only 14 patients. The models are fitting substantial numbers of parameters, and the covariance parameters may be estimated quite inaccurately, especially in the case of the second and third models. The sample size may also be inadequate for the likelihood ratio tests to be reliable. Under these circumstances, the observed covariances used in the empirical approach may also be unreliable, and our results suggest that this may be the case. Certainly, the more liberal significance levels which have been produced with the empirical approach could not be considered any more reliable than the model-based results. Therefore, with such small sample sizes we would advise use of the model-based standard errors and significance tests.

Some analysts might therefore decide on the above evidence to adopt the model-based approach to the presentation of results and to use the structured-by-treatment model, because this gives the highest likelihood. The conclusions would then be that treatments A and C produce a significant elevation in LVETs compared with treatments B and D; that the results from treatment A are at most weakly correlated with results from the other three treatments, which in turn are quite highly correlated; and that there is a substantial carry-over from treatment C to the observation in the following period.

A rather different view can be taken though with respect to the inclusion of the carry-over term in the model. The conclusion about the significance of the carry-over is strongly model-dependent, and the observed significance level in the chosen model is not extreme ($p = 0.03$). The trial design, with reasonable washout periods, makes physical carry-over of drugs unlikely, although there is always a theoretical possibility that any treatment may produce a disease-modifying effect. Even if carry-over does occur, the simple carry-over model used may be inappropriate (see Section 7.11). We think it is advisable, therefore, to go on to consider models in which carry-over is omitted.

The use of the same three covariance structures as before, but without carry-over in the model, is summarised in Table 7.10. Many of the comments made on the previous models still apply. The patterns of covariances are similar, and structuring by period or by treatment produces a significant increase in the likelihoods compared with the compound symmetry pattern. All of the models give similar estimates of the treatment differences which do not involve treatment C. The estimates involving treatment C are similar for the compound symmetry model and for the structured-by-treatment model, but are substantially different

Table 7.10　Comparison of covariance pattern models for the four-way, cross-over trial, without carry-over effects.

Error model	Period/ treatment	Variances ($\times 10^2$)	Correlations	-2 Log likelihood (no of parameters)	Estimated treatment differences (model SE) (empirical SE)
Compound symmetry	1–4	36	$\begin{pmatrix} 1 & & & \\ 0.49 & 1 & & \\ 0.49 & 0.49 & 1 & \\ 0.49 & 0.49 & 0.49 & 1 \end{pmatrix}$	544.6 (2)	A − B　43.1 (16.1) (20.4) A − C　23.1 (16.3) (15.1) A − D　70.4 (16.3) (14.5) B − C −20.0 (16.3) (12.9) B − D　27.3 (16.3) (15.0) C − D　47.3 (16.3)　(8.6)
General (across periods)	1 2 3 4	74 22 37 13	$\begin{pmatrix} 1 & & & \\ 0.75 & 1 & & \\ 0.44 & 0.61 & 1 & \\ 0.43 & 0.43 & 0.58 & 1 \end{pmatrix}$	528.2 (10)	A − B　45.9 (11.8) (13.0) A − C　47.9 (11.8) (10.8) A − D　70.5 (11.6) (10.3) B − C　2.0 (11.8) (10.6) B − D　24.6 (11.6) (12.2) C − D　22.6 (11.2)　(4.6)
General (across treatments)	A B C D	21 61 34 28	$\begin{pmatrix} 1 & & & \\ 0.20 & 1 & & \\ 0.39 & 0.82 & 1 & \\ 0.13 & 0.56 & 0.76 & 1 \end{pmatrix}$	527.4 (10)	A − B　43.1 (22.0) (21.0) A − C　24.4 (15.7) (14.4) A − D　69.0 (17.4) (16.4) B − C −18.7 (12.2) (11.6) B − D　25.8 (17.6) (16.4) C − D　44.6 (10.5)　(8.9)

with structuring by period. In view of the highest likelihood being obtained when structuring by treatment, and the consistency of the estimates when compared with the compound symmetry model, this model might be tentatively accepted.

As noted in the corresponding model when carry-over was included, the correlations involving treatment A are relatively low. This treatment is also the one producing the highest measurements. This may therefore increase further the plausibility of this model, suggesting as it does a different and more effective form of action.

Interpretation of both the model-based standard errors and the empirical standard errors is difficult. Structuring by period yields the smallest standard errors, despite not producing the best fit as indicated by the likelihoods. The empirical standard errors are generally lower than the model-based standard errors, with the first and third models giving similar values, as would follow from the treatment estimates being similar. These standard errors are also similar to the model-based standard errors from Model 3, again giving indirect support to the structured-by-treatment model.

We think therefore that there is a case for basing our inferences on the model without carry-over, with the covariances being structured by treatment. These results indicate, however, that choosing the model from which to present findings is not always straightforward. Decisions concerning the inclusion of carry-over terms may be based primarily on how the trial was designed, and in particular on the adequacy of washout periods. The choice between different covariance pattern models may be influenced by consideration of the likelihoods. However, statistical tests need not be the only factor determining model choice. The validity of the assumptions relating to a model are also important. For example, if we believe that periods and treatments are unlikely to have varying correlations based on past experience, or if the trial is too small to give precise covariance estimates, then a compound symmetry structure possibly may be the one of choice.

It will always be a cause for concern when different models give qualitatively different conclusions, and this is always a greater danger when the data are relatively sparse. The different conclusions which could be reached with different models using the data set we have considered here is therefore perhaps less surprising than the fact that we are able to present a strong case for a relatively complicated model. In larger studies it should be easier to compare models, and the inferences are not so likely to be strongly model dependent.

The above example makes the point that treating a multi-period, cross-over trial as repeated measures data with a covariance pattern which is structured by treatment provides an additional approach to analysis which can be informative. For analyses which are conducted in the pharmaceutical industry for drug registration purposes, the requirement to specify the analysis plan in the trial protocol may be restrictive. This is likely to cause the compound symmetry model to be the one of choice for a primary analysis, perhaps with the choice of the empirical estimator for the standard errors, so that a more complicated 'true' covariance structure can be partially reflected. It should be acceptable though to specify a secondary analysis which is structured by treatment, so that the interrelationships of responses to different treatments can be explored.

SAS code and output

The following code uses the same variable names specified at the end of Section 7.4. The code is given below initially for the compound symmetry model, with inclusion of carry-over effects, and specifying the use of the empirical estimators for the standard errors, by use of the EMPIRICAL option.

```
PROC MIXED EMPIRICAL; CLASS treat period patient carry;
TITLE 'COMPOUND SYMMETRY USING EMPIRICAL';
MODEL lvet=treat period carry/ DDFM=SATTERTH;
REPEATED period/ SUBJECT=patient TYPE=CS;
LSMEANS treat carry/DIFF PDIFF;
```

The code for the other covariance pattern models is identical except for the REPEATED statement.

```
TITLE 'STRUCTURED BY PERIOD USING EMPIRICAL';
REPEATED period/ SUBJECT=patient TYPE=UN;

TITLE 'STRUCTURED BY TREATMENT USING EMPIRICAL';
REPEATED treat/ SUBJECT=patient TYPE=UN;
```

Model-based standard errors are obtained by omission of the EMPIRICAL option, and the changes to remove carry-over effects are obvious. The following output is for the above code, where structuring by treatment is employed.

```
            STRUCTURED BY TREATMENT USING EMPIRICAL
                  Class Level Information
        Class     Levels  Values
        TREAT          4  1 2 3 4
        PERIOD         4  1 2 3 4
        PATIENT       14  1 2 3 4 5 6 7 8 9 10 11 12 13
                          14
        CARRY          4  1 2 3 4

              REML Estimation Iteration History
      Iteration  Evaluations    Objective      Criterion
             0            1   446.65094012
             1            2   423.46380445    0.02010791
             2            1   417.92341564    0.01043261
             3            1   415.11025471    0.00468615
             4            1   413.88418353    0.00154375
             5            1   413.49757907    0.00028985
             6            1   413.42928421    0.00002064
             7            1   413.42475153    0.00000023
             8            1   413.42470371    0.00000000
                  Convergence criteria met.
```

```
Covariance Parameter Estimates (REML)
```

Cov Parm	Subject	Estimate
UN(1,1)	PATIENT	1681.0233851
UN(2,1)	PATIENT	322.44529238
UN(2,2)	PATIENT	6360.2519018
UN(3,1)	PATIENT	390.80269939
UN(3,2)	PATIENT	3891.8999325
UN(3,3)	PATIENT	3429.5166903
UN(4,1)	PATIENT	-205.2595044
UN(4,2)	PATIENT	2555.1482991
UN(4,3)	PATIENT	2411.6523293
UN(4,4)	PATIENT	2531.9805687

```
Model Fitting Information for LVET
```

Description	Value
Observations	56.0000
Res Log Likelihood	-248.984
Akaike's Information Criterion	-258.984
Schwarz's Bayesian Criterion	-268.127
-2 Res Log Likelihood	497.9670
Null Model LRT Chi-Square	33.2262

```
Model Fitting Information for LVET
```

Description	Value
Null Model LRT DF	9.0000
Null Model LRT P-Value	0.0001

```
Tests of Fixed Effects
```

Source	NDF	DDF	Type III F	Pr>ChiSq	Pr>F	
TREAT	3	8.33	49.99	16.66	0.0001	0.0007
PERIOD	3	2.59	85.41	28.47	0.0001	0.0168
CARRY	3	12.9	33.23	11.08	0.0001	0.0007

```
Least Squares Means
```

Effect	TREAT	CARRY	LSMEAN	Std Error	DF	t	Pr>|t|
TREAT	1		388.46464898	10.19795839	9	38.09	0.0001
TREAT	2		344.63812147	20.76278820	11.1	16.60	0.0001
TREAT	3		380.15038167	17.71498177	20	21.46	0.0001
TREAT	4		322.09604889	13.05355892	8.97	24.67	0.0001
CARRY		1	338.37672590	10.21795495	5.77	33.12	0.0001
CARRY		2.	361.92171120	16.64885665	34.9	21.74	0.0001
CARRY		3	392.39293873	16.62964129	26.7	23.60	0.0001
CARRY		4	342.65782519	11.76421063	9.98	29.13	0.0001

```
Differences of Least Squares Means
```

Effect	TREAT	CARRY	_TREAT	_CARRY	Difference	Std Error	DF	t	Pr>|t|
TREAT	1		2		43.82652751	22.24063376	10.4	1.97	0.0759
TREAT	1		3		8.31426731	19.53692552	15.4	0.43	0.6763
TREAT	1		4		66.36860009	17.47253130	7.81	3.80	0.0055
TREAT	2		3		-35.51226020	11.58509868	7.48	-3.07	0.0168
TREAT	2		4		22.54207258	15.60090755	9.46	1.44	0.1808
TREAT	3		4		58.05433278	10.66474238	13.1	5.44	0.0001

```
CARRY    1           2    -23.54498530 12.19533278  17.2  -1.93  0.0702
CARRY    1           3    -54.01621283 12.11100097   5.12 -4.46  0.0063
CARRY    1           4     -4.28109929  7.71265733   1.82 -0.56  0.6395
CARRY    2           3    -30.47122752  6.56171016   0.53 -4.64  0.2787
CARRY    2           4     19.26388601 14.19044171  11.8   1.36  0.1999
CARRY    3           4     49.73511353 13.51124296   9.57  3.68  0.0046
```

7.8   ANALYSIS OF BINARY DATA

The earlier sections in this chapter have considered trials where the response variable was assumed to be normally distributed. Although this type of data is almost certainly the most common, binary data is by no means unusual. We have seen in Chapter 3 how the mixed model approach can be extended from normally distributed data to binary data using generalised linear mixed models. The approach described there can be used for any of the designs considered earlier in this chapter, but we will restrict ourselves to examining the simplest, and probably most common, cross-over design: the AB/BA design.

We utilise an example presented by Jones and Kenward (1989) with deletion of some observations in the second treatment period. The sample comes from safety data from a trial on cerebrovascular insufficiency. The response variable was whether an electrocardiogram was assessed by a cardiologist to be normal (1) or abnormal (0). The modified data are presented in Table 7.11.

In a fixed effects analysis, those subjects with a missing observation do not contribute to the evaluation of treatment effects. Testing the null hypothesis of no treatment difference can be performed most powerfully using Prescott's Test (Prescott, 1981). This requires reorganisation of the data in Table 7.11 to the form in Table 7.12 where the rows can be thought of as representing a 'change score' from period 1 to period 2, with values -1, 0 or 1.

Values of this change score are then compared between the treatment sequence groups. This is undertaken most appropriately using an exact trend test (which is equivalent to a permutation t test). Within SAS this can be achieved using an EXACT option within PROC FREQ. It yields a p-value of 0.33 (the corresponding asymptotic test gives $p = 0.22$).

A problem with the above approach is that it is based purely on significance testing. It does not yield an estimate of the magnitude of the treatment effect. The

Table 7.11   Data from an AB/BA trial on cerebrovascular deficiency. Outcomes 0 and 1 correspond to abnormal and normal ECG readings with deleted observations denoted by ●.

Sequence	Outcomes						
	(0, 0)	**(0, 1)**	**(0, ●)**	**(1, 0)**	**(1, 1)**	**(1, ●)**	**Total**
AB	11	1	2	6	27	3	50
BA	12	5	2	5	23	3	50
Total	23	6	4	11	50	6	100

Table 7.12   Data from Table 7.11 reorganised in
the form for the application of Prescott's Test.

Sequence	Change score			Total
	−1	0	1	
AB	1	38	6	45
BA	5	35	5	45
Total	6	73	11	90

use of a mixed model allows us to both recover the between-patient information,
and to obtain meaningful estimates of the magnitude of the treatment effect.

As discussed in Sections 3.2.4 and 3.3.2, problems caused by uniform random
effect categories are likely to arise if patients are fitted as random, since there
are only two observations per patient. This can be avoided by using instead a
covariance pattern model with a compound symmetry structure. The SAS code
and most relevant parts of the output are presented at the end of this section.

The correlation parameter estimate is 0.51 and indicates that observations on
the same patient are quite strongly correlated.

The estimate of the period effect (0.18) is similar to its standard error (0.20) and
is therefore clearly non-significant. It is not negligible, however, and we repeat our
recommendation that the period effect should always be included in the model.

The treatment estimate of −0.33 with 95% confidence limits from −0.72 to
0.05 corresponds to the estimate on the logistic scale. By exponentiating these
figures we obtain a point estimate and 95% confidence intervals for the odds ratio
(i.e. a ratio of proportion normal to proportion abnormal in the two treatment
groups). This gives an estimated odds ratio of 0.71, with 95% confidence limits of
0.48 to 1.06.

Note that the above confidence interval was obtained using the empirical
variance estimator, which is the default in the SAS procedure (PROC GENMOD)
used to perform the analysis. If model-based variance estimators are used, the
95% confidence interval is slightly wider at 0.47 to 1.10.

Although we should not be unduly influenced by single examples, we note
that the significance level obtained for the treatment effect in the above analysis
was $p = 0.09$ using the empirical variance estimator and $p = 0.12$ with the
model-based variance estimator, compared with $p = 0.33$ with Prescott's Test.
The ability of the model to recover information from patients with incomplete data,
together with its capacity to provide meaningful estimates of treatment effects,
should make it the preferred option. This latter feature means that it may also be
preferred when data are complete. When data are complete, treatment estimates
can also be obtained via a bivariate logistic model proposed by Jones and Kenward
(1989). They differ, however, from those provided by the mixed model approach.
The odds ratio which they produce is conditional on different responses being
observed in the two treatment periods. Thus, they report an odds ratio of 0.0385

on the complete data from which our example was derived. Senn (1993) has also suggested a method based on applying ordinal logistic regression techniques to the data as configured in Table 7.12. This method yields yet another different estimate of a treatment effect in the form of an odds ratio. Our initial examination of this method suggests that it is less powerful than the alternatives considered and we do not consider it further here. The estimate from the mixed model provides, we believe, a natural, comprehensible estimate, and is our recommendation for all cases where confidence interval estimates are required for the magnitude of the treatment effect. Note, however, that it is based on asymptotic theory, and the values from small studies must be regarded as approximate.

SAS code and output

```
PROC GENMOD;
CLASS treat period patient;
MODEL outcome/one=period treat/ ERROR=B;
REPEATED SUBJECT=patient/ WITHIN=period TYPE=CS MODELSE CORRW;
```

```
                    The GENMOD Procedure
                      Model Information
            Description                  Value
            Data Set                     WORK.A
            Distribution                 BINOMIAL
            Link Function                LOGIT
            Dependent Variable           OUTCOME
            Dependent Variable           ONE
            Observations Used            190
            Number Of Events             125
            Number Of Trials             190
            Missing Values               10

                   Class Level Information
        Class     Levels  Values
        TREAT          2  A B
        PERIOD         2  1 2
        PATIENT      100  1 2 3 4 5 6 7 8 9 10 11 12 13
                          14 15 16 17 18 19 20 21 22 23
                          24 25 26 27 28 29 30 31 32 33
                          34 35 36 37 38 39 40 41 42 43
                          44 45 46 47 48 49 50 51 52 53
                          54 55 56 57 58 59 60 61 62 63
                          64 65 66 67 68 69 70 71 72 73
                          74 75 76 77 78 79 80 81 82 83
                          84 85 86 87 88 89 90 91 92 93
                          94 95 96 97 98 99 100

                     Parameter Information
          Parameter      Effect     TREAT   PERIOD
          PRM1           INTERCEPT
          PRM2           PERIOD              1
          PRM3           PERIOD              2
```

```
            PRM4                TREAT     A
            PRM5                TREAT     B

        Criteria For Assessing Goodness Of Fit
    Criterion               DF          Value       Value/DF
    Deviance                187       242.0829        1.2946
    Scaled Deviance         187       242.0829        1.2946
    Pearson Chi-Square      187       189.9881        1.0160
    Scaled Pearson X2       187       189.9881        1.0160
    Log Likelihood            .      -121.0414             .

        Analysis Of Initial Parameter Estimates
    Parameter   DF Estimate Std Err ChiSquare Pr>Chi
    INTERCEPT    1   0.8128   0.2752    8.7228 0.0031
    PERIOD    1  1   0.1146   0.3077    0.1388 0.7095
    PERIOD    2  0   0.0000   0.0000         .      .
    TREAT     A  1  -0.4233   0.3082    1.8863 0.1696
    TREAT     B  0   0.0000   0.0000         .      .
    SCALE        0   1.0000   0.0000         .      .
```

NOTE: The scale parameter was held fixed.

As in the example considered in Section 6.4, the above output will always be produced, but it is irrelevant.

```
                GEE Model Information
    Description                     Value
    Correlation Structure           Exchangeable
    Within-Subject Effect           PERIOD (2 levels)
    Subject Effect                  PATIENT (100 levels)
    Number of Clusters              100
    Clusters With Missing Values    10
    Correlation Matrix Dimension    2
    Maximum Cluster Size            2
    Minimum Cluster Size            1

            Working Correlation Matrix
                        COL1      COL2
            ROW1      1.0000    0.5097
            ROW2      0.5097    1.0000

        Analysis Of GEE Parameter Estimates
        Empirical Standard Error Estimates

            Empirical  95% Confidence Limits
    Parameter   Estimate Std Err   Lower  Upper       Z   Pr>|Z|
    INTERCEPT     0.7011  0.2494  0.2124 1.1899  2.8116   0.0049
    PERIOD    1   0.1802  0.1984 -0.2086 0.5690  0.9085   0.3636
    PERIOD    2   0.0000  0.0000  0.0000 0.0000  0.0000   0.0000
    TREAT     A  -0.3348  0.1987 -0.7241 0.0546 -1.685    0.0920
    TREAT     B   0.0000  0.0000  0.0000 0.0000  0.0000   0.0000
    Scale         1.0012       .       .      .       .        .
```

NOTE: The scale parameter for GEE estimation was
computed as the square root of the normalized Pearson's
chi-square.

```
              Analysis Of GEE Parameter Estimates
              Model-Based Standard Error Estimates
                      Model  95% Confidence Limits
      Parameter    Estimate  Std Err   Lower  Upper       Z   Pr>|Z|
      INTERCEPT      0.7011   0.2468  0.2175 1.1848  2.8413   0.0045
      PERIOD   1     0.1802   0.2172 -0.2455 0.6059  0.8297   0.4067
      PERIOD   2     0.0000   0.0000  0.0000 0.0000  0.0000   0.0000
      TREAT    A    -0.3348   0.2175 -0.7611 0.0916 -1.539    0.1238
      TREAT    B     0.0000   0.0000  0.0000 0.0000  0.0000   0.0000
      Scale          1.0012        .       .      .       .        .
```

7.9 ANALYSIS OF CATEGORICAL DATA

Apart from the case of binary data, response variables which are purely categorical, without an underlying scale, are extremely rare. We will therefore only consider data on short ordered scales in this section. Variables classified as none, mild, moderate and severe will arise in a variety of contexts.

To illustrate techniques, we will again take an example from Jones and Kenward (1989) and delete five observations from the second treatment period. The example is a placebo controlled trial of a treatment for primary dysmenorrhoea. Thirty patients entered the trial and in each treatment period the amount of relief obtained was recorded as none or minimal (1), moderate (2) and complete (3). The data as we analyse them are summarised in Table 7.13.

Taking a fixed effects approach, a test of significance is most readily obtained using methods based on the analysis of an appropriate contingency table. A simple but inefficient way of producing such a contingency table would be to categorise the changes in the outcome variable from the first treatment period

Table 7.13 Data from an AB/BA trial on a treatment for primary dysmenorrhoea (A: placebo; B: high dose analgesic).

Sequence	(1, 1)	(1, 2)	(1, 3)	(1, •)	(2, 1)	(2, 2)	(2, 3)	(2, •)	(3, 1)	(3, •)	Total
AB	2	3	5	1	1	1	2	1	0	0	16
BA	3	2	0	0	1	0	1	1	4	2	14
Total	5	5	5	1	2	1	3	2	4	2	30

Sequence	Change Score					Total
	−2	−1	0	1	2	
AB	5	5	3	1	0	14
BA	0	3	3	1	4	11
Total	5	8	6	2	4	25

to the second as 'worse', 'no change' and 'better', and to tabulate this variable against the treatment sequence. The significance of the treatment effect could then be determined from this 3×2 contingency table, as in Prescott's Test. However, this configuration does not use the information that observations of 'none' and 'complete' in the two treatment periods represent a larger difference than between 'none' and 'moderate', or 'moderate' and 'complete'. If we arbitrarily assign numbers of 1, 2 and 3 to the outcome categories, we can generate a 5×2 contingency table based on the change scores. The 'obvious' approach is to then apply a permutation t test (test for trend) to this table to assess the significance of the treatment effect. Application to the change scores presented in Table 7.13 gives $p = 0.005$.

In applying this test it should be appreciated that the scores of -2, -1, 0, $+1$ and $+2$ are arbitrary. They should not be taken to imply that the difference between 'none' and 'complete' is twice as large as the difference between 'none' and 'moderate', nor that the difference between 'none' and 'moderate' is the same as the difference between 'moderate' and 'complete'. For this reason, some statisticians may wish to replace the change scores with ranks and apply an (exact) Wilcoxon Rank Sum Test. The choice will rarely make any practical difference, but it is clearly good practice to make this choice prior to analysis, rather than reporting the more favourable result! Note that the situation becomes more complicated when there are more than three categories for the outcome variable. Analysis could still be based on the change scores, but there would be an implicit strong assumption about the meaning of the intervals between the categories. Without such strong assumptions, many of the categories of change would be indistinguishable from each other and a simplified 5×2 contingency table would result. Such an example is presented by Senn (1993).

The mixed model approach with random patient effects and fixed period and treatment effects, based on carrying out ordinal logistic regression, cannot be undertaken using SAS. There is, however, a macro developed by Lipsitz *et al.* (1994) which allows this model to be fitted (see Section 9.1). The code for this model is not provided but can be found on [www.med.ed.ac.uk/phs/mixed]. Using the more robust empirical variance estimator (see Section 2.4.3), the coefficient for the treatment effect, on the logistic scale, is -1.9 with a standard error of 0.6 ($p = 0.0007$). By exponentiating, we obtain an estimate of the odds ratio of 0.15, with 95% confidence limits of $0.05–0.44$. The interpretation of the odds ratio in this situation is that the estimated odds of being in a favourable outcome category when treated with placebo compared to the analgesic is 0.15, whether favourable is defined as complete relief or moderate/complete relief.

7.10 USE OF RESULTS FROM RANDOM EFFECTS MODELS IN TRIAL DESIGN

Once a cross-over trial has been analysed and reported, there is a subsequent question to consider. What has the present study taught us about the trial design to be used in future studies of this condition? The factors we will wish to take into consideration are the sizes of the residual and patient variance components and the dropout rates at various stages during cross-over. If between-subject variability is large compared with residual variation, a cross-over design may be vastly more efficient than a corresponding parallel group study. On the other hand, the analysis of a cross-over trial requires more assumptions than a parallel group design, and if between-subject variability is relatively small, a parallel group study may be the design of choice. If a multi-period, cross-over trial experiences a substantial dropout in the later phases of the cross-over, or if the required duration of each treatment is long, than an incomplete block design may be considered.

Of course, the most basic way in which we can use data from one cross-over trial to plan a succeeding trial is to use the estimate of the residual variance in standard sample size formulae (see, for example, Senn 1993, p. 217). If there has been a sequence of similar trials, then a weighted average of the estimates of the residual variance would provide a more robust figure.

The more interesting question is whether a cross-over design should be used at all. This can usually be assessed satisfactorily by comparing the standard error of the treatment differences in the cross-over trial with the standard errors which would be expected from a comparable parallel group study. We reported such information for the trial described in Section 7.5, and we now consider another example.

7.10.1   Example

We consider the results from the oral mouthwash trial, summarised in Table 7.3. The estimate of the patient variance component is 0.029 and that of the residual variance is 0.066. Thus, the estimated residual variance in a parallel group trial is 0.095, the sum of these two variance components. The expected standard error of the treatment difference in a parallel group trial with 34 patients per group (giving the comparable number of treatment periods to the cross-over trial) is

$$\sqrt{\frac{0.095}{34} + \frac{0.095}{34}} = 0.075.$$

This compares with the comparable standard error of 0.065 in the data analysed. To achieve a similar standard error would require 45 patients in each arm of a parallel group study. There is therefore a trade-off to be made between the advantages of each design. The advantages of the parallel group design are

- six weeks in the trial per patient compared with 15 weeks;
- simpler administration;
- simpler analysis; and
- absence of assumptions concerning carry-over.

The advantage of the cross-over design is its greater efficiency:

- 34 patients (68 treatment periods) versus 90 patients.

A deciding factor between the alternatives in examples such as this will often be the availability of an adequate number of patients for the parallel group study.

7.11   GENERAL POINTS

This chapter has demonstrated that random effects models can have advantages over fixed effects models in the context of cross-over trials. In balanced situations, with normally distributed data, the results of both analyses will generally be identical. In unbalanced situations, however, the random effects models will lead to smaller standard errors of the estimates of treatment differences. If the degree of imbalance is slight (e.g. few missing observations in a balanced design) and/or if the patient variance component is large compared with the residual variance component, this reduction in the size of the standard error will be modest. It should be remembered though that some clinical trials are very expensive to conduct, and even a small gain in statistical efficiency may be equivalent to the recruitment of one or two additional costly patients.

It is tempting, therefore, to recommend routine use of the random effects model. There are situations, however, where the methods may not be sufficiently robust. This is of particular concern when we are dealing with non-normal data. The methods are based on asymptotic theory, and we are not aware of sufficient research to quantify the biases which may occur with small samples. The capability of the random effects model to summarise treatment effects on binary data in the form of odds ratios is an attractive feature, but at the present time it is perhaps prudent to exercise caution in its use if sample sizes are fairly small.

In the case of normally distributed data, there is somewhat greater experience, and the methods may be used with more confidence, although there are circumstances where we would still urge caution. Although cross-over designs are sometimes used in trials with a relatively large sample size (particularly in Phase III trials), they are also widely used in Phase I and Phase II trials, where typically there will be a small number of subjects. In the latter situation, we must be aware that the estimates of the variance components are likely to be imprecise, and the asymptotic theory on which the mixed model approach is based may not hold. In particular, the estimates of the fixed effect standard errors are based on the assumption that the variance components are known. Whenever a fixed effect is estimated from two (or more) error strata, it is known that the standard errors are biased downwards to some extent. This will occur in cross-over trials which are unbalanced by design (Section 7.5), or because of missing values (Sections 7.3 and 7.4), and in which mixed models are fitted. The size of this bias is usually small, is related to the accuracy with which the variance components have been estimated, and to the size of the variance components relative to the residual. Thus, if the number of subjects is very limited; say, less than 10 and the patient variance component is less than the residual, it may be safer to treat the subject effects as fixed.

Several of the examples presented in this chapter have demonstrated that the benefits of using a random effects model are much more pronounced in models where carry-over is being estimated. We have already alluded briefly to the fact that the use of such models may be inadvisable. Senn (1993) has expressed eloquently the arguments against the inclusion of simple carry-over terms and we find them compelling.

Senn summarises the case against adjusting for carry-over as

- The simple carry-over model has been developed without reference to phar-macological or biological models.
- It does not provide a useful approximation to reality.
- It leads to more complicated estimation procedures which are more difficult to describe and understand.
- Usually the adjusted estimators have higher variance than the unadjusted ones.
- Although the adjusted estimators will be unbiased if simple carry-over applies; in practice, if carry-over occurs, then they will be biased and it is perfectly possible that this bias will be larger than it is for unadjusted estimators.
- The most serious objection, however, is that the use of such approaches encourages the erroneous belief that the validity of estimates obtained from cross-over trials does not depend on adequate washout having taken place.

We end this section by highlighting the potential for the use of covariance pattern models in the analysis of cross-over trials. The nature of the cross-over with repeated observations on the same subject leads naturally to the consideration of a 'standard' repeated measures approach, with the covariance pattern structured by the visits. The use of this approach to structure by treatment is perhaps not immediately obvious, but in multi-treatment trials, this may well have much to offer, with subsets of similar treatments producing greater correlations than those with very different modes of action. Experimentation with such plausible covariance structures can provide a greater insight into the data. Importantly, the treatment effect estimates and their standard errors will also be more appropriate if the best covariance structure is modelled.

8

Other Applications of Mixed Models

In this chapter the use of mixed models in a variety of situations is considered. In Chapters 5, 6 and 7 we covered three different types of data structure: hierarchical; repeated measures and crossed. Designs with a combination of these features can also arise and some of these are considered in Sections 8.1–8.4. In Section 8.5 the matched case control study data is considered, and in Section 8.6 a covariance pattern model is used to allow treatment groups to have different variances in a simple between-patient study. The examples in Sections 8.7–8.13 have arisen from consultancy work and have a variety of structures.

8.1 TRIALS WITH REPEATED MEASUREMENTS WITHIN VISITS

Sometimes repeated measurements occur within visits in cross-over or repeated measures trials. For example, bioequivalance trials often record several blood or urine measurements at each visit within a cross-over design. Studies in cardiology sometimes involve exercise tests where repeated measurements are made throughout the test at each visit. Cross-over trials in asthma may involve a series of lung function measurements made after a 'challenge' designed to provoke an asthma attack.

When the data are complete at each visit, a simple approach would be to calculate summary statistics for each visit (e.g. area under the curve, maximum value or time to maximum value) and to analyse these derived variables using methods suggested for ordinary repeated measures or cross-over data (see Chapters 6 and 7). This approach has the advantage of simplicity and gives straightforward interpretation. It cannot, however, test the treatment·reps interaction (reps are repeated measurements within visits) or always overcome problems caused by missing data.

When there are missing data the use of summary statistics may not be satisfactory and a mixed model is often more appropriate. If visits and reps occur at fixed time intervals, a covariance pattern model can be used to structure the covariances by visits and by the reps. Alternatively, if the visits and/or reps occur at irregular intervals or if it is of interest to model the relationship of the response with time, then a random coefficients model can be used instead.

8.1.1   Covariance pattern models

There are several ways in which the covariance of the data can be modelled when measurements are taken across both visits and reps. In this section we will present five of the more plausible options. In models for ordinary repeated measures trials (considered in Chapter 6) the overall variance matrix, $\mathbf{V}$, had a block diagonal form with zero correlations between observations on different patients:

$$
\mathbf{V} = \begin{pmatrix}
V_1 & 0 & 0 & 0 & 0 & 0 & 0 & 0 & 0 \\
0 & V_2 & 0 & 0 & 0 & 0 & 0 & 0 & 0 \\
0 & 0 & V_3 & 0 & 0 & 0 & 0 & 0 & 0 \\
0 & 0 & 0 & V_4 & 0 & 0 & 0 & 0 & 0 \\
0 & 0 & 0 & 0 & V_5 & 0 & 0 & 0 & 0 \\
0 & 0 & 0 & 0 & 0 & V_6 & 0 & 0 & 0 \\
0 & 0 & 0 & 0 & 0 & 0 & V_7 & 0 & 0 \\
0 & 0 & 0 & 0 & 0 & 0 & 0 & V_8 & 0 \\
0 & 0 & 0 & 0 & 0 & 0 & 0 & 0 & V_9
\end{pmatrix},
$$

where the $\mathbf{V}_i$ were blocks of covariances for observations on the ith patient. We will again use this form for $\mathbf{V}$, but now there are more ways in which the $\mathbf{V}_i$ can be structured. We will illustrate a variety of possible structures assuming a dataset with three visits and three reps per visit (nine observations per patient), which leads to each $\mathbf{V}_i$ being a 9×9 sub-matrix.

Constant covariances   A very simple structure for $\mathbf{V}_i$ would assume a constant correlation between all observations on the same patient regardless of the visit or rep number such that

$$
V_i =
\begin{array}{cc}
\begin{matrix} 1 & 1 & 1 & 2 & 2 & 2 & 3 & 3 & 3 \\ 1 & 2 & 3 & 1 & 2 & 3 & 1 & 2 & 3 \end{matrix} & \begin{matrix} \textbf{Visit} \\ \textbf{Rep} \end{matrix} \\[4pt]
\begin{pmatrix}
\sigma^2 & \theta & \theta & \theta & \theta & \theta & \theta & \theta & \theta \\
\theta & \sigma^2 & \theta & \theta & \theta & \theta & \theta & \theta & \theta \\
\theta & \theta & \sigma^2 & \theta & \theta & \theta & \theta & \theta & \theta \\
\theta & \theta & \theta & \sigma^2 & \theta & \theta & \theta & \theta & \theta \\
\theta & \theta & \theta & \theta & \sigma^2 & \theta & \theta & \theta & \theta \\
\theta & \theta & \theta & \theta & \theta & \sigma^2 & \theta & \theta & \theta \\
\theta & \theta & \theta & \theta & \theta & \theta & \sigma^2 & \theta & \theta \\
\theta & \theta & \theta & \theta & \theta & \theta & \theta & \sigma^2 & \theta \\
\theta & \theta & \theta & \theta & \theta & \theta & \theta & \theta & \sigma^2
\end{pmatrix} &
\begin{matrix} 1 & 1 \\ 2 & 1 \\ 3 & 1 \\ 1 & 2 \\ 2 & 2 \\ 3 & 2 \\ 1 & 3 \\ 2 & 3 \\ 3 & 3 \end{matrix}
\end{array}
\qquad (A)
$$

where

θ = covariance between observations on same patient,

σ^2 = residual variance.

Extra covariance for observations at the same visit The pattern above is perhaps over-simplistic as it takes no account of the possibility that observations taken at the same visit are more highly correlated than observations taken at different visits. A simple way to account for this would be to parameterise $\mathbf{V}_i$ with a different covariance for observations on the same visit,

$$
\mathbf{V}_i = \begin{pmatrix}
\sigma^2 & \theta_v & \theta_v & \theta & \theta & \theta & \theta & \theta & \theta \\
\theta_v & \sigma^2 & \theta_v & \theta & \theta & \theta & \theta & \theta & \theta \\
\theta_v & \theta_v & \sigma^2 & \theta & \theta & \theta & \theta & \theta & \theta \\
& & & & & & & & \\
\theta & \theta & \theta & \sigma^2 & \theta_v & \theta_v & \theta & \theta & \theta \\
\theta & \theta & \theta & \theta_v & \sigma^2 & \theta_v & \theta & \theta & \theta \\
\theta & \theta & \theta & \theta_v & \theta_v & \sigma^2 & \theta & \theta & \theta \\
& & & & & & & & \\
\theta & \theta & \theta & \theta & \theta & \theta & \sigma^2 & \theta_v & \theta_v \\
\theta & \theta & \theta & \theta & \theta & \theta & \theta_v & \sigma^2 & \theta_v \\
\theta & \theta & \theta & \theta & \theta & \theta & \theta_v & \theta_v & \sigma^2
\end{pmatrix},
\tag{B}
$$

where

θ = covariance between observations on different visits,

θ_v = covariance between observations at the same visit,

σ^2 = residual variance.

A covariance pattern structured by visits Alternatively, it is possible that the correlation between observations is different for each pair of visits leading to

$$
\mathbf{V}_i = \begin{pmatrix}
\sigma_1^2 & \theta_{11} & \theta_{11} & \theta_{12} & \theta_{12} & \theta_{12} & \theta_{13} & \theta_{13} & \theta_{13} \\
\theta_{11} & \sigma_1^2 & \theta_{11} & \theta_{12} & \theta_{12} & \theta_{12} & \theta_{13} & \theta_{13} & \theta_{13} \\
\theta_{11} & \theta_{11} & \sigma_1^2 & \theta_{12} & \theta_{12} & \theta_{12} & \theta_{13} & \theta_{13} & \theta_{13} \\
& & & & & & & & \\
\theta_{12} & \theta_{12} & \theta_{12} & \sigma_2^2 & \theta_{22} & \theta_{22} & \theta_{23} & \theta_{23} & \theta_{23} \\
\theta_{12} & \theta_{12} & \theta_{12} & \theta_{22} & \sigma_2^2 & \theta_{22} & \theta_{23} & \theta_{23} & \theta_{23} \\
\theta_{12} & \theta_{12} & \theta_{12} & \theta_{22} & \theta_{22} & \sigma_2^2 & \theta_{23} & \theta_{23} & \theta_{23} \\
& & & & & & & & \\
\theta_{13} & \theta_{13} & \theta_{13} & \theta_{23} & \theta_{23} & \theta_{23} & \sigma_3^2 & \theta_{33} & \theta_{33} \\
\theta_{13} & \theta_{13} & \theta_{13} & \theta_{23} & \theta_{23} & \theta_{23} & \theta_{33} & \sigma_3^2 & \theta_{33} \\
\theta_{13} & \theta_{13} & \theta_{13} & \theta_{23} & \theta_{23} & \theta_{23} & \theta_{33} & \theta_{33} & \sigma_3^2
\end{pmatrix},
\tag{C1}
$$

where

θ_{ij} = covariance between observations at visits i and j,

σ_i^2 = residual variance at visit i (this may be parameterised as $\theta_{ii} + \sigma^2$).

This matrix, in fact, has a general covariance pattern structured by visits.

Another alternative would be to use a different covariance pattern for the correlations between visits. For example, by using a Toeplitz pattern the correlation between observations will depend on the separation of the visits and has the form

$$
V_i = \begin{pmatrix}
\sigma^2 & \theta_1 & \theta_1 & \theta_2 & \theta_2 & \theta_2 & \theta_3 & \theta_3 & \theta_3 \\
\theta_1 & \sigma^2 & \theta_1 & \theta_2 & \theta_2 & \theta_2 & \theta_3 & \theta_3 & \theta_3 \\
\theta_1 & \theta_1 & \sigma^2 & \theta_2 & \theta_2 & \theta_2 & \theta_3 & \theta_3 & \theta_3 \\
\theta_2 & \theta_2 & \theta_2 & \sigma^2 & \theta_1 & \theta_1 & \theta_2 & \theta_2 & \theta_2 \\
\theta_2 & \theta_2 & \theta_2 & \theta_1 & \sigma^2 & \theta_1 & \theta_2 & \theta_2 & \theta_2 \\
\theta_2 & \theta_2 & \theta_2 & \theta_1 & \theta_1 & \sigma^2 & \theta_2 & \theta_2 & \theta_2 \\
\theta_3 & \theta_3 & \theta_3 & \theta_2 & \theta_2 & \theta_2 & \sigma^2 & \theta_1 & \theta_1 \\
\theta_3 & \theta_3 & \theta_3 & \theta_2 & \theta_2 & \theta_2 & \theta_1 & \sigma^2 & \theta_1 \\
\theta_3 & \theta_3 & \theta_3 & \theta_2 & \theta_2 & \theta_2 & \theta_1 & \theta_1 & \sigma^2
\end{pmatrix}, \tag{C2}
$$

where

$\theta_i =$ covariance between observations separated by one visit,

$\sigma^2 =$ residual variance at visit.

A covariance pattern structured by reps   It is also possible that correlation between observations at the same visit differs depending on the rep number. A structure assuming constant correlation between observations on different visits (as in (B)) but a different correlation for each pair of reps at the same visit is

$$
V_i = \begin{pmatrix}
\sigma_1^2 & \theta_{12} & \theta_{13} & \theta & \theta & \theta & \theta & \theta & \theta \\
\theta_{12} & \sigma_2^2 & \theta_{23} & \theta & \theta & \theta & \theta & \theta & \theta \\
\theta_{13} & \theta_{23} & \sigma_3^2 & \theta & \theta & \theta & \theta & \theta & \theta \\
\theta & \theta & \theta & \sigma_1^2 & \theta_{12} & \theta_{13} & \theta & \theta & \theta \\
\theta & \theta & \theta & \theta_{12} & \sigma_2^2 & \theta_{23} & \theta & \theta & \theta \\
\theta & \theta & \theta & \theta_{13} & \theta_{23} & \sigma_3^2 & \theta & \theta & \theta \\
\theta & \theta & \theta & \theta & \theta & \theta & \sigma_1^2 & \theta_{12} & \theta_{13} \\
\theta & \theta & \theta & \theta & \theta & \theta & \theta_{12} & \sigma_2^2 & \theta_{23} \\
\theta & \theta & \theta & \theta & \theta & \theta & \theta_{13} & \theta_{23} & \sigma_3^2
\end{pmatrix}, \tag{D1}
$$

where

$\theta =$ covariance between observations on different visits,

$\theta_{ij} =$ covariance between observations on reps i and j (at the same visit),

$\sigma_i^2 =$ residual variance at visit i (this may be parameterised as $\theta_{ii} + \sigma^2$).

This matrix, in fact, has a general covariance pattern structured by reps (within visits).

Alternatively, a different pattern from the general pattern could be considered. For example, a first-order autoregressive pattern allowing the correlation between observations to decrease exponentially depending on the separation of the reps would have the form

$$
\mathbf{V}_i =
\begin{pmatrix}
\sigma^2 & \rho\sigma^2 & \rho^2\sigma^2 & \theta & \theta & \theta & \theta & \theta & \theta \\
\rho\sigma^2 & \sigma^2 & \rho\sigma^2 & \theta & \theta & \theta & \theta & \theta & \theta \\
\rho^2\sigma^2 & \rho\sigma^2 & \sigma^2 & \theta & \theta & \theta & \theta & \theta & \theta \\
 & & & & & & & & \\
\theta & \theta & \theta & \sigma^2 & \rho\sigma^2 & \rho^2\sigma^2 & \theta & \theta & \theta \\
\theta & \theta & \theta & \rho\sigma^2 & \sigma^2 & \rho\sigma^2 & \theta & \theta & \theta \\
\theta & \theta & \theta & \rho^2\sigma^2 & \rho\sigma^2 & \sigma^2 & \theta & \theta & \theta \\
 & & & & & & & & \\
\theta & \theta & \theta & \theta & \theta & \theta & \sigma^2 & \rho\sigma^2 & \rho^2\sigma^2 \\
\theta & \theta & \theta & \theta & \theta & \theta & \rho\sigma^2 & \sigma^2 & \rho\sigma^2 \\
\theta & \theta & \theta & \theta & \theta & \theta & \rho^2\sigma^2 & \rho\sigma^2 & \sigma^2
\end{pmatrix},
\tag{D2}
$$

where

θ = covariance between observations on different visits,

$\rho^{|i-j|}$ = correlation between observations on reps i and j,

σ^2 = residual variance.

Extra covariance for observations on the same reps It is also possible that there is additional correlation between observations on the same reps. Adding this feature to structure (B) above we obtain the structure

$$
\mathbf{V}_i =
\begin{pmatrix}
\sigma^2 & \theta_v & \theta_v & \theta_r & \theta & \theta & \theta_r & \theta & \theta \\
\theta_v & \sigma^2 & \theta_v & \theta & \theta_r & \theta & \theta & \theta_r & \theta \\
\theta_v & \theta_v & \sigma^2 & \theta & \theta & \theta_r & \theta & \theta & \theta_r \\
 & & & & & & & & \\
\theta_r & \theta & \theta & \sigma^2 & \theta_v & \theta_v & \theta_r & \theta & \theta \\
\theta & \theta_r & \theta & \theta_v & \sigma^2 & \theta_v & \theta & \theta_r & \theta \\
\theta & \theta & \theta_r & \theta_v & \theta_v & \sigma^2 & \theta & \theta & \theta_r \\
 & & & & & & & & \\
\theta_r & \theta & \theta & \theta_r & \theta & \theta & \sigma^2 & \theta_v & \theta_v \\
\theta & \theta_r & \theta & \theta & \theta_r & \theta & \theta_v & \sigma^2 & \theta_v \\
\theta & \theta & \theta_r & \theta & \theta & \theta_r & \theta_v & \theta_v & \sigma^2
\end{pmatrix},
\tag{E1}
$$

where

θ = covariance between observations at different visits and reps,

θ_v = covariance between observations at the same visit but different rep,

θ_r = covariance between observations at the same rep but different visits,

σ^2 = residual variance.

We can also make this structure slightly more complex by assuming that each pair of reps has a different correlation:

$$
V_i = \begin{pmatrix}
\sigma_{11}^2 & \theta_v & \theta_v & \theta_{12} & \theta & \theta & \theta_{13} & \theta & \theta \\
\theta_v & \sigma_{11}^2 & \theta_v & \theta & \theta_{12} & \theta & \theta & \theta_{13} & \theta \\
\theta_v & \theta_v & \sigma_{11}^2 & \theta & \theta & \theta_{12} & \theta & \theta & \theta_{13} \\
\\
\theta_{12} & \theta & \theta & \sigma_{22}^2 & \theta_v & \theta_v & \theta_{23} & \theta & \theta \\
\theta & \theta_{12} & \theta & \theta_v & \sigma_{22}^2 & \theta_v & \theta & \theta_{23} & \theta \\
\theta & \theta & \theta_{12} & \theta_v & \theta_v & \sigma_{22}^2 & \theta & \theta & \theta_{23} \\
\\
\theta_{13} & \theta & \theta & \theta_{23} & \theta & \theta & \sigma_{33}^2 & \theta_v & \theta_v \\
\theta & \theta_{13} & \theta & \theta & \theta_{23} & \theta & \theta_v & \sigma_{33}^2 & \theta_v \\
\theta & \theta & \theta_{13} & \theta & \theta & \theta_{23} & \theta_v & \theta_v & \sigma_{33}^2
\end{pmatrix}, \tag{E2}
$$

where

θ = covariance between observations at different visits and reps,

θ_v = covariance between observations at the same visit but at different reps,

θ_{ij} = covariance between observations at the same rep at visits i and j,

σ_{ii}^2 = residual variance at visit i.

Choosing the covariance pattern

It is clear from the options given above that there is a huge amount of flexibility in choosing how to define the covariance structure of the data. If the covariance structure itself is of particular interest it may be worthwhile to experiment with several different structures, building up a model by starting with a very simple structure such as (A) and then using likelihood ratio tests (see Section 6.2.2) to determine whether more complex structures lead to significant improvements. However, if interest lies only with, say, comparing two treatment groups or their interaction with visit or rep, estimates of means and standard errors from a fairly simple structure such as (B), which allows for correlation between observations on the same visit, are likely to differ little from those obtained using a more complex structure.

We note that the range of covariance structures available may be limited by the software and care is sometimes needed to ensure that covariance parameters are not confounded.

8.1.2   Example

This data was a three-period, cross-over trial taken from Jones and Kenward (1989) to compare the effects of three treatments on systolic blood pressure. Treatments A and B are 20 mg and 40 mg doses of an active drug and treatment C is a placebo. There were 12 patients and 10 measurements were made at each visit. These were taken at 30 and 15 minutes before treatment, and at 15, 30, 45, 60, 75, 90, 120 and 240 minutes after treatment. We assume that the objective is to assess whether there are any post-treatment differences between treatments and to test whether these differences are constant over the reps. Since the data are complete, calculating summary statistics (e.g. AUC, minimum or maximum value, time to maximum value) and analysing them as cross-over data would form a simple strategy, although the treatment·rep interaction could not then be tested. However, here, for illustration, we will analyse the raw data using covariance pattern models. Measurements at 15 minutes pre-treatment will be taken as baseline values.

Although the main interest in this example lies with the comparison of treatments, for illustration we will use likelihood ratio tests to determine an appropriate covariance structure for the data. The models tested and the resulting values of $-2\log(L)$ are shown in Table 8.1. The covariance structures relate to the example structures defined above. Initially, Model 1 was used to test treatment·visit, treatment·rep and carry-over effects to determine whether they could be omitted from the models. None of these effects was significant; however, treatment·rep effects were retained in the models so that mean treatment·rep profiles could be produced. The fixed effects included in the models were therefore baseline (15 min pre-treatment), treatment, visit, rep and treatment·rep effects.

Models 1 and 2 are compared to establish whether there is extra correlation between observations on the same rep. The likelihood ratio statistic is $1879.79 - 1876.84 = 2.95\ \chi_1^2$ $(p = 0.09)$. Therefore, we do not have strong statistical evidence that there is any additional correlation across the reps and structures taking account of this are not considered further.

Model 3, with a general structure across visits, is compared with Model 1 to determine whether different correlations are justified for each pair of visits. On comparison with Model 1 we obtain $\chi_4^2 = 1.70$ $(p = 0.19)$, which indicates no significant improvement over Model 1, which uses the same correlation between all visits.

Model 4 tests whether the correlation between observations within the same visit decreases as reps become more widely separated. The structure fitted is similar to (D2) but not quite the same due to the limitations of PROC MIXED. Its form will be shown in the results. The same number of parameters are used as in Model 2 and therefore the log likelihoods can be compared directly. Model 4 has a higher

Table 8.1 Values of $-2\log(L)$ for all models.

Model	Covariance structure	$-2\log(L)$	Parameters
1	B	1879.79	3
2	E1	1876.84	4
3	C1	1878.09	7
4	D2	1866.98	3
5	D2, by treatment	1861.68	9

likelihood and we therefore conclude that the first-order autoregressive structure is the more appropriate and that correlation does decrease with the separation of the reps.

Model 5 is similar to Model 4 but fits a separate covariance matrix for each treatment. The likelihood ratio statistic on comparison with Model 4 is $\chi_6^2 = 5.30$ ($p = 0.51$). This is non-significant, so there is no evidence that the covariances differ between treatments and we can choose to base our conclusions on results from Model 4.

Results from Model 4

The exact form of structure (D2) could not be fitted using PROC MIXED because the REPEATED statement can only be used to fit a single covariance pattern. It was necessary to use a RANDOM statement to give the constant covariance between observations on the same patient. The covariance parameter estimates obtained were

covariance between observations on same patient = 46.100
autoregressive correlation coefficient   = 0.323
residual   = 55.600

Thus, the variance matrix block for observations on the same patient, V_i has the form

$$
V_i =
\begin{pmatrix}
46 & 46 & 46 & 46 & 46 & 46 & 46 & 46 & 46 \\
46 & 46 & 46 & 46 & 46 & 46 & 46 & 46 & 46 \\
46 & 46 & 46 & 46 & 46 & 46 & 46 & 46 & 46 \\
 & & & & & & & & \\
46 & 46 & 46 & 46 & 46 & 46 & 46 & 46 & 46 \\
46 & 46 & 46 & 46 & 46 & 46 & 46 & 46 & 46 \\
46 & 46 & 46 & 46 & 46 & 46 & 46 & 46 & 46 \\
 & & & & & & & & \\
46 & 46 & 46 & 46 & 46 & 46 & 46 & 46 & 46 \\
46 & 46 & 46 & 46 & 46 & 46 & 46 & 46 & 46 \\
46 & 46 & 46 & 46 & 46 & 46 & 46 & 46 & 46
\end{pmatrix}
$$

$$
+ 56
\begin{pmatrix}
1 & 0.32 & 0.32^2 & 0 & 0 & 0 & 0 & 0 & 0 \\
0.32 & 1 & 0.32 & 0 & 0 & 0 & 0 & 0 & 0 \\
0.32^2 & 0.32 & 1 & 0 & 0 & 0 & 0 & 0 & 0 \\
 & & & & & & & & \\
0 & 0 & 0 & 1 & 0.32 & 0.32^2 & 0 & 0 & 0 \\
0 & 0 & 0 & 0.32 & 1 & 0.32 & 0 & 0 & 0 \\
0 & 0 & 0 & 0.32^2 & 0.32 & 1 & 0 & 0 & 0 \\
 & & & & & & & & \\
0 & 0 & 0 & 0 & 0 & 0 & 1 & 0.32 & 0.32^2 \\
0 & 0 & 0 & 0 & 0 & 0 & 0.32 & 1 & 0.32 \\
0 & 0 & 0 & 0 & 0 & 0 & 0.32^2 & 0.32 & 1
\end{pmatrix}.
$$

Note that the diagonal blocks are in fact 8×8 with autoregressive correlations ranging between 0.32 and 0.32^8. However, due to space limitations, their general form is illustrated here using 3×3 blocks.

The results for the treatment effects were

Effect	Difference (SE)	*t* test DF	*p*-value
A − B	−3.18 (1.46)	261	0.03
A − C	3.05 (1.61)	234	0.06
B − C	6.24 (1.52)	261	0.0001

Thus, the 40 mg treatment (B) produces significantly higher systolic blood pressure than the 20 mg treatment (A) and the placebo (C). The treatment·rep interaction was non-significant ($p = 0.35$) and therefore we can be reasonably confident in reporting treatment effects over all reps. The standard errors are 'model-based' and are thus calculated from the covariance pattern parameters estimated by the model. However, if we had assumed a simple pattern (e.g. Model A or B) without testing more complex patterns, it might have been preferable to use the 'empirical' estimates which would have taken more account of the observed covariance of the data.

SAS code and output

Variables

```
pat   = patient,
visit = visit,
treat = treatment (A = 20 mg dose, B = 40 mg dose, C = placebo),
rep   = rep,
sbp   = post-treatment systolic blood pressure (mmHg),
pre   = pre-treatment systolic blood pressure (mmHg),
time  = time since dose (minutes).
```

The following statements were used in each model to specify the fixed effects. The RANDOM and REPEATED statements to specify the covariance pattern are then shown for each model. Note that fitting effects as random leads to compound symmetry covariance structures within each effect specified. For example, in Model 1 a compound symmetry structure is fitted within patients and within patient·visits. The full code and output is given for Model 4 on which the conclusions were based.

```
PROC MIXED; CLASS pat visit treat rep;
MODEL sbp=pre treat visit rep treat*rep/ DDFM=SATTERTH;

Model 1: RANDOM pat pat*visit;
Model 2: RANDOM pat pat*visit pat*rep;
```

Model 3: RANDOM visit/ SUB=pat TYPE=un;
Model 4: RANDOM pat;
        REPEATED rep/ SUB=pat*visit TYPE=AR(1) R;
Model 5: RANDOM pat/ GROUP=treat;
        REPEATED rep/SUB=pat*visit TYPE=AR(1) R GROUP=treat;

Model 4 code and output

```
PROC MIXED; CLASS pat visit treat rep;
MODEL sbp=pre treat visit rep treat*rep/ DDFM=SATTERTH;
RANDOM pat;
REPEATED rep/ SUB=pat*visit TYPE=AR(1) R;
LSMEANS treat/ PDIFF DIFF;
LSMEANS treat*rep;
```

REML Estimation Iteration History

Iteration	Evaluations	Objective	Criterion
0	1	1525.2552760	
1	2	1387.2896067	0.00000136
2	1	1387.2886516	0.00000000

Convergence criteria met.

R Matrix for PAT*VISIT 1 1

Row	COL1	COL2	COL3	COL4	COL5
1	55.64300023	17.98436791	5.81272555	1.87873038	0.60722424
2	17.98436791	55.64300023	17.98436791	5.81272555	1.87873038
3	5.81272555	17.98436791	55.64300023	17.98436791	5.81272555
4	1.87873038	5.81272555	17.98436791	55.64300023	17.98436791
5	0.60722424	1.87873038	5.81272555	17.98436791	55.64300023
6	0.19626088	0.60722424	1.87873038	5.81272555	17.98436791
7	0.06343346	0.19626088	0.60722424	1.87873038	5.81272555
8	0.02050232	0.06343346	0.19626088	0.60722424	1.87873038

R Matrix for PAT*VISIT 1 1

COL6	COL7	COL8
0.19626088	0.06343346	0.02050232
0.60722424	0.19626088	0.06343346
1.87873038	0.60722424	0.19626088
5.81272555	1.87873038	0.60722424
17.98436791	5.81272555	1.87873038
55.64300023	17.98436791	5.81272555
17.98436791	55.64300023	17.98436791
5.81272555	17.98436791	55.64300023

Covariances in the **R** matrix are calculated as 55.6×0.323^d, where $d =$ separation of reps.

Covariance Parameter Estimates (REML)

Cov Par m	Subject	Estimate
PAT		46.05975898
AR(1)	PAT*VISIT	0.32320989
Residual		55.64300023

```
                    Model Fitting Information for SBP
              Description                         Value
              Observations                      288.0000
              Res Log Likelihood                -933.487
              Akaike's Information Criterion     -936.487
              Schwarz's Bayesian Criterion      -941.834
              -2 Res Log Likelihood             1866.975
              Covariance Parameter Estimates    (REML)
```

```
                      Tests of Fixed Effects
              Source      NDF    DDF   Type III F    Pr>F
              PRE           1    250      12.02     0.0006
              TREAT         2    250       8.57     0.0003
              VISIT         2    250       3.44     0.0336
              REP           7    250       3.29     0.0023
              TREAT*REP    14    250       1.11     0.3511
```

```
                          Least Squares Means
Level                    LSMEAN       Std Error    DDF      T      Pr>|T|
TREAT 1              105.56846134    2.23012358    3.79   47.34    0.0001
TREAT 2              108.75042943    2.20836254    3.71   49.24    0.0001
TREAT 3              102.51444256    2.24436461    3.99   45.68    0.0001
TREAT*REP 1 2       103.63096134    2.92905863    8.77   35.38    0.0001
TREAT*REP 1 3       105.29762801    2.92905863    9.53   35.95    0.0001
TREAT*REP 1 4       101.54762801    2.92905863    9.52   34.67    0.0001
TREAT*REP 1 5       105.46429468    2.92905863    9.52   36.01    0.0001
TREAT*REP 1 6       108.79762801    2.92905863    9.52   37.14    0.0001
TREAT*REP 1 7       106.96429468    2.92905863    9.52   36.52    0.0001
TREAT*REP 1 8       106.29762801    2.92905863    9.53   36.29    0.0001
TREAT*REP 1 9       106.54762801    2.92905863    8.77   36.38    0.0001
TREAT*REP 2 2       100.94834610    2.91252440    8.71   34.66    0.0001
TREAT*REP 2 3       103.61501277    2.91252440    9.47   35.58    0.0001
TREAT*REP 2 4       107.19834610    2.91252440    9.47   36.81    0.0001
TREAT*REP 2 5       110.19834610    2.91252440    9.47   37.84    0.0001
TREAT*REP 2 6       111.69834610    2.91252440    9.47   38.35    0.0001
TREAT*REP 2 7       114.44834610    2.91252440    9.47   39.30    0.0001
TREAT*REP 2 8       113.03167943    2.91252440    9.47   38.81    0.0001
TREAT*REP 2 9       108.86501277    2.91252440    8.71   37.38    0.0001
TREAT*REP 3 2       101.58735922    2.93991595    9.17   34.55    0.0001
TREAT*REP 3 3        99.92069256    2.93991595    9.89   33.99    0.0001
TREAT*REP 3 4        98.92069256    2.93991595    9.89   33.65    0.0001
TREAT*REP 3 5       102.58735922    2.93991595    9.89   34.89    0.0001
TREAT*REP 3 6       104.08735922    2.93991595    9.89   35.40    0.0001
TREAT*REP 3 7       103.33735922    2.93991595    9.89   35.15    0.0001
TREAT*REP 3 8       103.75402589    2.93991595    9.89   35.29    0.0001
TREAT*REP 3 9       105.92069256    2.93991595    9.17   36.03    0.0001
```

```
                    Differences of Least Squares Means
Level 1    Level 2    Difference     Std Error   DDF      T     Pr>|T|
TREAT 1    TREAT 2    -3.18196809   1.45517590   261   -2.19   0.0297
TREAT 1    TREAT 3     3.05401878   1.61197367   234    1.89   0.0594
TREAT 2    TREAT 3     6.23598688   1.51944096   261    4.10   0.0001
```

8.1.3   Random coefficients models

Random coefficients models can also be utilised for analysing data with repeated measurements made within visits. This type of model would be more appropriate if greatest interest centred on explaining the relationship of a measurement with time. However, time is now measured across both visits and across reps and it is therefore possible to consider fitting slopes across both of these time scales. Alternatively, slopes can be fitted across just one time scale, i.e. against either visit time or rep time. In many applications one of these simpler models will be of greatest relevance. We will now describe three possible models. Model 1 fits slopes against visit time only, Models 2(a) and 2(b) fit slopes against rep time only, and Model 3 fits slopes against both visit time and rep time. The SAS code required to fit each model will be supplied following Section 8.1.4.

Model 1 — Modelling response against visit time

In some examples only the relationship of the response with visit time may be of interest. For example, this might be the case for a trial where patients make varying numbers of visits to a hospital clinic at unevenly spaced intervals during the course of a treatment with only two replicates of measurements per visit. In this situation, a slope against visit time can be fitted and reps can be taken as categorical fixed effects. To allow the slopes to vary randomly between the patients, patient and patient·tvisit (tvisit = visit time) are fitted as random coefficients. The patient coefficients will represent the intercepts and the patient·tvisit coefficients the slopes of separate regression lines for each patient. The average slopes will be determined by the fixed effects tvisit and treatment·tvisit. The random coefficients cause the regression lines (each made up of an intercept and slope) to be compared between treatments against an appropriate background of between-patient variability. It is unlikely that the estimation of slopes against visit will be of interest in a cross-over trial so this model will be of greatest use for analysing repeated measures designs.

Random coefficients	Fixed effects
Patient	Baseline
Patient·tvisit	Treatment
	Tvisit
	Rep
	Treatment·tvisit
	Treatment·rep

Note that tvisit represents the actual times of the visits (e.g. in weeks) rather than the visit number. However, in clinical trials with evenly spaced visits, use of the visit number will often suffice as the visit time.

Model 2 — Modelling response against rep time

In some applications the slope against against rep time is of greatest interest. To allow the slopes to vary randomly between the patients, patient and patient·trep (trep = rep time) are fitted as random coefficients. Additionally, to allow the slopes to vary randomly within the patients, patient·visit and patient·visit·trep (trep = rep time) are also fitted as random coefficients. Thus a separate regression line is fitted for each patient at each visit. In repeated measures trials slopes will be appropriately compared between treatments against a background of between-patient variability, and in crossover trials against a background of within-patient variability. Slope effects can also be compared between visits and this is always against a background of within-patient variability.

Random coefficients	Fixed effects
Patient	Baseline
Patient·trep	Treatment
Patient·visit	Visit
Patient·visit·trep	Trep
	Treatment·visit
	Visit·trep
	Treatment·trep

Visit effects are still included in the model but as categorical fixed effects. In a cross-over study fixed carryover effects can also be included if required. Note that the trep represents the actual times of the reps (e.g. in minutes) rather than the rep number (denoted by rep in Model 1).

Model 3 — Modelling response against rep time and visit time

This model can be used when both the slopes against visit time and against rep time are of interest. Separate slopes are fitted across visit time and rep time. To allow slopes to be compared between treatments against an appropriate background of between-patient variability, patient, patient·trep and patient·tvisit are fitted as random coefficients. As with Model 2(a), this model is appropriate for comparing slopes between treatments in repeated measures designs only.

We do not suggest a corresponding model for cross-over trials because it is unlikely that the estimation of slopes against visit will be sensible in a cross-over trial.

Random coefficients	Fixed effects
Patient	Baseline
Patient·tvisit	Treatment
Patient·trep	Tvisit
	Trep
	Treatment·tvisit
	Treatment·trep

Non-linear models

We have in the above section only considered linear relationships with time. However, the models can also be adapted to fit non-linear relationships if required as discussed in Section 6.5.

Choice of model

The models we have introduced each fit different fixed effects so cannot be compared using likelihood ratio tests. However, a statistical comparison of the models is not really relevant here since the choice of model should depend on which model best answers the required questions. For this reason, statistical comparisons between random coefficients models and covariance pattern models are also not helpful, since the two approaches are designed to answer different questions.

8.1.4   Example: Random coefficients models

We again consider the three-period, cross-over trial introduced in Section 8.1.2. A random coefficients model might be chosen instead of a covariance pattern model if interest were principally focused on modelling the relationship of systolic blood pressure across either visits or reps. In this trial it is likely that comparing the rep slopes between treatments will be of greatest interest and therefore only Model 2 will be considered. This model examines whether the rates of change of systolic blood pressure during each phase of the cross-over differs between treatments. It was found that taking the log of rep time gave a more linear relationship with systolic blood pressure. Thus, log(rep time) is used in place of rep time in the model.

The covariance parameter estimates are shown in Table 8.2. The positive diagonal terms in the two covariance matrices indicate that there is additional random variation in the regression lines both between patients and within patients over the set of visits (i.e. the regression lines differ to a greater extent than would

Table 8.2 Covariance parameter estimates.

	Covariance parameters	
Covariances for patient and patient·trep coefficients	$\begin{pmatrix} 63.6 & \\ -8.7 & 2.96 \end{pmatrix}$	
Covariances for patient· visit and patient·visit·trep coefficients	$\begin{pmatrix} 182.5 & \\ -33.8 & 6.48 \end{pmatrix}$	
Residual	42.9	

Note: Here, trep = log (rep time).

Table 8.3 Rep slope estimates (standard errors).

Rep slopes		
A	1.41	(1.22)
B	4.11	(1.22)
C	2.04	(1.22)
Treatment differences in rep slope		
A − B	−2.70	(1.58)
A − C	−0.63	(1.58)
B − C	2.07	(1.58)

be expected as a result of the residual variation). The negative covariance term is not surprising because of the expected negative correlation between estimates of slopes and intercepts in regression analysis. It is hard to assess the relative sizes of the covariance parameters, since the patient·trep and patient·visit·trep coefficients involve continuous effects. However, a comparison of the patient, patient·visit and residual variances indicates that most of the variation in the data is occurring at the patient·visit level.

The results for the treatment and treatment·rep effects are shown in Table 8.3. The rep slope estimates are positive for each treatment showing an increase in blood pressure following the administration of treatment. The overall test of the treatment.slope interaction was not significant ($p = 0.22$), and neither were any of the pairwise comparisons of rep slopes between treatments. Note that identical standard errors are obtained here for each of the rep slopes because the data were complete.

SAS code and output

Variables

```
pat    = patient,
sbp    = post-treatment systolic blood pressure (mmHg),
pre    = pre-treatment systolic blood pressure (mmHg),
```

treat   = treatment (A=20 mg dose, B=40 mg dose, C=placebo),
visit   = visit number,
tvisit = visit time,
rep   = rep,
trep   = log(time since dose in minutes).

The SAS code and output is shown for Model 2, which was used to analyse this example. Following this, the code required for the other models described in Section 8.1.3 (Models 1 and 3) is shown.

```
PROC MIXED; CLASS pat treat visit;
TITLE 'MODEL 2B. SLOPES ACROSS REP TIME - RANDOM SLOPES FOR
PATIENTS AND PATIENT.VISITS';
MODEL sbp=pre treat visit trep treat*visit treat*trep/
DDFM=SATTERTH;
RANDOM int trep / SUB=pat TYPE=UN;
RANDOM int trep / SUB=pat*visit TYPE=UN;
LSMEANS treat/ DIFF PDIFF;
ESTIMATE 'A, REP SLOPE' trep 1 treat*trep 1 0 0;
ESTIMATE 'B, REP SLOPE' trep 1 treat*trep 0 1 0;
ESTIMATE 'C, REP SLOPE' trep 1 treat*trep 0 0 1;
ESTIMATE 'A-B, REP SLOPE' treat*trep 1 -1 0;
ESTIMATE 'A-C, REP SLOPE' treat*trep 1 0 -1;
ESTIMATE 'B-C, REP SLOPE' treat*trep 0 1 -1;
```

The ESTIMATE statements are required to estimate the slope effects by treatment and their differences here. This is because the LSMEANS statement can only be used with CLASS variables.

The MIXED Procedure

Class Level Information

Class	Levels	Values
PAT	12	1 2 3 4 5 6 7 8 9 10 11 12
TREAT	3	A B C
VISIT	3	1 2 3

REML Estimation Iteration History

Iteration	Evaluations	Objective	Criterion
0	1	1563.3973083	
1	2	1435.0630982	0.00001352
2	1	1435.0529938	0.00000006
3	1	1435.0529499	0.00000000

Convergence criteria met.

Covariance Parameter Estimates (REML)

Cov Parm	Subject	Estimate
UN(1,1)	PAT	63.63064236
UN(2,1)	PAT	-8.66755753
UN(2,2)	PAT	2.96457217
UN(1,1)	PAT*VISIT	182.46284857

```
UN(2,1)     PAT*VISIT     -33.84496969
UN(2,2)     PAT*VISIT       6.48423735
Residual                   42.90008785
```

UN(1,1) and UN(2,2) are the variance component estimates for the patient and patient·trep random coefficients, and UN(2,1) is the covariance between the two random coefficients. Likewise, the second UN(1,1) and UN(2,2) are the variance component estimates for the patient·visit and patient·visit·trep random coefficients, and UN(2,1) is their covariance.

```
              Model Fitting Information for SBP
        Description                          Value

        Observations                       288.0000
        Res Log Likelihood                 -970.235
        Akaike's Information Criterion     -977.235
        Schwarz's Bayesian Criterion       -989.893
        -2 Res Log Likelihood             1940.469
        Null Model LRT Chi-Square           128.3444
        Null Model LRT DF                     6.0000
        Null Model LRT P-Value                0.0000
```

```
                  Tests of Fixed Effects
        Source      NDF     DDF    Type III F     Pr>F

        PRE          1     25.9       10.29       0.0035
        TREAT        2     21.1        0.58       0.5676
        VISIT        2     16.5        3.22       0.0662
```

```
                  Tests of Fixed Effects
        Source      NDF     DDF    Type III F     Pr>F

        TREP          1      11        9.61       0.0101
        TREAT*VISIT   4     18.4       0.58       0.6796
        TREP*TREAT    2      22        1.61       0.2234
```

```
                  ESTIMATE Statement Results
  Parameter          Estimate     Std Error      DF      t      Pr>|t|

  A, REP SLOPE       1.41140499   1.22072925    31.3    1.16    0.2564
  B, REP SLOPE       4.11180728   1.22072925    31.3    3.37    0.0020
  C, REP SLOPE       2.04107228   1.22072925    31.3    1.67    0.1045
  A-B, REP SLOPE    -2.70040229   1.57678929    22     -1.71    0.1008
  A-C, REP SLOPE    -0.62966729   1.57678929    22     -0.40    0.6935
  B-C, REP SLOPE     2.07073501   1.57678929    22      1.31    0.2026
```

The first three estimates give the rep slopes corresponding to each of the treatment groups. The last three estimates are of the differences between the slopes for each treatment and are the slope differences shown in Table 8.3.

```
                     Least Squares Means
  Effect  TREAT      LSMEAN      Std Error      DF      t      Pr>|t|

  TREAT    A     105.41969606   2.29665699    14.5    45.90    0.0001
  TREAT    B     108.71032749   2.26491307    14.1    48.00    0.0001
  TREAT    C     102.70330978   2.31736153    14.8    44.32    0.0001
```

```
                   Differences of Least Squares Means
Effect  TREAT  _TREAT    Difference    Std Error    DF      t    Pr>|t|
TREAT   A      B        -3.29063143   1.79205310   14.9  -1.84   0.0863
TREAT   A      C         2.71638628   1.98304161   17.9   1.37   0.1877
TREAT   B      C         6.00701770   1.87030685   16.2   3.21   0.0054
```

SAS code for Model 1

```
PROC MIXED; CLASS pat time treat;
TITLE 'MODEL 1. SLOPES ACROSS VISIT TIME ONLY';
MODEL sbp=pre treat tvisit time treat*tvisit treat*time/
DDFM=SATTERTH;
RANDOM INT tvisit / SUB=pat TYPE=UN;
LSMEANS treat/ DIFF PDIFF;
ESTIMATE 'A, VISIT SLOPE' tvisit 1 treat*tvisit 1 0 0;
ESTIMATE 'B, VISIT SLOPE' tvisit 1 treat*tvisit 0 1 0;
ESTIMATE 'C, VISIT SLOPE' tvisit 1 treat*tvisit 0 0 1;
ESTIMATE 'A-B, VISIT SLOPE' treat*tvisit 1 -1 0;
ESTIMATE 'A-C, VISIT SLOPE' treat*tvisit 1 0 -1;
ESTIMATE 'B-C, VISIT SLOPE' treat*tvisit 0 1 -1;
```

SAS code for Model 3

```
PROC MIXED; CLASS pat treat;
TITLE 'MODEL 3. SLOPES ACROSS VISIT TIME AND REP TIME';
MODEL sbp=pre treat tvisit trep treat*tvisit treat*trep/
DDFM=SATTERTH;
RANDOM INT tvisit trep / SUB=pat TYPE=UN;
LSMEANS treat/ DIFF PDIFF;
ESTIMATE 'A, VISIT SLOPE' tvisit 1 treat*tvisit 1 0 0;
ESTIMATE 'B, VISIT SLOPE' tvisit 1 treat*tvisit 0 1 0;
ESTIMATE 'C, VISIT SLOPE' tvisit 1 treat*tvisit 0 0 1;
ESTIMATE 'A, REP SLOPE' trep 1 treat*trep 1 0 0;
ESTIMATE 'B, REP SLOPE' trep 1 treat*trep 0 1 0;
ESTIMATE 'C, REP SLOPE' trep 1 treat*trep 0 0 1;
ESTIMATE 'A-B, VISIT SLOPE' treat*tvisit 1 -1 0;
ESTIMATE 'A-C, VISIT SLOPE' treat*tvisit 1 0 -1;
ESTIMATE 'B-C, VISIT SLOPE' treat*tvisit 0 1 -1;
ESTIMATE 'A-B, REP SLOPE' treat*trep 1 -1 0;
ESTIMATE 'A-C, REP SLOPE' treat*trep 1 0 -1;
ESTIMATE 'B-C, REP SLOPE' treat*trep 0 1 -1;
```

8.2   MULTI-CENTRE TRIALS WITH REPEATED MEASUREMENTS

It is not uncommon for clinical trials to record measurements over several visits and also to recruit patients from several centres. The hypertension study introduced in Section 1.3 in fact has this structure. A mixed model can then be used to allow treatment effects to vary randomly across centres, and also to fit a covariance pattern for the repeated measurements. As in ordinary multi-centre studies, treatments can be allowed to vary randomly across centres by fitting centre and centre·treatment effects as random and inference is then wider and can be applied to the population of centres (see Chapter 5). The treatment standard errors will be increased compared with models omitting centre·treatment effects or fitting them as fixed, whenever the centre·treatment variance component is positive. Time and treatment·time effects can also be allowed to vary randomly between centres by fitting centre·time and centre·treatment·time effects as random. Alternatively, if a 'local' interpretation is required, interactions with centre effects can be omitted or taken as fixed. If the interaction terms involving centre are omitted, retaining centre effects as random in the model will still allow any additional information on treatments available from the centre stratum to be recovered. Covariance patterns for the repeated observations occurring on the same patients can be constructed in the same way as described for ordinary repeated measures trials and compared using likelihood ratio tests (see Section 6.2.2).

8.2.1 Example: Multi-centre hypertension trial

The multi-centre hypertension trial introduced in Section 1.3 had visits at 2, 4, 6 and 8 weeks post treatment. We will initially analyse the primary endpoint, DBP using the following mixed model:

Random effects	Fixed effects
Centre	Treatment
Centre·treatment	Time
Centre·time	Treatment·time
Centre·treatment·time	Baseline
Patient	

Note that a compound symmetry structure for the repeated measurements is obtained here by fitting patients as random.

The resulting variance component estimates were

Effect	Variance component
Centre	4.4
Centre·treat	0.5
Visit·centre	1.0
Visit·centre·treat	0.5
Patient	35.0
Residual	34.3

These values indicate that most of the variation occurs at the patient and residual level. The very small variance components for treatment·centre, visit·centre and visit·centre·treatment effects will influence the results very little and could be removed from the analysis in order to simplify the model. The small centre·treatment variance component indicates that treatments are varying hardly at all across the centres. Thus, results can still be related to the population of centres with some confidence, even when centre·treatment effects are removed. However, retaining the centre effects as random allows additional information on treatments to be recovered from the centre error stratum. The fixed treatment·time interaction was not significant and we will omit it from the model.

8.2.2   Covariance pattern models

We can also consider fitting covariance patterns to the repeated measurements, although in many situations a constant compound symmetry covariance (achieved above by fitting patients as random) will be sufficient. As in ordinary repeated measures trials, models using different covariance patterns can be compared using likelihood ratio tests (see Section 6.2.2). The six models below were fitted to the hypertension data. In each model, centre effects are fitted as random, and treatment, time and baseline effects are fitted as fixed. The other terms which were considered in the initial model but found to be negligible have been omitted. The covariance parameters estimated in each model are shown in Table 8.4 and values of $-2\log(L)$ in Table 8.5.

The comparison of the covariance pattern models is quite similar to that presented in Section 6.3, where the effects of centre were not considered. Model 2 has the same number of covariance parameters as Model 1 and a direct comparison of their likelihoods can be made. Model 1 has the highest likelihood, therefore we have no evidence that correlations decay exponentially as visits become further apart. The likelihood ratio tests comparing Models 3–6 with Model 1 are all highly significant, indicating that all these models are preferable to Model 1. Models 3 and 4 are not nested and therefore cannot be compared using a likelihood ratio test. Model 4 is nested within Model 5 and the likelihood ratio test statistic is $6.78 \sim \chi_6^2$. This is not significant and therefore Model 5 can be rejected. Models 3 and 4 are nested within Model 6 and the likelihood ratio statistics for comparisons with Model 6 are $22.47 \sim \chi_6^2$ and $27.92 \sim \chi_8^2$, which are both highly significant. Since Model 6 is a significant improvement over both Models 3 and 4 we might choose to base our conclusions on it. However, we note that in practice it may be difficult to justify experimentation with so many structures, particularly in regulatory trials where it is necessary to specify analysis methods in the protocol.

The treatment effects from Model 6 are given in Table 8.6 with model-based standard errors. They indicate that treatment C produces a significantly lower DBP than treatment A. Ideally, we would like additionally to calculate empirical standard errors to reassure ourselves further that the covariance pattern fitted by Model 6 is close to the observed covariance in the data. However, an error message occurred when these standard errors were requested using *PROC MIXED*. Model 6 in Section 6.3 used an identical covariance pattern to Model 6 in this section but

Table 8.4 Variance parameters from covariance pattern models.

Model	Period	Variance components Centre	Residual	Correlation
1. Compound symmetry	1 – 4	4.8	71.4	0.52
2. First-order autoregressive	1 2 3 4	5.1	70.8	$\begin{pmatrix} 1 & & & \\ 0.53 & 1 & & \\ 0.53^2 & 0.53 & 1 & \\ 0.53^3 & 0.53^2 & 0.53 & 1 \end{pmatrix}$
3. Compound symmetry with separate covariances	1 – 4 A 1 – 4 B 1 – 4 C	4.6	78.7 63.5 72.6	0.49 0.35 0.61
4. Toeplitz	1 2 3 4	4.8	71.3	$\begin{pmatrix} 1 & & & \\ 0.54 & 1 & & \\ 0.45 & 0.54 & 1 & \\ 0.42 & 0.45 & 0.54 & 1 \end{pmatrix}$
5. Unstructured	1 2 3 4	4.5	73.2 66.8 80.0 66.4	$\begin{pmatrix} 1 & & & \\ 0.49 & 1 & & \\ 0.44 & 0.57 & 1 & \\ 0.42 & 0.47 & 0.56 & 1 \end{pmatrix}$
6. Toeplitz with separate covariances	1 A 2 A 3 A 4 A	4.7	78.5	$\begin{pmatrix} 1 & & & \\ 0.54 & 1 & & \\ 0.45 & 0.54 & 1 & \\ 0.46 & 0.45 & 0.54 & 1 \end{pmatrix}$
	1 B 2 B 3 B 4 B		63.4	$\begin{pmatrix} 1 & & & \\ 0.38 & 1 & & \\ 0.29 & 0.38 & 1 & \\ 0.40 & 0.29 & 0.38 & 1 \end{pmatrix}$
	1 C 2 C 3 C 4 C		72.1	$\begin{pmatrix} 1 & & & \\ 0.68 & 1 & & \\ 0.60 & 0.68 & 1 & \\ 0.43 & 0.60 & 0.68 & 1 \end{pmatrix}$

Table 8.5 Values of $-2\log(L)$ for covariance pattern models.

Model	$-2\log(L)$	Covariance Parameters
1. Compound symmetry	7483.21	3
2. First-order autoregressive	7492.70	3
3. Compound symmetry separate treatment covariances	7454.48	7
4. Toeplitz	7459.93	5
5. Unstructured	7453.15	11
6. Toeplitz separate treatment covariances	7432.01	13

Table 8.6   Treatment effect estimates from Model 6 and corresponding estimates from section 6.3 without centre effects.

Difference	Mean difference	Model-based SE
Model 6 including centre effects		
A − B	1.26	0.93
A − C	3.01	1.02
B − C	1.75	0.95
Model 6 from Section 6.3 excluding centre effects		
A − B	1.25	0.99
A − C	3.04	1.08
B − C	1.79	0.98

did not include centre effects. This model gave similar treatment effect results but with larger standard errors, indicating that fitting centre effects as random had lead to some recovery of extra information on the treatments.

SAS code and output

Variables

pat   = patient,
centre = centre,
visit   = visit,
treat   = treatment (A,B,C),
dbp   = post-treatment diastolic blood pressure (mmHg),
dbp1   = pre-treatment diastolic blood pressure (mmHg).

The first three statements below were used in all the models. The RANDOM and REPEATED statements to set the covariance structure are then shown for each model. The full code and output is given only for Model 6 on which the conclusions were based.

```
PROC MIXED NOCLPRINT; CLASS visit centre pat treat;
MODEL dbp=dbp1 treat visit/ DDFM=SATTERTH;
LSMEANS treat/ DIFF PDIFF;

Model 1:   RANDOM centre;
           REPEATED visit/ SUB=pat TYPE=CS GROUP=treat;
Model 2:   RANDOM centre;
           REPEATED visit/ SUB=pat TYPE=AR(1);
Model 3:   RANDOM centre;
           REPEATED visit/ SUB=pat TYPE=CS GROUP=treat;
Model 4:   RANDOM centre;
           REPEATED visit/ SUB=pat TYPE=TOEP;
```

```
Model 5:   RANDOM centre;
           REPEATED visit/ SUB=pat TYPE=UN;
Model 6:   RANDOM centre;
           REPEATED visit/ SUB=pat TYPE=TOEP GROUP=treat;
```

Model 6 — code and output

```
PROC MIXED NOCLPRINT; CLASS visit centre pat treat;
MODEL dbp=dbp1 treat visit centre/ DDFM=SATTERTH;
RANDOM centre;
REPEATED visit/ SUB=pat TYPE=TOEP GROUP=treat;
LSMEANS treat/ DIFF PDIFF;
```

```
               REML Estimation Iteration History
        Iteration  Evaluations      Objective     Criterion
               0            1     5820.4854789
               1            3     5438.1764142     0.00008952
               2            1     5437.9173028     0.00000111
               3            1     5437.9142630     0.00000000
                  Convergence criteria met.
```

```
               Covariance Parameter Estimates (REML)
        Cov Parm   Subject   Group        Estimate
        CENTRE                           4.74659770
        Variance   PAT       TREAT A    78.46134293
        TOEP(2)    PAT       TREAT A    41.62047546
        TOEP(3)    PAT       TREAT A    35.01930026
        TOEP(4)    PAT       TREAT A    35.50724749
        Variance   PAT       TREAT B    63.42633075
        TOEP(2)    PAT       TREAT B    24.22830512
        TOEP(3)    PAT       TREAT B    18.01833013
        TOEP(4)    PAT       TREAT B    24.50943924
        Variance   PAT       TREAT C    72.10747355
        TOEP(2)    PAT       TREAT C    48.66750102
        TOEP(3)    PAT       TREAT C    42.51507349
        TOEP(4)    PAT       TREAT C    30.66580862
```

```
               Model Fitting Information for DBP
        Description                          Value
        Observations                       1092.000
        Res Log Likelihood                 -3716.01
        Akaike's Information Criterion     -3729.01
        Schwarz's Bayesian Criterion       -3761.44
        -2 Res Log Likelihood              7432.011
```

```
               Covariance Parameter Estimates (REML)
                  Tests of Fixed Effects
        Source     NDF    DDF    Type III F    Pr>F
        DBP1         1    229        29.44     0.0001
        TREAT        2    168         4.39     0.0139
        VISIT        3    455        13.20     0.0001
```

```
                     Least Squares Means
     Level              LSMEAN     Std Error      DDF          T      Pr>|T|
     TREAT A        93.33986196   0.85769795     31.6     108.83     0.0001
     TREAT B        92.07724166   0.77533744     21.3     118.76     0.0001
     TREAT C        90.32930450   0.86925944     32.5     103.92     0.0001

                  Differences of Least Squares Means
   Level 1    Level 2     Difference      Std Error    DDF      T      Pr>|T|
   TREAT A    TREAT B     1.26262030     0.93432463    168    1.35     0.1784
   TREAT A    TREAT C     3.01055746     1.01988812    171    2.95     0.0036
   TREAT B    TREAT C     1.74793716     0.95214117    165    1.84     0.0682
```

8.3   MULTI-CENTRE CROSS-OVER TRIALS

Multi-centre trials are most frequently used for demonstrating the effectiveness of a drug in its later stages of development and usually have between-patient designs. However, occasionally multi-centre cross-over trials will be encountered. Although we do not show an example of this design, we suggest how a mixed model can be applied.

As in ordinary multi-centre trials 'global' estimates can be obtained by allowing treatment effects to vary randomly across centres. This is achieved by fitting centre and centre·treatment effects as random. Additionally, period effects can be allowed to vary randomly across the centres by fitting centre·period effects as random. A 'local' model (Model A) where results relate only to the centres sampled can be obtained by fitting interactions with centre effects as fixed. If the interaction terms involving centre can safely be removed from the model, then there can be an advantage in setting the centre effects to random to allow any additional information on treatments and periods to be recovered from the between-centre error stratum (Model B).

The fixed and random effects that are fitted in global and local models are listed below. Fitting patients as random gives a constant correlation for observations on the same patient. Alternatively, a more complex covariance pattern can be fitted by using a covariance pattern model (see Section 7.7).

Global model

Random	**Fixed**
Centre	Treatment
Centre·treatment	Period
Centre·period	Baseline
Patient	

Local model (A)

Random	Fixed
Patient	Treatment
	Period
	Centre
	Centre·treatment
	Centre·period
	Baseline

Local model (B)

Random	Fixed
Centre	Treatment
Patient	Period
	Baseline

8.4 HIERARCHICAL MULTI-CENTRE TRIALS AND META-ANALYSIS

Sometimes centres within a multi-centre trial, or trials within a meta-analysis, may be grouped in some way; for example, by country or continent. Such trials can then be described as having a double hierarchical structure. Although we do not consider an example with this design, we suggest how a mixed model can be applied. The double hierarchical structure can be taken into account in the analysis by modelling both hierarchies and their interactions with treatment effects as random. For example, in a multi-centre trial with centres grouped by country a 'global' model taking into account possible random variation in the treatment effect between centres and countries would be

Random	Fixed
Centre	Treatment
Centre·treatment	Baseline
Country	
Country·treatment	

This model will allow treatment effect results to be related with more confidence to the potential population of centres and countries, and 'shrunken' estimates of treatment effects can be obtained for each country and centre. Note that it is not necessary in the mixed model to specify formally that countries are nested within centres provided each centre is numbered individually. However, if centres are numbered separately within each country, specifying centre(country) in the software will usually avoid the need to create a variable with separate numbers for each centre (this is the case with `PROC MIXED`). In a meta-analysis with trials grouped by country, trial effects would be substituted for centre effects in the above model. There are also other possibilities for double hierarchical structures in which this type of analysis can be used. For example, trials within a meta-analysis may each have multi-centre designs and the above model can be fitted with trial effects substituted for country effects.

8.5   MATCHED CASE–CONTROL STUDIES

In a matched case–control study a group of subjects who have a particular disease or outcome (cases) are compared with a group of subjects who do not have the disease or outcome (controls). Each case is matched to one or more controls using one or more factors that are known to be connected with the disease; for example, age and sex are often used. We will refer to sets of matched subjects as 'matched sets'. The primary objective in a case–control analysis is to determine which factors (not used in matching) differ between the case and control groups. However, in doing so it is important to allow for the matched nature of the data. This can be achieved by taking candidate risk factors as outcome variables and by fitting matched sets as a random effect.

The design of the matched case–control study has similarities with that of the cross-over trial. In the cross-over trial the treatment effects are 'crossed' with patient effects (i.e. each patient may receive several treatments). In the matched case–control study, group effects (i.e. whether case or control) are crossed with matched set effects (each matched set may contain cases and controls). The effect of fitting matched sets as random in a case–control study is similar to that of fitting patients as random in a cross-over study. Results will be identical to an analysis fitting matched sets as a fixed effect when there are the same number of controls for every case and the matched set variance component is positive (when it is negative and set to zero, the mean group differences will be identical but their standard errors will differ). In an analysis fitting matched sets as fixed, information is completely lost on group effects in any matched sets which either contain only a case or only controls (although matched sets containing two or more controls but no cases can contribute information if the model fits at least one effect in addition to the group effect). Additionally, in a fixed effects analysis of binary data (which can be performed using conditional logistic regression), matched sets whose members all have identical outcomes (i.e. are uniform) do not contribute information to the analysis. This loss of information does not occur when matched sets are fitted as random because information is then 'recovered' from the matched sets error stratum.

Another fixed effects approach sometimes used is to fit the matching variables (e.g. sex and age) as covariates but to otherwise ignore the matching. However, this can sometimes cause a bias in the group estimates if the matching variables are associated with group (i.e. case or control). This is most likely to occur when there are an uneven number of controls per case causing the groups to be unbalanced. Results may also be misleading if the relationship with quantitative matching variables such as age is non-linear.

8.5.1 Example

This study was carried out by the Scottish Cot Death Trust which aimed to interview the parents of every sudden infant death syndrome (SIDS) baby in Scotland during 1992–1995 (Brooke *et al.*, 1997). The parents of two matched control babies born immediately before and after each case at the same hospital were also interviewed. As with most interview studies not all parents agreed to participate and this caused some of the matched sets to be incomplete. A summary of the content of the matched sets is given below for the interviewed subjects. Only 65% of the matched sets had their full complement of subjects.

Matched set content	Number of matched sets (%)
ABB	108 (65)
AB	28 (17)
A	11 (7)
BB	12 (7)
B	7 (4)

Note: A = case, B = control.

8.5.2 Analysis of a quantitative variable

We consider analysing a social deprivation score (`depcat`) which is measured on a scale of one to seven and is derived from post codes using information given in the 1991 Census (Carstairs and Morris, 1991). The distribution of `depcat` by group is shown in Figure 8.1. Although the score is an ordered categorical variable we analyse it using a normal mixed model and then check that this is a reasonable approximation by examining residual plots.

Results from a random effects model are compared with results from three alternative fixed effects approaches. In Model 1, group (case or control) is fitted as fixed and matched set as random. Model 2 allows for the matching by fitting matched sets as fixed. Model 3 fits the matching variables, age and season (at death/interview), as fixed effects. Season is categorised as Summer (June–August), Winter (December–February) or Spring/Autumn (March–May, September–November). Model 4 simply treats the data as unmatched.

The results are shown in Table 8.7. From all the models we would conclude that SIDS is associated with deprivation, since the cases on average have increased

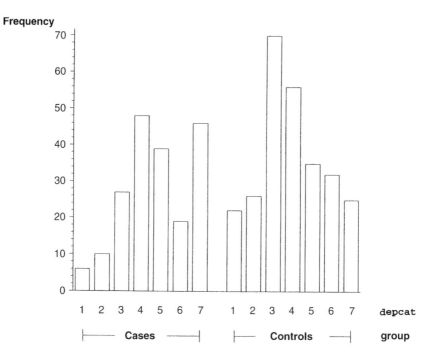

Figure 8.1   Histogram of deprivation category (`depcat`).

Table 8.7   Results from analyses of deprivation category (`depcat`).

Model	Fixed effects	Random effects	Method
1	Group	Matched set	REML
2	Group, matched set	—	OLS
3	Group, age, season	—	OLS
4	Group	—	OLS

Variance components

Model	Matched set	Residual
1	0.96	1.86
2	—	1.87
3	—	2.83
4	—	2.82

Group difference and SE (Cases–controls)

Model	
1	0.84 (0.13)
2	0.88 (0.14)
3	0.82 (0.16)
4	0.82 (0.16)

depcat scores. The residual estimates in Models 3 and 4 are greatly increased over those in Models 1 and 2, indicating that for analyses of depcat it is important to allow for matching.

Model 1 fits matched sets as random and the positive variance component shows that depcat scores are positively correlated within the matched sets. This is likely to be because matching is carried out within hospitals, which each have different catchment areas reflecting different levels of deprivation.

Model 2 takes account of the matching by fitting matched set as fixed. The group estimate differs from that in Model 1 and its standard error is larger. This is mainly because Model 2 does not use information from the matched set error stratum, and the 30 matched sets which either contain just cases or just controls are effectively lost from the analysis.

In Model 3 adjustment is made for the two matching variables, age and season of case death. Neither of these is significant ($p = 0.76$ and 0.55, respectively) and the group estimates are the same as in Model 4, which ignores matching.

8.5.3 Check of model assumptions

Since depcat is an ordered categorical variable, we should check model assumptions to ensure that a normal mixed model is reasonable. The residuals from Model 1 are plotted against their predicted values in Figure 8.2 and appear evenly

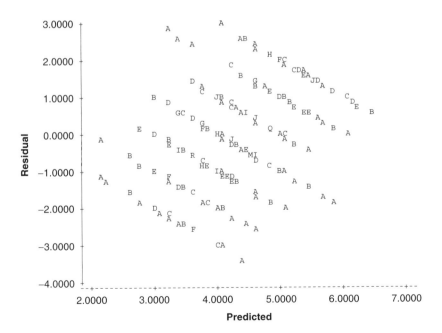

Figure 8.2 Plots of residuals vs. predicted values; A = 1 obs, B = 2 obs, etc.

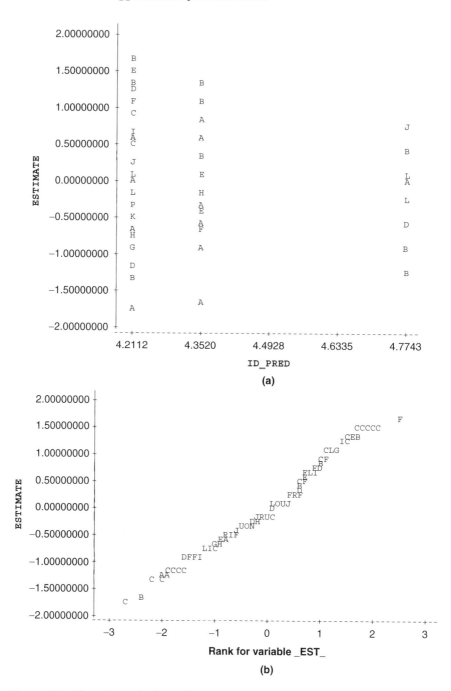

Figure 8.3 Plots of matched set effects: (a) vs. their predicted values; (b) Normal Plot; A = 1 obs, B = 2 obs, etc.

distributed. Note that the diagonal pattern is caused by the fact that the score is an ordered categorical variable.

The random matched set effects from Model 1 are plotted against their predicted values in Figure 8.3. These appear reasonably evenly distributed although there is a hint that observations with lower predicted values may have higher variances. To check the effects further we use a normal plot. This plot produces a fairly straight line, indicating that the matched set effects are approximately normally distributed.

8.5.4 Analysis of binary variables

Here, we consider analysing two binary variables recorded in the cot death study—the sex of the infant and whether the infant slept in a cot. Now that the analysis variable is binary, several uniform categories (see Section 3.2.4) are likely to occur for the matched set effects (i.e. there will be several matched sets in which all subjects have the same response, e.g. are of the same sex). In the random effects model this may cause bias in the matched sets variance component and we will therefore reparameterise the model as a covariance pattern model with a compound symmetry covariance structure (see Sections 3.2.4 and 3.3.2). This approach is usually preferable to a conditional logistic regression model (see Section 3.1.8), where information is lost on all tied matched sets (i.e. any sets with all members answering 'no' or all members answering 'yes') and on any matched sets which either contain only a case or only controls.

A similar set of models is considered to those used for analysing depcat (see Table 8.8) except that Model 2 is fitted using conditional logistic regression (see Section 3.1.8) and conditions the data on the matched sets. We will consider the results obtained for sex and sleeping in a cot separately.

Sex

The within-matched-set correlation parameter was 0.00 in Model 1. This indicates, not surprisingly, that control babies were no more likely to be of the same sex as the case baby than of the opposite sex. The zero correlation parameter causes the results from Model 1 to be the same as Model 4 where matching is completely ignored. We should mention that since Model 1 was reparameterised as a covariance pattern model, a negative correlation estimate will not be set to zero by default. When a negative variance correlation occurs it would be advisable to remove the effect of matched sets from the analysis by fitting Model 4 (effectively setting the correlation parameter to zero).

All models show statistically significant group effects, with a greater chance of SIDS in male infants. However, the group estimate in Model 2 is lower than in the other models and the standard error higher. This is because information from the matched set error stratum is completely omitted from matched sets with all cases

Table 8.8   Results from analyses of sex of the infant and sleeps in cot.

Model	Fixed effects	Random effects	Method
1	Group	Matched set	GLMM
2	Group, (matched set)	—	Conditional LR
3	Group, age, season	—	GLM
4	Group	—	GLM

Group effect on logit scale (cases–controls) and odds ratio (cases ÷ controls)

| | Male | | Sleeps in cot | |
Model	On logit scale	OR	On logit scale	OR
1	0.78 (0.19)	2.18 (1.50, 3.17)	−0.67 (0.19)	0.51 (0.35, 0.74)
2	0.61 (0.21)	1.84 (1.22, 2.78)	−0.33 (0.25)	0.72 (0.44, 1.16)
3	0.77 (0.20)	2.15 (1.47, 3.15)	−1.01 (0.27)	0.37 (0.22, 0.62)
4	0.78 (0.19)	2.18 (1.50, 3.17)	−0.65 (0.22)	0.52 (0.34, 0.80)

Variance parameters for Model 1

Parameter	Male	Sleeps in cot
Within matched set correlation	0.00	0.29
Dispersion	1.00	1.00

or all controls, or with the same sex for all members. Age and season were not significantly related to sex in Model 3 ($p = 0.21$ and 0.76) and this leads to only slight differences in the results compared with Model 4.

Sleeps in cot

The within-matched-set correlation parameter is now positive and this causes the results to differ between Models 1 and 4. It indicates that control infants were more likely to sleep in a cot if the case infant did and vice versa. This is not surprising since the infants were matched by age and older babies are more likely to sleep in cots.

All of the models except Model 2 show a statistically significant group effect, indicating that the risk of SIDS is less in those infants who were sleeping in a cot compared with infants who slept elsewhere (in carrycots, Moses baskets, prams, etc.). However, the group effect differs widely between the four models. This is mainly because sleep place is associated with age and each model allows for this association in a different way. In Models 1 and 2 the exact matching is taken into account and the results only differ because information from the matched set error stratum is completely omitted in Model 2 from the matched sets containing all cases or all controls, or all with the same sleep place. The age effect was highly significant in Model 3 ($p = 0.0001$). However, this model fits age as a quantitative variable and therefore does not allow for any non-linear effects (i.e. it only allows the proportion of babies sleeping in a cot to increase linearly with age on the logistic scale). Model 4 did not fit age at all and is clearly inappropriate given the influence that age has on sleep place. At first sight it is therefore surprising that the standard error in Model 4 is smaller than that in Model 3. This has occurred because the pattern of 'dropouts' from the study has resulted in cases and controls being unbalanced for age, with a resultant increase in the standard error in Model 3.

Analysis of both variables demonstrates the value of the mixed model approach. The standard errors of the group effect are minimised, as we might expect, because the data are being utilised more fully than is the case with any of the fixed effects approaches.

SAS code and output

Depcat analyses

Variables

group   = group (A = cases, B = controls),
id   = matched set,
depcat = deprivation score (1–7, 7 = most deprived),
age   = age (weeks),
seas   = 1 = Winter, 2 = Spring/Autumn, 3 = Summer.

Full code and output are given for Model 1. SAS code only is given for Models 2–4.

Model 1 and model checking

```
PROC MIXED DATA=a NOCLPRINT; CLASS group id;
MODEL depcat= group/ PREDICTED PM DDFM=SATTERTH;
RANDOM id/ SOLUTION;
ESTIMATE 'A-B' group 1 -1;
ID id group;
MAKE 'PREDMEANS' OUT=predm NOPRINT;
MAKE 'PREDICTED' OUT=resid NOPRINT;
MAKE 'SOLUTIONR' OUT=solut NOPRINT;

PROC PLOT DATA=resid; PLOT _resid_*_pred_;
TITLE 'RESIDUALS AGAINST PREDICTED VALUES';

DATA resid; SET resid;
KEEP _pred_ _resid_ patient id group;
PROC SORT; BY id group;

DATA solut; SET solut;
idx=id*1; * Obtain numeric id variable;
drop id;

DATA est(KEEP=id _est_); SET solut;
id=idx;
OUTPUT est;

DATA _est_; MERGE predm est; BY id;
PROC MEANS NOPRINT; BY id; id _est_;
VAR _pred_; OUTPUT OUT=est MEAN=id_pred N=freq;
```

```
PROC PLOT DATA=est; PLOT _est_*id_pred;
TITLE 'MATCHED SET EFFECTS AGAINST THEIR PREDICTED VALUES';
PROC RANK OUT=norm NORMAL=TUKEY DATA=_est_; VAR _est_; RANKS
  rank;
PROC PLOT DATA=norm; PLOT _est_*rank;
TITLE 'MATCHED EFFECTS - NORMAL PLOT';
```

Note that although ID is a numeric variable the MAKE 'SOLUTIONR' creates a dataset with ID as a character variable and additional code is necessary to convert it back to numeric. However, the MAKE 'PREDMEANS' and MAKE 'PREDICTED' statements output ID as numeric as expected!

REML Estimation Iteration History

Iteration	Evaluations	Objective	Criterion
0	1	945.54086980	
1	2	906.01304204	0.00000000

Convergence criteria met.

Covariance Parameter Estimates (REML)

Cov Parm	Estimate
ID	0.95755938
Residual	1.86019231

Model Fitting Information for DEPCAT

Description	Value
Observations	461.0000
Res Log Likelihood	-874.799
Akaike's Information Criterion	-876.799
Schwarz's Bayesian Criterion	-880.928
-2 Res Log Likelihood	1749.599

Tests of Fixed Effects

Source	NDF	DDF	Type III F	Pr>F
GROUP	1	260	40.19	0.0001

ESTIMATE Statement Results

Parameter	Estimate	Std Error	DF	t	Pr>|t|
A-B	0.84461985	0.13323438	315	6.34	0.0001

NB. Plots are printed in the main text.

Model 2

```
PROC GLM; CLASS group id;
MODEL depcat= group id;
ESTIMATE 'A-B' group 1 -1;
```

Model 3

```
PROC GLM; CLASS group id seas;
MODEL depcat= group age seas;
ESTIMATE 'A-B' group 1 -1;
```

Model 4

```
PROC GLM; CLASS group;
MODEL depcat= group;
ESTIMATE 'A-B' group 1 -1;
```

Binary analyses

SAS code is given below for the analyses of 'sleeps in cot'. Identical code is used for the sex analyses with 'sex' replacing 'sleepn1' in the MODEL statement. Again, full code and output are given for Models 1–3 to illustrate the use of different procedures and macros to fit each type of model. SAS code only is given for Model 4 because the output is similar to that from Model 3.

Variables

grp   = group (1 = cases, 0 = controls),
id   = matched set,
one   = 1 for all observations,
sleepn1 = sleeps in cot at night (1 = yes, 0 = no).

Model 1

```
PROC GENMOD DATA=a; CLASS id;
MODEL sleepn1/one = grp/ DIST=B;
REPEATED SUB=id/ TYPE=CS CORRW MODELSE;
```

```
                     The GENMOD Procedure
                      Model Information
               Description              Value

               Data Set                WORK.A
               Distribution            BINOMIAL
               Link Function           LOGIT
               Dependent Variable      SLEEPN1
               Dependent Variable      ONE
               Observations Used       421
               Number Of Events        175
               Number Of Trials        421
               Missing Values          56

                    Class Level Information
          Class   Levels   Values
          ID        201    1  2  3  4  5  6  7  8  9  10  11  12  13
                           14 15 16 17 18 19 20 22 23 24
                           25 26 27 28 29 30 31 32 33 34
                           35 36 37 38 39 40 41 42 43 44
                           45 46 47 48 49 51 52 53 54 55
                           56 57 58 59 61 63 64 65 66 67
                           68 69 70 71 72 73 74 75 76 77
                           78 79 80 81 82 83 84 85 86 87
```

```
88 89 90 91 92 93 94 95 96 97
98 99 100 101 102 103 104 105
106 107 108 109 110 112 113
114 115 116 117 118 120 121
122 123 124 125 126 127 128
129 130 131 132 133 134 135
136 137 138 139 140 141 142
143 144 145 146 147 148 149
150 151 152 153 154 155 156
157 158 159 160 161 162 163
164 165 166 167 168 169 170
171 172 173 174 175 176 177
178 179 180 181 182 183 184
185 186 187 188 189 190 191
192 193 194 195 196 197 198
199 200 201 202 203 205 206
207 300
```

Parameter Information

Parameter	Effect
PRM1	INTERCEPT
PRM2	GRP

Criteria For Assessing Goodness Of Fit

Criterion	DF	Value	Value/DF
Deviance	419	562.1235	1.3416
Scaled Deviance	419	562.1235	1.3416
Pearson Chi-Square	419	421.0000	1.0048
Scaled Pearson X2	419	421.0000	1.0048
Log Likelihood	.	-281.0617	.

Analysis Of Initial Parameter Estimates

Parameter	DF	Estimate	Std Err	ChiSquare	Pr>Chi
INTERCEPT	1	-0.1238	0.1208	1.0496	0.3056
GRP	1	-0.6527	0.2153	9.1942	0.0024
SCALE	0	1.0000	0.0000	.	.

NOTE: The scale parameter was held fixed.

The above results correspond to a GLM analysis (that would be obtained without using a REPEATED statement) and can be ignored.

GEE Model Information

Description	Value
Correlation Structure	Exchangeable
Subject Effect	ID (201 levels)
Number of Clusters	201
Clusters With Missing Values	56
Correlation Matrix Dimension	3
Maximum Cluster Size	3
Minimum Cluster Size	0

```
                    Working Correlation Matrix

                         COL1      COL2      COL3
             ROW1      1.0000    0.2858    0.2858
             ROW2      0.2858    1.0000    0.2858
             ROW3      0.2858    0.2858    1.0000
```

This matrix has dimension 3×3 because there are up to three subjects per matched set.

```
              Analysis Of GEE Parameter Estimates

                 Empirical Standard Error Estimates

                     Empirical 95% Confidence Limits
  Parameter   Estimate   Std Err     Lower      Upper       Z    Pr>|Z|

  INTERCEPT   -0.0888    0.1386    -0.3604    0.1829   -.6403   0.5220
  GRP         -0.6715    0.1905    -1.0449   -0.2982   -3.525   0.0004
  Scale        1.0069       .          .         .        .       .
```

NOTE: The scale parameter for GEE estimation was computed as the square root of the normalized Pearson's chi-square.

```
              Analysis Of GEE Parameter Estimates
                Model-Based Standard Error Estimates

                       Model  95% Confidence Limits
  Parameter   Estimate   Std Err     Lower      Upper      Z    Pr>|Z|

  INTERCEPT   -0.0888    0.1347    -0.3527    0.1752  -.6589   0.5099
  GRP         -0.6715    0.1862    -1.0366   -0.3065  -3.606   0.0003
  Scale        1.0069       .          .         .       .       .
```

The standard error given above is the model-based estimate and is obtained by using the MODELSE option.

Model 2

By using the TIES=DISCRETE option and STRATA < matched set >, the PHREG procedure can be used to perform a conditional logistic regression analysis. Note that the procedure will produce the same results here, regardless of whether group is fitted as the dependent variable with sleep place as the independent variable, or the other way around.

```
PROC PHREG DATA=a NOSUMMARY;
MODEL time*sleepn1(0)=grp/ TIES=DISCRETE;
STRATA id;
```

```
                     The PHREG Procedure

      Data Set: WORK.A
      Dependent Variable: TIME
      Censoring Variable: SLEEPN1
      Censoring Value(s): 0
      Ties Handling: DISCRETE
```

```
               Testing Global Null Hypothesis: BETA=0
               Without       With
Criterion      Covariates    Covariates    Model Chi-Square

-2 LOG L       163.116       161.248       1.868 with 1 DF (p=0.1717)
Score             .             .          1.847 with 1 DF (p=0.1742)
Wald              .             .          1.833 with 1 DF (p=0.1758)
```

```
             Analysis of Maximum Likelihood Estimates
               Parameter   Standard     Wald        Pr >      Risk
Variable   DF  Estimate    Error      Chi-Square  Chi-Square  Ratio

GRP         1  -0.332214   0.24538    1.83298     0.1758      0.717
```

Model 3

```
PROC GENMOD DATA=a;
MODEL sleepn1/one = grp seas age/ DIST=B;
```

```
                    The GENMOD Procedure
                     Model Information
               Description              Value

               Data Set                 WORK.A
               Distribution             BINOMIAL
               Link Function            LOGIT
               Dependent Variable       SLEEPN1
               Dependent Variable       ONE
               Observations Used        421
               Number Of Events         175
               Number Of Trials         421
               Missing Values           56
```

```
             Criteria For Assessing Goodness Of Fit
       Criterion            DF        Value      Value/DF
       Deviance             417     419.0846      1.0050
       Scaled Deviance      417     419.0846      1.0050
       Pearson Chi-Square   417     428.0713      1.0265
       Scaled Pearson X2    417     428.0713      1.0265
       Log Likelihood        .     -209.5423        .
```

```
             Analysis Of Parameter Estimates
       Parameter   DF   Estimate   Std Err   ChiSquare   Pr>Chi
       INTERCEPT    1    -2.5500    0.4393    33.6967     0.0001
       GRP          1    -1.0051    0.2693    13.9298     0.0002
       SEAS         1     0.1366    0.1797     0.5778     0.4472
       AGE          1     0.1587    0.0175    82.4572     0.0001
       SCALE        0     1.0000    0.0000       .          .
```

```
NOTE:  The scale parameter was held fixed.
```

Model 4

```
PROC GENMOD DATA=a;
MODEL sleepn1/one = grp/ DIST=B;
```

8.6 DIFFERENT VARIANCES FOR TREATMENT GROUPS IN A SIMPLE BETWEEN-PATIENT TRIAL

In a simple between-patient trial the treatment groups will sometimes have different variances. Allowing for this can produce more appropriate standard errors for treatment estimates and also the variance values themselves may aid the understanding of the different treatment mechanisms. Often, the possibility of different treatment variances is not considered when choosing an analysis model. However, it can easily be allowed for in a mixed model by structuring the residual matrix to have a different variance for each treatment group. Observations will remain uncorrelated (provided no random effects are specified) and thus the model retains many of the features of a fixed effects analysis. Consider the example data used to illustrate the notation in Section 2.1.

Centre	Treatment	Baseline systolic BP	Post-treatment systolic BP
1	A	178	176
1	A	168	194
1	B	196	156
1	B	170	150
2	A	165	150
2	B	190	160
3	A	175	150
3	A	180	160
3	B	175	160

If different variances are allowed for the treatment groups, the variance matrix would have the form (assuming no random effects are fitted):

$$
\mathbf{V} = \mathbf{R} = \begin{pmatrix}
\sigma_A^2 & 0 & 0 & 0 & 0 & 0 & 0 & 0 & 0 \\
0 & \sigma_A^2 & 0 & 0 & 0 & 0 & 0 & 0 & 0 \\
0 & 0 & \sigma_B^2 & 0 & 0 & 0 & 0 & 0 & 0 \\
0 & 0 & 0 & \sigma_B^2 & 0 & 0 & 0 & 0 & 0 \\
0 & 0 & 0 & 0 & \sigma_A^2 & 0 & 0 & 0 & 0 \\
0 & 0 & 0 & 0 & 0 & \sigma_B^2 & 0 & 0 & 0 \\
0 & 0 & 0 & 0 & 0 & 0 & \sigma_A^2 & 0 & 0 \\
0 & 0 & 0 & 0 & 0 & 0 & 0 & \sigma_A^2 & 0 \\
0 & 0 & 0 & 0 & 0 & 0 & 0 & 0 & \sigma_B^2
\end{pmatrix}.
$$

8.6.1   Example

We now consider the hypertension trial introduced in Section 1.3 as a simple between-patient trial. The final DBP measurement is analysed and centre effects are ignored. A model with separate variances for each treatment group (Model 2) is compared with a standard model which assumes a constant variance over all treatments (Model 1). Both models fit baseline and treatment effects as fixed. The results obtained are given in Table 8.9.

Model 2 indicates that treatment A may be more variable than treatments B and C. A likelihood ratio test is used to test whether this model is significantly better than Model 1. This gives $2 \times (1036.15 - 1031.72) = 8.86 \sim \chi_2^2$ (two DF are used because Model 2 uses two additional variance parameters) and shows that Model 2 is a significant improvement ($p = 0.012$). Thus, it is unlikely that the differences in the treatment variances have occurred by chance. The standard errors in Model 2 reflect the different treatment variances. The standard error has become larger for $A - C$, and smaller for $B - C$.

Including random centre and centre·treatment effects

Different treatment variances can be modelled at the residual level even when random effects are included. We illustrate this by adding random centre and centre·treatment effects (Model 3). The standard errors are increased over those in Models 1 and 2 because treatment effects are assumed to vary randomly across the centres. As in Model 2 the standard errors reflect the different residual variances for treatments.

Table 8.9   Results from Models 1–3.

		Variance estimates			
Model	**Treatment**	**Residual variance**	**Centre variance component**	**Centre·treatment variance component**	**Log (L)**
1	$A - C$	79.8	—	—	−1036.15
2	A	104.3	—	—	−1031.72
	B	56.5			
	C	76.9			
3	A	92.4	5.76	4.43	−1022.47
	B	44.5			
	C	66.7			

	Fixed effects (SEs)			
Model	**Baseline**	**A − B**	**A − C**	**B − C**
1	0.30 (0.11)	1.01 (1.29)	3.04 (1.28)	2.04 (1.31)
2	0.31 (0.11)	1.00 (1.29)	3.05 (1.36)	2.05 (1.19)
3	0.30 (0.11)	1.23 (1.40)	2.88 (1.47)	1.65 (1.31)

SAS code and output

Variables

```
pat     = patient,
centre = centre,
treat   = treatment (A = 20 mg dose, B = 40 mg dose, C = placebo),
dbp     = post-treatment DBP (mmHg),
dbp1    = pre-treatment DBP (mmHg).
```

Output is only given for Model 2 since it is the main model we are illustrating here.

Model 1

```
PROC MIXED; CLASS CENTRE TREAT;
MODEL dbp = dbp1 treat/ S DDFM=SATTERTH;
LSMEANS treat/ DIFF PDIFF CL;
```

Model 2

```
PROC MIXED NOCLPRINT; CLASS centre treat pat;
MODEL dbp = dbp1 treat/ S DDFM=SATTERTH;
REPEATED /SUBJECT=pat GROUP=treat;
LSMEANS treat/ DIFF PDIFF CL;
```

The REPEATED statement causes the covariance matrix to be blocked by patients. Because there is only one observation per patient there will no correlation between the observations. The GROUP option causes a separate variance parameter to be used for each treatment group. Note that the S option in the MODEL statement is an abbreviation for the SOLUTION option.

```
              REML Estimation Iteration History

     Iteration    Evaluations       Objective     Criterion

         0               1     1550.3451721
         1               2     1541.4770205    0.00000000
              Convergence criteria met.


            Covariance Parameter Estimates (REML)

     Cov Parm    Subject    Group         Estimate

      DIAG        PAT      TREAT A     104.31429964
      DIAG        PAT      TREAT B      56.45174700
      DIAG        PAT      TREAT C      76.88222838
```

The DIAG covariance terms are the residual variances for each treatment group.

```
            Model Fitting Information for DBP

     Description                          Value

     Observations                       288.0000
     Res Log Likelihood                 -1031.72
     Akaike's Information Criterion     -1034.72
```

```
Schwarz's Bayesian Criterion      -1040.19
-2 Res Log Likelihood             2063.434
Null Model LRT Chi-Square            8.8682
Null Model LRT DF                    2.0000
Null Model LRT P-Value               0.0119
```

Solution for Fixed Effects

Effect	TREAT	Estimate	Std Error	DF	t	Pr>|t|
INTERCEPT		56.79606542	11.21271093	281	5.07	0.0001
DBP1		0.30850753	0.10834425	281	2.85	0.0047
TREAT	A	3.04531878	1.36109984	191	2.24	0.0264
TREAT	B	2.04936832	1.19334356	183	1.72	0.0876
TREAT	C	0.00000000	.	.	.	.

Tests of Fixed Effects

Source	NDF	DDF	Type III F	Pr>F
DBP1	1	281	8.11	0.0047
TREAT	2	197	2.74	0.0668

Least Squares Means

Effect	TREAT	LSMEAN	Std Error	DF	t	Pr>|t|	Alpha	Lower	Upper
TREAT	A	91.57266902	1.02154215	98.6	89.64	0.0001	0.05	89.5456	93.5997
TREAT	B	90.57671856	0.78106477	92.4	115.97	0.0001	0.05	89.0256	92.1279
TREAT	C	88.52735024	0.90020525	93.5	98.34	0.0001	0.05	86.7399	90.3148

Differences of Least Squares Means

Effect	TREAT	_TREAT	Difference	Std Error	DF	t	Pr>|t|	Alpha	Lower	Upper
TREAT	A	B	0.99595046	1.28679289	181	0.77	0.4400	0.05	-1.5431	3.5350
TREAT	A	C	3.04531878	1.36109984	191	2.24	0.0264	0.05	0.3606	5.7301
TREAT	B	C	2.04936832	1.19334356	183	1.72	0.0876	0.05	-0.3052	4.4039

Model 3

```
PROC MIXED; CLASS centre treat pat;
MODEL dbp = dbp1 treat/ S DDFM=SATTERTH;
RANDOM centre centre*treat;
REPEATED /SUBJECT=pat GROUP=treat;
LSMEANS treat/ DIFF PDIFF CL;
```

8.7 ESTIMATING VARIANCE COMPONENTS IN AN ANIMAL PHYSIOLOGY TRIAL

The purpose of this experiment was to calculate variance components for breathing measurements in rabbits and to use them to design a future clinical trial to compare two treatments. One hundred breaths were measured on four rabbits on each of four days. This gave rise to four potential sources of variation—rabbits, days, rabbit·day interaction and residual (between breaths).

Note that the variability of individual breaths is given by the sum of the variance components, $\sigma_b^2 + \sigma_d^2 + \sigma_{bd}^2 + \sigma_r^2$. A random effects model was fitted to inspiration time (seconds) with each of these effects taken as random. The resulting variance components were

Source	Variance component
Rabbit (σ_b^2)	0.00231
Day (σ_d^2)	0.00247
Rabbit·day (σ_{bd}^2)	0.00442
Residual (σ_r^2)	0.00318

The positive rabbit component indicates that, not surprisingly, inspiration time varies between rabbits. The positive day and rabbit·day components show there is variation between days and additional variation within rabbits on different days. Thus, the rabbits' breathing is sensitive to the trial environment on each day. Factors such as time of day, time since last meal, the experimenter, differences in trial set-up, could all have contributed to this variation. However, we should bear in mind that there were only three DF available for estimating the rabbit and day components (σ_b^2 and σ_d^2) and these estimates are therefore only approximate.

8.7.1 Sample size estimation for a future experiment

We seek to design a study to compare two treatments which are expected to influence breathing in rabbits. Two possibilities are a parallel group study, where each rabbit receives only one treatment, or a cross-over design where each rabbit will receive each treatment. For ethical reasons the number of rabbits should be kept as low as possible, as should the number of days for which each rabbit is studied. We set a maximum of 10 days in the study for each rabbit, but as many as 100 breaths could easily be sampled on each day.

Initially, we must consider the mean difference in inspiration time between the two treatments, Δ, which we wish the study to be able to detect. We must also set the significance level α, and the power β. Then, for either design, the standard sample size estimation method can be used to estimate the required number of rabbits, number of days and the number of breaths per session. We require

$$\Delta = (Z_{(1-\alpha)/2} + Z_\beta) \times \mathrm{SE}(t_1 - t_2).$$

A range of possible values for the number of rabbits per treatment (n_t), the number of days (n_d) and the number of breaths (n_b) can be calculated from which the preferred design can be chosen. If the relative costs of using rabbits, days and breaths were specified, we could determine the optimal design. In practice, though, it is usually easier to select the design by eyeballing the range of possible designs which satisfy the power requirements. In the context of this design, there is a clear hierarchy in the design priorities. The first priority is to minimise the number of rabbits, then the number of days per rabbit, and then the number of breaths per day.

In the first instance we consider setting $\Delta = 0.10$, $\alpha = 0.05$ and $\beta = 0.1$.

Between-rabbit design

The variation of the mean treatment difference in inspiration duration is given by

$$\text{var}(t_1 - t_2) = 2\{\sigma_b^2/n_t + \sigma_d^2/n_d + \sigma_{bd}^2/(n_t \times n_d) + \sigma_r^2/(n_t \times n_d \times n_b)\},$$

and from this we obtain

$$\Delta = (Z_{\alpha/2} + Z_{1-\beta}) \times \{2[\sigma_b^2/n_t + \sigma_d^2/n_d + \sigma_{bd}^2/(n_t \times n_d) + \sigma_r^2/(n_t \times n_d \times n_b)]\}^{1/2}.$$

The approach which we recommend to determine the final design is to calculate Δ for a range of values for n_t, n_d and n_b, accepting only those which yield $\Delta < 0.10$. A SAS program to undertake this calculation is given at the end of this section. The smallest number of rabbits to give a viable design is 14 rabbits per treatment with nine days of observation and 25 breaths per day. By increasing the number of rabbits to 18 per treatment, the number of days per rabbit can be reduced to eight.

Within-rabbit design

In this design each treatment is received by each of the n_t rabbits for n_d days. Again n_b breaths are measured on each day. The variation of the mean treatment difference in inspiration duration on one treatment is now given by

$$\text{var}(t_1 - t_2) = 2 \times \{\sigma_d^2/n_d + \sigma_{bd}^2/(n_t \times n_d) + \sigma_r^2/(n_t \times n_d \times n_b)\},$$

$$\Delta = (Z_{\alpha/2} + Z_{1-\beta}) \times \{2 \times (\sigma_d^2/n_d + \sigma_{bd}^2/(n_t \times n_d) + \sigma_r^2/(n_t \times n_d \times n_b))\}^{1/2}.$$

As before, Δ is calculated for a range of values for n_t, n_d and n_b. We find in this instance that none of the designs satisfies the study requirements. This occurs because the requirement to limit rabbits to 10 days of study means that a maximum of five days are allowed per treatment. Thus, however large n_t is set, $\text{var}(t_1 - t_2) > 2\sigma_d^2/n_d$, and $\Delta > (Z_{\alpha/2} + Z_{1-\beta}) \times (2\sigma_d^2/n_d)^{1/2} = 0.102$.

Thus, with these design requirements we could only undertake a between-rabbit design with 14 rabbits per treatment group. This could cause the design requirements to be rethought, and we illustrate the method below with Δ changed to 0.15.

The following table lists for the between-rabbit design, combinations of the numbers of rabbits, days and breaths which satisfy $\Delta < 0.15$. The table is structured so that for any specified number of rabbits, the number of days is the minimum possible to give $\Delta < 0.15$, and then for that number of rabbits and days, the number of breaths is also the minimum to satisfy $\Delta < 0.15$.

Rabbits per group (n_t)	Days (n_d)	Breaths (n_b)	Δ
4	8	5	0.14816
6	5	5	0.14839
8	4	5	0.14957
10	4	5	0.14315
12	4	5	0.13871
14	4	5	0.13545
16	3	10	0.14972
18	3	5	0.14823
20	3	5	0.14665
—	—	—	—
—	—	—	—
100	3	5	0.13471

Since our priority in this study is to minimise the number of rabbits, we would prefer the design with four rabbits per group, studied for eight days each, with five breaths measured per day. If our priorities were different we could alternatively consider six rabbits studied for five days, eight rabbits studied for four days, each with five breaths per day, or even 16 rabbits studied for three days with 10 breaths per day.

With the requirement that $\Delta < 0.15$, it is now also possible to use the within-rabbit design. The following table, constructed along the lines of the previous table, provides alternative designs.

Rabbits (n_t)	Days (n_d)	Breaths (n_b)	Δ
2	5	5	0.14495
4	4	5	0.14008
6	4	5	0.13194
8	3	5	0.14743
10	3	5	0.14439
12	3	5	0.14233
14	3	5	0.14084
—	—	—	—
—	—	—	—
48	3	5	0.13433
50	3	5	0.13422

We could therefore opt for only two rabbits studied for five days per treatment with five breaths per day. Although this does satisfy the sample size requirement as specified, it is intuitively unappealing to conduct such a small study and the design with four rabbits studied for four days per treatment might be preferred.

Note that in the above tables we have only considered intervals of two in the number of rabbits. We could, of course, fill in the gaps in the regions which we are considering implementing. We should bear in mind, however, the limited accuracy of the estimates of the variance components on which our calculations are based. Apparently precise sample size calculations could be misleading. Therefore, we recommend that the sample size calculations be viewed as establishing ball park figures for the size of the study, and helping in determining the most appropriate type of design.

We also recommend that some form of sensitivity analysis is performed before a trial design is finalised. This can have two dimensions. One of these is how the design changes with the choice of Δ, α and β. We have seen above that a between-rabbit design would be essential with our first choice of figures, while a within-subject design would be preferable when Δ was larger. The second dimension is the sensitivity of the design to the values used for the estimates of variance components. Sometimes quite small changes can modify the design appreciably. The wisest design might not be the one which is optimal with the initial choice of parameters, but the one which performs well over a wider range of possible parameter values.

SAS code and output

Variables

rabbit = rabbit number
ins   = inspiration time (seconds)
day   = day number

Analysis model

```
PROC MIXED; CLASS rabbit day;
MODEL ins=;
RANDOM day rabbit day*rabbit;
```

```
              The MIXED Procedure
             Class Level Information
          Class     Levels    Values
          RABBIT       4      1 2 3 4
          DAY          4      1 2 3 4

         REML Estimation Iteration History
    Iteration   Evaluations      Objective      Criterion
           0             1      -5543.937520
           1             2      -7197.315049    0.00014476
```

```
                2          1     -7197.943785     0.00002574
                3          1     -7198.048272     0.00000138
                4          1     -7198.053437     0.00000001
                    Convergence criteria met.
```

```
              Covariance Parameter Estimates (REML)
```

Cov Parm	Ratio	Estimate	Std Error	Z	Pr>|Z|
DAY	0.77477066	0.00246726	0.00296963	0.83	0.4061
RABBIT	0.72607826	0.00231220	0.00284513	0.81	0.4164
RABBIT*DAY	1.38772764	0.00441923	0.00209679	2.11	0.0351
Residual	1.00000000	0.00318451	0.00011555	27.56	0.0001

```
              Model Fitting Information for INS
```

Description	Value
Observations	1535.000
Variance Estimate	0.0032
Standard Deviation Estimate	0.0564
REML Log Likelihood	2189.375
Akaike's Information Criterion	2185.375
Schwarz's Bayesian Criterion	2174.704
-2 REML Log Likelihood	-4378.75

Sample size estimation for between-rabbit trial

```
DATA a; SET a;
DO rabbit = 2 TO 100 BY 2;
DO day = 1 TO 10;
DO breath = 5 TO 100 BY 5;
OUTPUT;
END;
END;
END;

DATA a; SET a;
d = (2*10.51*(0.00231/rabbit + 0.00247/day +
    0.00442/(day*rabbit) + 0.00318/(breath*day*rabbit)))**0.5;
*IF d<0.10; * for delta<0.10;
IF d<0.15;
PROC SORT; BY rabbit day breath;

DATA a; SET a; BY rabbit;
IF FIRST.rabbit;
PROC PRINT NOOBS; VAR rabbit day breath d;
```

Output appears in the main text.

Sample size estimation for within-rabbit trial

```
DATA a; SET a;
DO rabbit=2 TO 50 by 2;
DO day=1 TO 10;
DO breath=5 TO 100 BY 5;
```

```
OUTPUT;
END;
END;
END;

DATA a; SET a;
d = (2*10.51*(0.00442/(day*rabbit)
    + 0.00247/day + 0.00318/(breath*day*rabbit)))**0.5;
IF d<0.15;
PROC SORT; BY rabbit day breath;

DATA a; SET a; BY rabbit;
IF first.rabbit;
PROC PRINT NOOBS; VAR rabbit day breath d;
```

Output appears in the main text.

8.8   INTER- AND INTRA-OBSERVER VARIATION IN FOETAL SCAN MEASUREMENTS

Ultrasound scans are often used during pregnancy to predict gestation (age of foetus). However, predictions made using ultrasound can be unreliable, particularly in the latter stages of pregnancy. An experiment to measure inter- and intra- observer variability in ultrasound measurements was carried out at an Edinburgh maternity hospital. Six radiologists participated in the experiment. Fifty-two women in the latter stages of pregnancy with a mean gestation of 29.9 (SD 3.2) weeks were each scanned by two of the radiologists selected at random. Both scans were carried out in the same session. Note that it would not have been ethical or, indeed, feasible to have used all six radiologists at each session.

A random effects model was fitted to the data with radiologist effects taken as random and subject (women) effects as fixed. Here, subjects are fitted as fixed because they each have a different gestation and therefore cannot be treated as a randomly distributed sample. The resulting variance components were

Radiologist 0.000,
Residual   0.287.

The zero radiologist component indicates that there was no systematic variability between the radiologists. This reassured the radiologists that although they often obtained different gestation predictions on the same women, none of them had a tendency to produce particularly high or low readings. The residual variance of 0.287 potentially incorporates several types of variability such as image variation, variation caused by foetus changing position, and radiologist measurement error. The residual value indicates that predicted gestations have a standard deviation of 0.54 weeks and hence a 95% confidence interval of

$\pm t_{47,0.975} \times 0.54 = \pm 2.01 \times 0.54 = \pm 1.09$ weeks (a t_{47} statistic is used because 47 is the residual DF). Note that this is the error encountered in measuring foetus size from which gestation is predicted. It does not incorporate the variability in foetal size that occurs naturally at given gestations. The standard error would be higher if this were taken into account.

SAS code and output

Variables

obs = observer ID,

pat = patient number,

gest = estimated gestation (weeks).

```
PROC MIXED; CLASS pat obs;
MODEL gest=pat/ DDFM=SATT;
RANDOM obs;
```

```
                    The MIXED Procedure
                  Class Level Information
           Class    Levels  Values
           PAT        52    1 2 3 4 5 6 7 8 9 10 11 12 13
                            14 15 16 17 18 19 20 21 22 23
                            24 25 26 27 28 29 30 31 32 33
                            34 35 36 37 38 39 40 41 42 43
                            44 45 46 47 48 49 50 51 52
           OBS         6    e i j n r s

                REML Estimation Iteration History
        Iteration  Evaluations     Objective      Criterion
              0             1     23.41422782
              1             1     23.41422782     0.00000000
                    Convergence criteria met.

           Covariance Parameter Estimates (REML)
                    Cov Parm       Estimate
                    OBS          0.00000000
                    Residual     0.28855385

             Model Fitting Information for GEST
           Description                       Value
           Observations                     104.0000
           Res Log Likelihood               -59.4919
           Akaike's Information Criterion    -61.4919
           Schwarz's Bayesian Criterion     -63.4432
           -2 Res Log Likelihood            118.9838

                  Tests of Fixed Effects
           Source    NDF   DDF   Type III F    Pr>F
           PAT        51    52        71.92    0.0001
```

8.9   COMPONENTS OF VARIATION AND MEAN ESTIMATES IN A CARDIOLOGY EXPERIMENT

In this experiment the heart wall thickness of 11 healthy dogs was measured using ultrasound scans. Each scan consisted of 20 thickness measurements taken over a single heartbeat cycle. Here, we consider the maximum thickness (mm) obtained over the cycle. This was not a carefully planned experiment and a varying number of scans and observers were used for each dog. Each dog had between two and six scans and each scan was assessed by between one and three observers (see Table 8.10). Although the data are not balanced, a random effects model can still be used to estimate variance components and to calculate an appropriate estimate for the mean and standard error of heart wall thickness. Taking the raw mean thickness and its standard error would be inappropriate in this study since greater weight would then be given to the dogs and observers who were used most frequently. A random effects model was fitted with dog, observer, dog·observer and scan effects taken as random. The variance component estimates were

Source	Variance component
Dog	0.294
Obs	0.083
Dog·obs	0.307
Scan	0.705
Residual	0.566

Table 8.10   Number of scans and observers per scan for each dog.

Dog	Number of scans	Minimum number of observers	Maximum number of observers	Mean thickness
Aussie	4	1	2	3.52
Corrie	2	3	3	3.78
Dance	4	1	3	4.27
Gem	2	1	3	5.53
Gus	2	1	1	3.87
Isla	2	2	3	6.54
Jenny	2	3	3	4.51
Jos	2	3	3	4.64
Midge	4	1	1	2.94
Mist	3	1	3	3.78
Tara	2	3	3	3.67

These indicate little systematic variation (bias) between observers. Not surprisingly, there is some additional variation occurring between the heart wall thicknesses of individual dogs. There is also some variation between the observers in their overall readings (i.e. over all scans) for each dog (dog·obs). However, most of the variation occurs at the scan and residual levels. Thus, there is variability between scans even after allowing for between-dog variability, and there is variability in readings made from the same scans by different observers (residual).

The overall mean maximum thickness estimate was 4.42 mm with a standard error of 0.33. This compares with a raw mean of 4.25 mm with standard error 0.18 (calculated taking the naive approach that all observations are independent). While the means are of a similar order, the raw mean standard error does not adequately take into account the variation from each source.

SAS code and output

Variables

dog = dog name,
obs = observer initials,
max = maximum heart wall thickness (mm),
scan = scan ID (dog in CLASS levels below is irrelevant).

```
PROC MIXED; CLASS dog obs scan;
MODEL max =/ SOLUTION;
RANDOM dog obs dog*obs scan;
```

```
                    The MIXED Procedure
                  Class Level Information
          Class     Levels  Values
          DOG          11   Aussie Corrie Dance Gem Gus
                            Isla Jenny Jos Midge Mist Tara
          OBS           3   AA BB CC
          SCAN         29   dog 100 dog 101 dog 102 dog
                            105 dog 106 dog 107 dog 108
                            dog 110 dog 111 dog 112 dog
                            113 dog 114 dog 115 dog 67
                            dog 77 dog 78 dog 79 dog 80
                            dog 83 dog 84 dog 85 dog 86
                            dog 90 dog 92 dog 95 dog 96
                            dog 97 dog 98 dog 99

           REML Estimation Iteration History
      Iteration  Evaluations     Objective     Criterion
              0            1   91.43150790
              1            3   78.57375156     0.00038608
```

```
            2           1   78.55749018   0.00000595
            3           1   78.55725348   0.00000000
                Convergence criteria met.

        Covariance Parameter Estimates (REML)
               Cov Parm         Estimate
               DOG              0.29385853
               OBS              0.08280518
               DOG*OBS          0.30701589
               SCAN             0.70474637
               Residual         0.56559602

            Model Fitting Information for MAX
        Description                        Value
        Observations                     55.0000
        Res Log Likelihood              -88.9013
        Akaike's Information Criterion  -93.9013
        Schwarz's Bayesian Criterion    -98.8738
        -2 Res Log Likelihood           177.8026

                Solution for Fixed Effects
        Effect       Estimate    Std Error   DF    t    Pr>|t|
        INTERCEPT   4.41571805   0.32975802   2   13.39  0.0055
```

8.10   CLUSTER SAMPLE SURVEYS

Sometimes data occur within clusters; for example, patients may be treated at particular hospitals or by particular GPs. In a cluster sample survey, clusters are usually sampled at random. Either all items (e.g. patients) within a cluster are then observed or, alternatively, a random sample of items is taken from each cluster (a two-stage cluster sample). Random variation between clusters can be allowed by fitting cluster effects as random in the analysis. The 'global' results obtained can then be related with some confidence to the population of clusters. Note that inference will be stronger than in multi-centre analyses where centres are rarely sampled randomly. The model can also be used to produce shrunken cluster estimates which help to prevent unrealistic estimates occurring due to chance variation when cluster sizes are small.

8.10.1   Example: Cluster sample survey

This was a cluster sample survey undertaken to determine the prevalence of a disease in animals taken from Thrusfield (1995) (neither disease nor animal species are specified in this reference). There were a total of 865 farms in the population of interest and 14 were sampled at random. All animals at these farms were assessed for the presence of the disease. The disease frequencies and prevalences are listed as follows:

Farm	Animals	Diseased animals	Prevalence
1	272	17	0.063
2	87	15	0.172
3	322	71	0.220
4	176	17	0.097
5	94	9	0.096
6	387	23	0.059
7	279	78	0.280
8	194	59	0.304
9	65	37	0.569
10	110	34	0.309
11	266	23	0.087
12	397	57	0.144
13	152	19	0.125
14	231	17	0.074
Total	3032	476	0.157

A random effects model is fitted to the data using pseudo-likelihood with farm effects taken as random. The data are analysed in binomial form. Since there will be a separate farm effect for each observation, the dispersion parameter is fixed at one to prevent the farm variance component from becoming incorporated into the dispersion parameter. There are no farms with a prevalance of zero. Therefore, there will be no uniform farm effects and we would not expect any bias in the farm variance component (see Section 3.2.4).

The farm variance component was 0.727 and indicates that, not surprisingly, disease rates vary by more than chance variation between the farms. The random effects model gave the logit (SE) for overall disease as -1.684 (0.236), which leads to an overall prevalance rate of 15.7% with 95% confidence limits of 10.0–23.6% (calculated by $-1.684 \pm t_{13,0.975} \times 0.236 = -1.684 \pm 2.160 \times 0.236$ and converting to rates by $(1 + \exp(-\text{logit}))^{-1}$).

This compares with the 'local' raw prevalance rate of 15.7(14.4,17.0)%, which has a much narrower confidence interval since it does not take the between-farm variation into account. (Note also that since there is no estimation of a variance component, the asymptotic normality of the estimate is used to calculate the 95% confidence limits from estimate $\pm 1.96 \times$ SE.) The latter estimate and standard error would only be applicable if the animals were (erroneously) assumed to be a random sample of all animals in the population. In contrast, the 'global' estimate obtained from the random effects model can legitimately be related to the potential population of farms.

SAS code and output

Variables

farm = farm ID,

dis = number of animals with the disease,

n   = number of animals at farm.

```
%GLIMMIX(DATA=a,
    STMTS=%STR(
    CLASS farm;
    MODEL dis/n=/ SOLUTION DDFM=SATTERTH;
    RANDOM farm;
    PARMS (0) (1.0)/ EQCONS=2;
    ),
    ERROR=B);
```

The PARMS statement is used to set the initial values of the variance parameters. The EQCONS = 2 option caused the second variance parameter (which here is the dispersion parameter) to be fixed at 1.0 throughout the analysis.

Class Level Information

Class	Levels	Values
FARM	14	1 2 3 4 5 6 7 8 9 10 11 12 13 14

Covariance Parameter Estimates

Cov Parm	Estimate
FARM	0.72665423

GLIMMIX Model Statistics

Description	Value
Deviance	0.8814
Scaled Deviance	0.8814
Pearson Chi-Square	0.8734
Scaled Pearson Chi-Square	0.8734
Extra-Dispersion Scale	1.0000

Parameter Estimates

Effect	Estimate	Std Error	DF	t	Pr>|t|
INTERCEPT	-1.6843	0.2355	13	-7.15	0.0001

8.11   SMALL AREA MORTALITY ESTIMATES

There is a need to assess health needs of particular populations to allocate medical resources effectively. We consider mortality rates in the 40–64 year age group in Edinburgh by post code area (e.g. EH6). These rates provide some measure of general health in the area and could conceivably have some value for health care planning. However, mortality rates are subject to random variation and may be inaccurate, particularly in small areas. This problem can be alleviated by obtaining shrunken mortality estimates using a random effects model with post code area taken as random. The data are analysed in binomial form with the dispersion parameter fixed at one. This prevents the area variance component from becoming incorporated into the dispersion parameter. The data could alternatively have been analysed in Bernoulli form (i.e. a separate observation for each person), although this would have lead to a very large dataset. If there had been no uniform area categories (i.e. with zero mortality) we would expect the results to have been identical regardless of the data form. Since one of the areas does has a mortality of zero, the Bernoulli analysis might have been marginally preferable since the (unfixed) dispersion parameter would then help to overcome any bias caused by random effects shrinkage (see Section 3.2.4). However, we found that the results changed only very slightly when the data were analysed in this form.

The overall mortality rate of 40–64 years olds in 1991 in Edinburgh was 1.29% (2845/220178). The variance component obtained for post code area is 0.0576, indicating that more variability occurs between the areas than expected by chance variation. This is not surprising since the areas are known to differ in terms of socio-economic factors and age distribution within the 40–64 range. The shrunken mortality estimates are listed in Table 8.11 in increasing order. Note that they are calculated by $\{1 + \exp(-(\text{intercept} + \text{logit}))\}^{-1}$.

The ranking of the areas is quite different between the raw and shrunken mortality rates. As expected, greatest shrinkage towards the overall mortality rate of 1.29% occurs in the areas with smaller populations. For example, 0% mortality is observed in area EH38 with a population of 93, but the shrunken rate of 1.19% is close to the overall rate of 1.29%. The area with the highest raw mortality rate, EH24, with a relatively small population of 386, has its raw rate of 2.85% shrunken to 1.64% and no longer ranks the highest. Any planning decisions based on the mortality rates would differ widely depending on which estimates were taken. Use of shrunken estimates is an important aid to overcoming the problem of extreme estimates that can occur when estimates are based on small populations.

It is possible to use t tests to determine whether the mortality estimates for each area are significantly different from the average. From the SAS output produced at the end of this section we find that post code areas EH1, EH11, EH16, EH6 and EH7 have significantly higher mortality rates ($p = 0.01, 0.0001, 0.0001, 0.04, 0.0001$, respectively), and areas EH10, EH12, EH14 and EH9 have significantly lower mortality rates ($p = 0.0003, 0.003, 0.05, 0.02$, respectively). If required, the estimated standard errors produced by the analysis can be used to calculate confidence intervals for the shrunken mortality rates.

Table 8.11 Mortality and shrunken mortality prediction of 40–64 year olds in Edinburgh by post code area.

Post code sector	Deaths aged 40–64	Population aged 40–64	Shrunken logit estimate	Raw mortality rate per 100	Shrunken mortality rate per 100	Original rank
EH10	60	8298	−0.442	0.72	0.81	3
EH9	38	4704	−0.316	0.81	0.92	7
EH12	99	11105	−0.300	0.89	0.94	9
EH53	19	2456	−0.271	0.77	0.96	4
EH39	20	2399	−0.233	0.83	1.00	8
EH26	53	5550	−0.214	0.95	1.02	15
EH45	23	2534	−0.195	0.91	1.04	11
EH14	122	11922	−0.186	1.02	1.05	16
EH54	114	10956	−0.170	1.04	1.07	17
EH35	5	693	−0.148	0.72	1.09	2
EH32	52	4941	−0.139	1.05	1.10	19
EH51	47	4459	−0.134	1.05	1.11	20
EH34	5	644	−0.126	0.78	1.11	5
EH29	9	999	−0.124	0.90	1.12	10
EH55	24	2294	−0.112	1.05	1.13	18
EH37	4	512	−0.105	0.78	1.14	6
EH31	7	769	−0.102	0.91	1.14	12
EH46	8	853	−0.101	0.94	1.14	14
EH42	25	2315	−0.094	1.08	1.15	21
EH3	52	4630	−0.089	1.12	1.16	22
EH38	0	93	−0.064	0.00	1.19	1
EH41	40	3416	−0.053	1.17	1.20	23
EH30	32	2678	−0.036	1.19	1.22	24
EH20	25	2087	−0.032	1.20	1.22	25
EH19	45	3704	−0.028	1.21	1.23	26
EH21	77	6258	−0.022	1.23	1.23	28
EH36	1	108	−0.019	0.93	1.24	13
EH49	56	4529	−0.016	1.24	1.24	29
EH18	9	732	−0.009	1.23	1.25	27
EH48	102	8130	−0.006	1.25	1.25	30
EH43	3	237	0.000	1.27	1.26	31
EH40	7	548	0.003	1.28	1.27	33
EH25	14	1099	0.004	1.27	1.27	32
EH44	12	906	0.019	1.32	1.29	35
EH33	42	3233	0.020	1.30	1.29	34
EH4	220	16382	0.058	1.34	1.34	36
EH2	3	134	0.068	2.24	1.35	52
EH15	85	6106	0.082	1.39	1.37	37
EH28	14	918	0.082	1.53	1.37	40
EH22	116	8297	0.089	1.40	1.38	38
EH47	113	7649	0.137	1.48	1.44	39
EH13	65	4251	0.152	1.53	1.47	41

Table 8.11 (*continued*).

Post code sector	Deaths aged 40–64	Population aged 40–64	Shrunken logit estimate	Raw mortality rate per 100	Shrunken mortality rate per 100	Original rank
EH5	73	4772	0.156	1.53	1.47	42
EH8	84	5480	0.162	1.53	1.48	43
EH52	78	5044	0.167	1.55	1.49	44
EH17	79	5084	0.171	1.55	1.49	45
EH27	12	569	0.191	2.11	1.52	51
EH6	121	7624	0.202	1.59	1.54	46
EH24	11	386	0.268	2.85	1.64	54
EH23	39	2031	0.287	1.92	1.67	49
EH16	147	7731	0.370	1.90	1.82	47
EH11	165	8604	0.383	1.92	1.84	48
EH7	143	7355	0.389	1.94	1.85	50
EH1	26	970	0.425	2.68	1.92	53

SAS code and output

Variables

pc   = post code area,
death = number of deaths in post code area,
pop   = population in post code area.

```
%GLIMMIX(DATA=a,
   STMTS=%STR(
   CLASS pc;
   MODEL death/pop=/SOLUTION DDFM=SATTERTH;
   RANDOM pc/ SOLUTION;
   PARMS (0) (1.0)/ EQCONS=2;
   ),
   ERROR=B);
                    Class Level Information
       Class    Levels
       PC        54
                         Values
EH1 EH10 EH11 EH12 EH13 EH14 EH15 EH16 EH17 EH18 EH19 EH2 EH20 EH21
EH22 EH23 EH
               Covariance Parameter Estimates
                    Cov
                    Parm       Estimate
                    PC       0.05754614

               GLIMMIX Model Statistics
       Description                   Value
       Deviance                     20.8820
       Scaled Deviance              20.8820
```

```
Pearson Chi-Square              20.4766
Scaled Pearson Chi-Square       20.4766
Extra-Dispersion Scale           1.0000
```

Parameter Estimates

Effect	Estimate	Std Error	DF	t	Pr>|t|
INTERCEPT	-4.3591	0.0419	44.9911	-104.1	0.0001

Random Effects Estimates

Effect	PC	Estimate	SE Pred	DF	t	Pr>|t|
PC	EH1	0.4250	0.1689	53.0000	2.52	0.0149
PC	EH10	-0.4418	0.1138	53.0000	-3.88	0.0003
PC	EH11	0.3825	0.0849	53.0000	4.51	0.0001
PC	EH12	-0.3002	0.0978	53.0000	-3.07	0.0034
PC	EH13	0.1519	0.1173	53.0000	1.30	0.2009
PC	EH14	-0.1863	0.0918	53.0000	-2.03	0.0474
PC	EH15	0.0816	0.1059	53.0000	0.77	0.4441
PC	EH16	0.3701	0.0884	53.0000	4.19	0.0001
PC	EH17	0.1713	0.1095	53.0000	1.56	0.1237
PC	EH18	-0.0092	0.1951	51.2958	-0.05	0.9624
PC	EH19	-0.0282	0.1302	53.0000	-0.22	0.8292
PC	EH2	0.0684	0.2285	27.5615	0.30	0.7667
PC	EH20	-0.0315	0.1552	53.0000	-0.20	0.8397
PC	EH21	-0.0215	0.1087	53.0000	-0.20	0.8442
PC	EH22	0.0894	0.0948	53.0000	0.94	0.3500
PC	EH23	0.2866	0.1429	53.0000	2.01	0.0501
PC	EH24	0.2678	0.2061	41.4435	1.30	0.2010
PC	EH25	0.0039	0.1802	53.0000	0.02	0.9829
PC	EH26	-0.2141	0.1209	53.0000	-1.77	0.0824
PC	EH27	0.1912	0.1969	49.4773	0.97	0.3362
PC	EH28	0.0822	0.1841	53.0000	0.45	0.6572
PC	EH29	-0.1242	0.1883	53.0000	-0.66	0.5123
PC	EH3	-0.0890	0.1234	53.0000	-0.72	0.4736
PC	EH30	-0.0363	0.1446	53.0000	-0.25	0.8029
PC	EH31	-0.1023	0.1964	49.9576	-0.52	0.6047
PC	EH32	-0.1386	0.1227	53.0000	-1.13	0.2635
PC	EH33	0.0202	0.1340	53.0000	0.15	0.8810
PC	EH34	-0.1256	0.2025	44.3836	-0.62	0.5381
PC	EH35	-0.1476	0.2010	45.7005	-0.73	0.4665
PC	EH36	-0.0194	0.2313	26.2532	-0.08	0.9336
PC	EH37	-0.1052	0.2081	39.8347	-0.51	0.6161
PC	EH38	-0.0635	0.2327	25.6122	-0.27	0.7872
PC	EH39	-0.2333	0.1577	53.0000	-1.48	0.1448
PC	EH4	0.0577	0.0761	53.0000	0.76	0.4518
PC	EH40	0.0033	0.2035	43.5278	0.02	0.9872
PC	EH41	-0.0533	0.1347	53.0000	-0.40	0.6939
PC	EH42	-0.0942	0.1533	53.0000	-0.61	0.5416
PC	EH43	0.0003	0.2218	30.9770	0.00	0.9988
PC	EH44	0.0194	0.1868	53.0000	0.10	0.9177
PC	EH45	-0.1951	0.1537	53.0000	-1.27	0.2100
PC	EH46	-0.1008	0.1930	53.0000	-0.52	0.6036
PC	EH47	0.1373	0.0960	53.0000	1.43	0.1584
PC	EH48	-0.0057	0.0987	53.0000	-0.06	0.9544

PC	EH49	-0.0163	0.1213	53.0000	-0.13	0.8937
PC	EH5	0.1560	0.1125	53.0000	1.39	0.1715
PC	EH51	-0.1340	0.1268	53.0000	-1.06	0.2953
PC	EH52	0.1669	0.1100	53.0000	1.52	0.1350
PC	EH53	-0.2714	0.1583	53.0000	-1.71	0.0923
PC	EH54	-0.1701	0.0940	53.0000	-1.81	0.0762
PC	EH55	-0.1118	0.1545	53.0000	-0.72	0.4726
PC	EH6	0.2019	0.0940	53.0000	2.15	0.0363
PC	EH7	0.3890	0.0894	53.0000	4.35	0.0001
PC	EH8	0.1619	0.1070	53.0000	1.51	0.1363
PC	EH9	-0.3155	0.1320	53.0000	-2.39	0.0205

8.12 ESTIMATING SURGEON PERFORMANCE

The performance of nine surgeons undertaking mastectomy operations at an Edinburgh hospital was recorded in terms of whether any post-operative complications arose (reported by Dixon *et al.*, 1996). However, some of the surgeons only performed a few operations and therefore their complication rates are likely to be unreliable. More appropriate shrunken estimates can be obtained by fitting a random effects model with surgeons taken as random. The data were analysed in binomial form with the dispersion parameter fixed at one.

The overall complication rate was 42.6%. The surgeon variance component was 1.296, indicating greater variability between surgeons than expected by chance. The raw and shrunken complication rates are given below. As in the previous example the shrunken rates are calculated by $\{1 + \exp(-(\text{intercept} + \text{logit}))\}^{-1}$.

Surgeon number	Number of complications	Number of operations	Shrunken logit estimate	Complication rate (%)	Shrunken complication rate (%)	Original rank
1	7	91	-1.957	8	9	1.0
3	3	15	-0.825	20	25	2.0
9	1	3	-0.179	33	38	3.0
5	2	5	-0.057	40	41	4.0
4	7	13	0.375	54	52	5.0
6	3	5	0.438	60	53	6.0
8	2	3	0.479	67	54	7.5
2	10	15	0.818	67	63	7.5
7	2	2	0.908	100	65	9.0

Although the ranking of the surgeons does not change, some of the shrunken rates have changed noticeably from the raw rates. For example, surgeon 7 has a raw complication rate of 100% based on only two operations, but his shrunken rate of 65% is more acceptable. However, even after shrinkage it is clear that complication rates differ widely between the surgeons. From the SAS output produced at the end of this section we find that surgeon 1 has a significantly lower complication rate than average ($p = 0.007$), but that no surgeon has a significantly higher rate than average. If required, the estimated standard errors produced by the analysis can be used to calculate confidence intervals for the complication rate of each surgeon.

SAS code and output

Variables

surg   = surgeon ID,

outcome = number of patients with post-operative complications,

n   = number of patients operated on by surgeon.

```
%GLIMMIX(DATA=a,
  STMTS=%STR(
  CLASS surg;
  MODEL outcome/n=/ SOLUTION DDFM=SATTHERTH;
  RANDOM surg/ SOLUTION;
  PARMS (0) (1.0)/ EQCONS=2;
  ),
  ERROR=B
  );
```

Class Level Information

Class	Levels	Values
SURG	9	1 2 3 4 5 6 7 8 9

Covariance Parameter Estimates

Cov Parm	Estimate
SURG	1.27790041

GLIMMIX Model Statistics

Description	Value
Deviance	2.6973
Scaled Deviance	2.6973
Pearson Chi-Square	2.0110
Scaled Pearson Chi-Square	2.0110
Extra-Dispersion Scale	1.0000

Parameter Estimates

Effect	Estimate	Std Error	DF	t	Pr>|t|
INTERCEPT	-0.3115	0.4668	8	-0.67	0.5234

Random Effects Estimates

Effect	SURG	Estimate	SE Pred	DF	t	Pr>|t|
SURG	1	-1.9572	0.5451	8	-3.59	0.0071
SURG	2	0.8180	0.6151	8	1.33	0.2202
SURG	3	-0.8247	0.6439	8	-1.28	0.2361

SURG	4	0.3750	0.6243	8	0.60	0.5646
SURG	5	-0.0570	0.7632	8	-0.07	0.9423
SURG	6	0.4377	0.7585	8	0.58	0.5798
SURG	7	0.9078	0.9142	8	0.99	0.3498
SURG	8	0.4789	0.8406	8	0.57	0.5845
SURG	9	-0.1785	0.8489	8	-0.21	0.8387

8.13   EVENT HISTORY ANALYSIS

Sometimes data on the exact times of a particular event (or events) are available on a group of patients. Examples of events would include: asthma attacks; epilepsy attacks; myocardial infarctions; hospital admissions. Often, occurrence (and non-occurrence) of an event is available on a regular basis (e.g. daily) and the data can then be thought of as having a repeated measures structure. An objective may be to determine whether any concurrent events or measurements have influenced the occurrence of the event of interest. For example, daily pollen counts may influence the risk of asthma attacks; high blood pressure may precede a myocardial infarction.

8.13.1   Example

We will consider data from a placebo controlled study of a treatment for eczema, which can cause itchiness. Thirty-four female subjects were asked to complete a diary recording the severity of their eczema and days of menstrual bleeding. This was done daily during a four-week, run-in period and for six months following treatment. We will consider analysing the occurrence of severe itchiness. To assess whether itchiness was related to the menstrual cycle, a covariance pattern model was fitted with average pre-treatment itchiness, treatment (active or placebo) and cycle (menstrual bleeding, y/n) taken as fixed effects, and with a compound symmetry covariance pattern to model the correlation between the repeated observations on each subject. Many of the subjects had not recorded their symptoms on every day or had stopped filling in their diaries before the end of the trial. However, such missing data did not pose problems in fitting a covariance pattern model.

This analysis produced a compound symmetry correlation parameter of 0.27 showing a moderate correlation between itchiness records on the same subjects. The cycle effect was significant ($p = 0.01$) and subjects were more likely to experience severe itchiness during menstrual bleeding. The cycle odds ratio and 95% confidence interval based on empirical standard errors was 1.41 (1.07,1.86). We suggest that the empirical standard errors are taken because more complex covariance patterns have not been explored and it is possible that the true covariance pattern is not compound symmetry. In this instance, the empirical standard errors are larger than the model-based standard errors, making this the conservative approach.

SAS code and output

Variables

`treat`	treatment (a = active, p = placebo)
`pat`	subject
`cycle`	menstrual bleeding (1 = yes, 2 = no)
`itch1`	proportion of days of severe itchiness pre-treatment
`one`	=1 for all observations

```
PROC GENMOD; CLASS pat treat cycle;
MODEL itch/one = itch1 treat cycle/ DIST=B;
REPEATED SUB=pat/ TYPE=CS CORRW MODELSE;
```

```
                    The GENMOD Procedure
                      Model Information

                Description              Value

                Data Set                 WORK.A
                Distribution             BINOMIAL
                Link Function            LOGIT
                Dependent Variable       ITCH
                Dependent Variable       ONE
                Observations Used        4883
                Number Of Events         1084
                Number Of Trials         4883
                Missing Values           42

                     Class Level Information
          Class      Levels  Values
          PAT            34   101 102 103 104 105 107 108
                             109 110 112 114 115 116 117
                             119 120 121 122 123 124 125
                             126 201 202 203 204 206 207
                             209 210 212 213 214 216
          TREAT           2   a p
          CYCLE           2   1 2

                     Parameter Information
             Parameter     Effect      TREAT   CYCLE
             PRM1          INTERCEPT
             PRM2          ITCH1
             PRM3          TREAT         a
             PRM4          TREAT         p
             PRM5          CYCLE                  1
             PRM6          CYCLE                  2

           Criteria For Assessing Goodness Of Fit
         Criterion           DF       Value    Value/DF
         Deviance           4879   5005.5344     1.0259
         Scaled Deviance    4879   5005.5344     1.0259
         Pearson Chi-Square 4879   4859.8497     0.9961
```

```
Scaled Pearson X2    4879      4859.8497      0.9961
Log Likelihood          .     -2502.7672          .
```

```
              Analysis Of Initial Parameter Estimates
     Parameter    DF    Estimate    Std Err   ChiSquare   Pr>Chi
     INTERCEPT    1     -1.7708     0.0651    740.1724    0.0001
     ITCH1        1      1.9283     0.1563    152.2795    0.0001
     TREAT   a    1      0.0393     0.0714      0.3037    0.5816
     TREAT   p    0      0.0000     0.0000          .         .
     CYCLE   1    1      0.3212     0.0834     14.8415    0.0001
     CYCLE   2    0      0.0000     0.0000          .         .
     SCALE        0      1.0000     0.0000          .         .
```

NOTE: The scale parameter was held fixed.

Note that these parameter estimates are those obtained from PROC GENMOD without using a REPEATED statement and therefore can be ignored.

```
                  GEE Model Information
     Description                         Value
     Correlation Structure               Exchangeable
     Subject Effect                      PAT (34 levels)
     Number of Clusters                  34
     Clusters With Missing Values        8
     Correlation Matrix Dimension        179
     Maximum Cluster Size                179
     Minimum Cluster Size                77
```

```
                    Working Correlation Matrix
          COL1      COL2      COL3      COL4      COL5     COL6      COL7   ETC
ROW1    1.0000    0.2656    0.2656    0.2656    0.2656   0.2656    0.2656
ROW2    0.2656    1.0000    0.2656    0.2656    0.2656   0.2656    0.2656
ROW3    0.2656    0.2656    1.0000    0.2656    0.2656   0.2656    0.2656
ROW4    0.2656    0.2656    0.2656    1.0000    0.2656   0.2656    0.2656
ROW5    0.2656    0.2656    0.2656    0.2656    1.0000   0.2656    0.2656
ROW6    0.2656    0.2656    0.2656    0.2656    0.2656   1.0000    0.2656
  .
  .
```

This is the within-subject correlation matrix, **P**, estimated by the GLMM with a single parameter to give the compound symmetry pattern. Note that the matrix is very large, because there are many observations per subject.

```
             Analysis Of GEE Parameter Estimates
             Empirical Standard Error Estimates

                    Empirical 95% Confidence Limits
Parameter   Estimate   Std Err    Lower     Upper        Z      Pr>|Z|
INTERCEPT    -1.8595    0.3705   -2.5856   -1.1334   -5.019     0.0000
ITCH1         2.0069    1.0377   -0.0269    4.0408    1.9340    0.0531
TREAT    a    0.0982    0.4338   -0.7520    0.9484    0.2265    0.8208
TREAT    p    0.0000    0.0000    0.0000    0.0000    0.0000    0.0000
```

CYCLE	1	0.3440	0.1398	0.0700	0.6180	2.4608	0.0139
CYCLE	2	0.0000	0.0000	0.0000	0.0000	0.0000	0.0000
Scale		1.0087	.	.	.	.	.

NOTE: The scale parameter for GEE estimation was computed as the square root of the normalized Pearson's chi-square.

Analysis Of GEE Parameter Estimates
Model-Based Standard Error Estimates

Parameter		Estimate	Std Err	Model 95% Confidence Limits Lower	Upper	Z	Pr>|Z|
INTERCEPT		-1.8595	0.3785	-2.6014	-1.1176	-4.913	0.0000
ITCH1		2.0069	0.8824	0.2775	3.7364	2.2745	0.0229
TREAT	a	0.0982	0.4371	-0.7584	0.9549	0.2248	0.8222
TREAT	p	0.0000	0.0000	0.0000	0.0000	0.0000	0.0000
CYCLE	1	0.3440	0.0763	0.1945	0.4935	4.5090	0.0000
CYCLE	2	0.0000	0.0000	0.0000	0.0000	0.0000	0.0000
Scale		1.0087	.	.	.	.	.

9

Software for Fitting Mixed Models

In this chapter we will look at mixed models software with a particular emphasis on the SAS package which has been used to analyse the majority of our examples. In Section 9.1 general information is given on some of the packages and programs that are available for fitting mixed models. Basic details on the use of the SAS procedure PROC MIXED for fitting normal mixed models are given in Section 9.2. In Section 9.3 we give details on using PROC GENMOD and the SAS macro GLIMMIX for fitting GLMMs.

9.1 PACKAGES FOR FITTING MIXED MODELS

SAS   The PROC MIXED procedure in SAS is to our knowledge the most versatile software available for fitting mixed models to normal data. It can be used to fit all types of mixed models (random effects, random coefficients and covariance pattern models) and more detail on its capabilities will be given in Section 9.2.

There is not yet a SAS procedure available to fit GLMMs. However, GLMMs with covariance patterns but no random effects can be fitted using the REPEATED option in PROC GENMOD introduced in Release 6.12 of SAS. A SAS macro known as GLIMMIX is also available for fitting all types of GLMMs. This may be obtained from the SAS web site www.sas.com.

There is no SAS procedure available for fitting models to categorical data. We have used a macro written by Stuart Lipsitz (Lipsitz *et al.*, 1994) designed for fitting covariance pattern models, which works by iteratively calling PROC LOGISTIC. It may be obtained from Web page [www.med.ed.ac.uk/phs/mixed].

MLWin   MLWin is a package that has been developed by the Multilevel Modelling Project at the Institute of Education in London. It fits mixed models (multilevel models are, in fact, mixed models) by using iterative generalised weighted least squares methods (see Section 2.2.1). It can be used to fit both normal mixed models and GLMMs, and macros are available for fitting categorical mixed models. However, the choice of covariance patterns available is much more limited than in PROC MIXED and we have found it less user friendly. Further information can be found on Web page www.ioe.ac.uk/multilevel.

Bayesian software   BUGS is a package dedicated to Bayesian analysis using the Gibbs sampler and has been developed by the Medical Research Council Biostatistics Unit in Cambridge. It can be used to fit random effects models to all types of data. The package can be obtained from Web page www.mrc-bsu.cam.ac.uk/bugs (it is free of charge at the time of writing).

Other software   Genstat (REML and VCOMPONENTS directives) and BMDP (3V and 5V procedures) contain software for fitting normal mixed models. Addresses for these packages are given in the reference section. Additionally, information on other lesser-known packages can be found on Web page www.ioe.ac.uk/multilevel/softgen.html. In particular, a Fortran program written by Donald Hedeker, MIXOR, is available for fitting random effects and random coefficients models to ordered categorical data (see Section 4.2). This software can be obtained free of charge from Web page www.ioe.ac.uk/multilevel/mixreg.html.

9.2   BASIC USE OF PROC MIXED

PROC MIXED is a SAS procedure for fitting normal mixed models. It can be used to fit any type of mixed model (random effects, random coefficients, covariance pattern or a combination). It has great flexibility and there are many options available for defining mixed models and for requesting output. By default, the REML estimation method is applied using Newton–Raphson iteration with Fisher scoring for the first iteration (see Section 2.2.4). Alternatively, a Bayesian analysis is available for random effects and random coefficients models (see Section 2.3). Documentation for Release 6.12 can be found in the manual *SAS/STAT Software: Changes and Enhancements through Release 6.12*. Another excellent text on using SAS to fit mixed models is *SAS System for Mixed Models* by Littell *et al.* (1996).

Our aim in this Section is to give a basic description of the most useful PROC MIXED statements and options to enable those not wishing to learn the procedure in depth to perform mixed model analyses. The presentation will assume a working knowledge of SAS.

9.2.1   Syntax

The general syntax and statements available in PROC MIXED are as follows:

```
PROC MIXED options;
CLASS <effect names>;
MODEL <dependent> = <fixed effects>/ options;
RANDOM <random effects>/ options;
LSMEANS <fixed effects>/ options;
```

```
ESTIMATE '<estimate label>' <estimate definition>/
                                       options;
CONTRAST '<contrast label>' <contrast definition>/
                                       options;
REPEATED <repeated effect>/ SUBJECT=<blocking effect>/
                                       options;
PARMS <grid search definition>/ options;
MAKE '<table>' OUT=<dataset>;
PRIOR <prior distribution>/ options;
ID <ID variables>;
WEIGHT <weighting variable>;
```

Simple example: random effects model

Data from the multicentre hypertension trial introduced in Section 1.3 will be used to illustrate the use of PROC MIXED without options to fit a simple random effects model. The code below fits a random effects model with pre-treatment DBP (dbp1) and treatment effects fixed, and centre and centre-treatment effects random.

```
PROC MIXED; CLASS centre treat;
MODEL dbp = dbp1 treat;
RANDOM centre centre*treat;
```

```
                    The MIXED Procedure
                 Class Level Information
       Class      Levels     Values
       CENTRE        29       1 2 3 4 5 6 7 8 9 11 12 13 14
                             15 18 23 24 25 26 27 29 30 31
                             32 35 36 37 40 41
       TREAT          3       A B C

           REML Estimation Iteration History
     Iteration  Evaluations      Objective     Criterion
             0            1     1550.3451721
             1            3     1533.6847949    0.00000322
             2            1     1533.6822800    0.00000000
                 Convergence criteria met.
```

The above table shows how quickly the algorithm has converged. The criterion is a measure of convergence and should be very close to zero. In this example convergence has been reached quickly. The objective function is -2 times the REML log likelihood, omitting the constant term. If this value is very large and negative, then it is likely that the covariance matrix, **V**, is singular, leading to an infinite likelihood. In this situation the results would be invalid and the model would need to be respecified, probably refitting certain random effects as fixed or removing them altogether. The 'evaluations' column gives the number of evaluations of the objective function carried out at each iteration. Occasionally, a message stating that the Hessian matrix is not positive definite will appear. In this situation it is possible that only a local maximum has been reached and a 'grid search' for other solutions might be advisable (see the PARMS statement).

```
Covariance Parameter Estimates (REML)

Cov Parm                      Estimate

CENTRE                        6.46282024
CENTRE*TREAT                  4.09615054
Residual                     68.36773938
```

This gives the variance parameter estimates. The variance component for centres is 6.46, and for the centre·treatment interaction 4.10. The residual estimate is 68.37.

```
Model Fitting Information for DBP

Description                              Value

Observations                           288.0000
Res Log Likelihood                     -1027.82
Akaike's Information Criterion         -1030.82
Schwarz's Bayesian Criterion          -1036.29
-2 Res Log Likelihood                  2055.639
```

This table gives information about the model fit. The `Res Log Likelihood` is the REML log likelihood. Following this are several statistics based on the likelihood value. Akaike's criteria ($L - q$, where L is the log likelihood and q is the number of variance components) and Schwartz's criteria ($L - q \log(N - p)/2$, where N = number of observations and p = number of fixed effects) can be used to make direct comparisons of the fit of different models. These are of greatest use for choosing between different covariance patterns (see Section 6.2).

```
Tests of Fixed Effects

Source    NDF    DDF    Type III F    Pr>F

DBP1       1     208         6.31    0.0128
TREAT      2      48         2.28    0.1131
```

Results from F tests are given for all fixed effects. The 'Type III' tests are the Wald tests described in Section 2.4.4. The denominator DF for the F tests is given by the residual DF or, if the fixed effect is contained, by the DF of the containing effect. Here, `DBP1` is tested using the residual DF of 208 and `TREAT` using the centre·treatment DF of 48 since it is contained within this effect. Other methods for calculating the denominator DF are available by using the `DDFM` option in the `MODEL` statement (see below).

9.2.2   PROC MIXED statement options

We will now give details on the use of each `PROC MIXED` statement.

PROC MIXED options

METHOD = \<method\>	Specifies estimation method
= REML	REML (default)
= ML	Maximum likelihood
= MINQUE	Minimum variance quadratic variance estimation. This is an estimation method that we have not covered. It is based on equating mean squares to their expected values and it is less computationally expensive than maximum likelihood. Further information can be found in Rao (1971, 1972). Searle *et al.* (1992) show that the solution is equivalent to that obtained using just one iteration with REML.
ASYCOV	Prints asymptotic covariance matrix of variance components.
EMPIRICAL	Empirical estimator of variance matrix of fixed effects, $\text{var}(\hat{\boldsymbol{\alpha}}) = (\mathbf{X}'\mathbf{V}^{-1}\mathbf{X})^{-1}(\mathbf{X}'\mathbf{V}^{-1}\text{cov}(\mathbf{y})\,\mathbf{V}^{-1}\mathbf{X})(\mathbf{X}'\mathbf{V}^{-1}\mathbf{X})^{-1}$, is used in place of $\mathbf{V}$ for all fixed effect variance estimates in covariance pattern models fitted using REPEATED statement (see Section 2.4.3).
NOCLPRINT	Suppresses printing of class levels. (This is useful when there are many categories.)

Example

```
PROC MIXED ASYCOV; CLASS centre treat;
MODEL dbp = dbp1 treat;
RANDOM centre centre*treat;
```

The output below occurs in addition to the output above. Use of the EMPIRICAL option is illustrated for repeated measures data in Section 6.3.

```
            Asymptotic Covariance Matrix of Estimates

    Cov Parm      Row        COVP1          COVP2          COVP3

    CENTRE          1   18.93134429    -9.68534402     1.90580654
    CENTRE*TREAT    2   -9.68534402    28.42105962   -12.21943311
    Residual        3    1.90580654   -12.21943311    42.52634677
```

This matrix gives an indication of the correlation between the variance components. The covariance values are only asymptotically normal and are only likely to be accurate when large numbers of DF have been used for estimating each variance component. The COVP1, COVP2 and COVP3 column headings relate to the centre, centre-treatment and residual variance components.

The `MODEL` statement

This statement is used to specify the dependent variable and the fixed effects in the model. Options are available for significance tests and for requesting specific output to be printed.

`MODEL <dependent> = <fixed effects>/ options;`

Options relating to the model specification

`NOINT`	Requests no intercept be included in the model.

Options relating to statistical tests

`CHISQ`	Carries out chi-squared tests in addition to the F tests of fixed effects.
`DDFM=<DF type>`	Selects DF type for F test denominator DF for all F tests.
`RESIDUAL`	Uses residual DF.
`CONTAIN`	If the effect of interest is contained within another effect, then the DF of the containing effect is used. If the effect is not contained, then residual DF is used. This is the default DF.
`BETWITHIN`	Assigns between-or within-subject DF to effects when `REPEATED` statement is used. Effects nested within the `SUBJECT` effect take the subject effect DF, others take the residual DF.
`SATTERTH`	Uses Satterthwaite's approximation to the true DF (see Section 2.4.4).

Options relating to output

`SOLUTION`	Solution for fixed effects is printed.
`CL`	Requests t-type confidence intervals for fixed effects given by `SOLUTION` option.
`ALPHA=<p>`	Specifies size of confidence intervals (default 0.05).
`E3`	Prints design matrix for all fixed effects.
`PREDICTED`	For each observation gives predicted values and residuals based on fixed and random effects ($\mathbf{X}\hat{\alpha} + \mathbf{Z}\hat{\beta}$ and $\mathbf{y} - \mathbf{X}\hat{\alpha} - \mathbf{Z}\hat{\beta}$). Approximate standard errors and 95% confidence intervals for each predicted value are also listed.
`PM`	For each observation gives predicted values and residuals based on fixed effects only ($\mathbf{X}\hat{\alpha}$ and $\mathbf{y} - \mathbf{X}\hat{\alpha}$). Approximate standard errors and 95% confidence intervals for each predicted value are also listed.

Example

```
PROC MIXED; CLASS centre treat;
MODEL dbp = dbp1 treat/ CL E3 PREDICTED PM SOLUTION
                                     DDFM=SATT;
RANDOM centre centre*treat;
```

Use of the SOLUTION and CL options

```
                    Solution for Fixed Effects

Parameter    Estimate    Std Error    DDF     T    Pr>|T|  Alpha   Lower   Upper

INTERCEPT  61.76382843  11.10659489   284   5.56  0.0001  0.05  39.9021 83.6255
DBP1        0.26890417   0.10703183   284   2.51  0.0125  0.05   0.0582  0.4796
TREAT A     2.92739048   1.37270937  25.6   2.13  0.0427  0.05   0.1036  5.7512
TREAT B     1.64147028   1.40487730  25.7   1.17  0.2534  0.05  -1.2482  4.5312
TREAT C     0.00000000      .          .      .      .      .      .       .
```

Use of the E3 option

```
           Type III Coefficients for DBP1

            Parameter          Row 1
            INTERCEPT            0
            DBP1                 1
            TREAT A              0
            TREAT B              0
            TREAT C              0

           Type III Coefficients for TREAT

            Parameter          Row 1       Row 2

            INTERCEPT            0           0
            DBP1                 0           0
            TREAT A              1           0
            TREAT B              0           1
            TREAT C             -1          -1
```

This shows that SAS has parameterised treatment effects by using differences from the last treatment (C).

Use of the DDFM=SATT option

```
              Tests of Fixed Effects

        Source  NDF  DDF  Type III F    Pr>F
        DBP1     1   284     6.31      0.0125
        TREAT    2    25     2.28      0.1230
```

Note that the DF now differ from the earlier analysis where the SATTERTH option was not used.

Use of the PM option

```
                  Predicted Means

      DBP Predicted SE Pred     L95      U95    Residual
   108.0000   91.5816  1.1571 89.3006 93.8627   16.4184
   116.0000   91.5816  1.1571 89.3006 93.8627   24.4184
    88.0000   91.5816  1.1571 89.3006 93.8627   -3.5816
    82.0000   93.1951  1.1404 90.9469 95.4432  -11.1951
    78.0000   94.2707  1.3165 91.6754 96.8660  -16.2707
```

```
98.0000    92.6573  1.1052 90.4783 94.8362    5.3427
88.0000    91.5816  1.1571 89.3006 93.8627   -3.5816
80.0000    92.6573  1.1052 90.4783 94.8362  -12.6573
82.0000    93.7329  1.2128 91.3419 96.1238  -11.7329
   .
   .
```

Use of the PREDICTED option

```
                      Predicted Values

        DBP Predicted SE Pred     L95     U95  Residual
   108.0000   91.9177  1.8098 88.3499 95.4855   16.0823
   116.0000   91.9177  1.8098 88.3499 95.4855   24.0823
    88.0000   91.9177  1.8098 88.3499 95.4855   -3.9177
    82.0000   93.5311  1.8168 89.9495 97.1128  -11.5311
    78.0000   94.6068  1.9431 90.7760 98.4375  -16.6068
    98.0000   92.9933  1.7890 89.4664 96.5202    5.0067
    88.0000   91.9177  1.8098 88.3499 95.4855   -3.9177
    80.0000   92.9933  1.7890 89.4664 96.5202  -12.9933
    82.0000   94.0690  1.8688 90.3847 97.7532  -12.0690
     .
     .
```

The RANDOM statement

This statement can be used to specify random effects and/or random coefficients, β, and the form of their variance matrix, **G**. Several RANDOM statements may be specified, although usually only one will be needed. When no RANDOM statement is included the results will be the same as those obtained using PROC GLM.

```
RANDOM <random effects>/ options;
```

Output options

CL   Requests t-type confidence intervals for each random effect estimate.

ALPHA=<p>   Specifies size of confidence interval (default is 0.05).

SOLUTION   Solution for random effects is printed.

Options corresponding to the G and V matrices

G	Prints **G**.
GCORR	Prints correlation matrix corresponding to **G**.
GI	Prints inverse of **G**.
V	Prints **V**.
VCORR	Prints correlation matrix corresponding to **V**.
VI	Prints inverse of **V**.
GDATA=<dataset>	Specifies a fixed **G** matrix.
RATIOS	Indicates that the **G** matrix given by GDATA is specified in terms of ratios to the residual.

Example

```
PROC MIXED; CLASS centre treat;
MODEL dbp = dbp1 treat/DDFM=SATTERTH;
RANDOM centre centre*treat/ V CL ALPHA=0.01 SOLUTION;
```

Use of the V option

V Matrix for Subject 1

Row	COL1	COL2	COL3	COL4	COL5
1	78.92671016	10.55897078	10.55897078	10.55897078	10.55897078
2	10.55897078	78.92671016	10.55897078	10.55897078	10.55897078
3	10.55897078	10.55897078	78.92671016	10.55897078	10.55897078
4	10.55897078	10.55897078	10.55897078	78.92671016	10.55897078
5	10.55897078	10.55897078	10.55897078	10.55897078	78.92671016
6	10.55897078	10.55897078	10.55897078	10.55897078	10.55897078
7	10.55897078	10.55897078	10.55897078	10.55897078	10.55897078
8	10.55897078	10.55897078	10.55897078	10.55897078	10.55897078
9	10.55897078	10.55897078	10.55897078	10.55897078	10.55897078
10	10.55897078	10.55897078	10.55897078	10.55897078	10.55897078
11	10.55897078	10.55897078	10.55897078	10.55897078	10.55897078
12	10.55897078	10.55897078	10.55897078	10.55897078	10.55897078
13	10.55897078	10.55897078	10.55897078	10.55897078	10.55897078
14	6.46282024	6.46282024	6.46282024	6.46282024	6.46282024
15	6.46282024	6.46282024	6.46282024	6.46282024	6.46282024
.					
.					

This matrix has a dimension equal to the number of observations in the dataset (unless the SUBJECT option is used in the RANDOM statement) and is hence very large in this example. Note that the SAS output wrongly states that the **V** matrix is only for Subject 1. The diagonal terms are equal to the sum of the variance components. Off-diagonal terms are equal to the sum of the centre and centre·treatment variance components, 10.6, when patients are at the same centre and receive the same treatment (the SAS dataset has been ordered so that the first 13 patients received treatment A at centre 1); to the centre variance component, 6.5, when patients are at the same centre but receive different treatments; and are zero when patients are at different centres.

Use of the SOLUTION option

Solution for Random Effects

Effect	CENTRE	TREAT	Estimate	SE Pred	DF	t	Pr>|t|
CENTRE	1		0.79661438	1.54552072	20.7	0.52	0.6117
CENTRE	2		-2.34405654	1.93373015	14.5	-1.21	0.2448
CENTRE	3		1.97599802	2.01060030	12.1	0.98	0.3449
.							

```
.
.
CENTRE          31                -4.79132011  1.56171590  21.5  -3.07  0.0057
.
.

CENTRE*TREAT   1    A    -0.46052816  1.68465564  2.36  -0.27  0.8067
CENTRE*TREAT   1    B     1.68999268  1.67788596  2.38   1.01  0.4049
CENTRE*TREAT   1    C    -0.72456847  1.69668457  2.3   -0.43  0.7061
CENTRE*TREAT   2    A    -0.83296961  1.89086285  1.56  -0.44  0.7129
CENTRE*TREAT   2    B    -1.18755089  1.86304412  1.66  -0.64  0.6004
.
.
```

The `Estimate` and `SE Pred` columns give the random effect estimates and their standard errors. The Wald *t* tests help to determine whether any centre is outlying. Here, centre 31 is a potential outlier.

Options for specifying the G matrix structure

`GROUP=<effect>` This causes a separate **G** matrix to be estimated within each group. For example, if patient effects were fitted as random in a repeated measures analysis and the `GROUP=treat` option were used, then a separate variance patient component would be estimated for each treatment group. If patients 1 and 3 received treatment A and patients 2 and 4 received treatment B, then the **G** matrix for the first four patients would have the form

$$\mathbf{G} = \begin{pmatrix} \sigma_A^2 & 0 & 0 & 0 \\ 0 & \sigma_B^2 & 0 & 0 \\ 0 & 0 & \sigma_A^2 & 0 \\ 0 & 0 & 0 & \sigma_B^2 \end{pmatrix}$$

The statement should be used cautiously when there are many group levels since a large number of parameters will then need to be estimated.

Non-diagonal G matrices   The **G** matrix is always diagonal in random effects models and the options below are not required. However, they are necessary for random coefficient models to allow the intercepts and slopes of the random coefficients to be correlated. They are also required for less usual covariance pattern models that have correlated random effects and hence a non-diagonal **G** matrix (see Section 2.1.4). The **G** matrix is blocked by the specified `SUBJECT` effect and the covariance pattern is specified by the `TYPE` options in a very similar way to the `REPEATED` statement (see below).

```
RANDOM <random effects or coefficients>/
  SUBJECT=<blocking effect>
  TYPE=<covariance pattern>;
```

Covariance patterns available for the TYPE= option will be listed under the REPEATED statement.

Random coefficients models   In these models the random coefficients on the same subject are assumed to be correlated (e.g. intercepts and slope effects are correlated for each subject in a repeated measures trial). This is achieved using a RANDOM statement of the following form,

```
RANDOM <random coefficients>/ SUBJECT=<subject effect>
                                            TYPE=UN;
```

For example, to model random coefficients for the quadratic effect of time in a repeated measures analysis, the code might be

```
RANDOM INT time time2/ SUBJECT=patient TYPE=UN;
```

This fits the random coefficients patient, patient·time and patient·time^2. Patient effects are specified by the INT (intercept) term.

Example
Examples of fitting random coefficients models are given in Section 6.6.

The LSMEANS statement

This statement calculates the least squares mean estimates of specified fixed effects.

```
LSMEANS <fixed effects>/ options;
```

Options

CL	Requests t-type confidence intervals for least squares means.
ALPHA=<p>	Specifies size of confidence interval (default is 0.05).
DF=<df>	Specifies DF for t test of hypothesis that the least squares means equal zero.
ADJUST=<type>	Requests adjusted p-values for multiple comparisons (see manual for further details)
BON	
DUNNETT	
SCHEFFE	
SIDAK	
TUKEY	
COV	Prints covariances of least squares means.
CORR	Prints correlations of least squares means.
DIFF	Prints differences between each pair of least squares means.
PDIFF	Prints p-values for comparisons between each pair of least squares means.

Example

```
PROC MIXED; CLASS centre treat;
MODEL dbp = dbp1 treat/ DDFM=SATTERTH;
RANDOM centre centre*treat;
LSMEANS treat/ CL ADJUST=BON COV CORR DIFF PDIFF;
```

Use of the CL, COV and CORR options in LSMEANS statement

Least Squares Means

Effect	TREAT	LSMEAN	Std Error	DF	t	Pr>|t|	Alpha
TREAT	A	92.34913293	1.10348051	52	83.69	0.0001	0.05
TREAT	B	91.06321274	1.14545866	51.4	79.50	0.0001	0.05
TREAT	C	89.42174246	1.11231265	58.7	80.39	0.0001	0.05

Least Squares Means

Lower	Upper	COV1	COV2	COV3	CORR1	CORR2	CORR3
90.1349	94.5634	1.22	0.30	0.29	1.00	0.23	0.23
88.7640	93.3624	0.30	1.31	0.29	0.23	1.00	0.23
87.1957	91.6477	0.29	0.29	1.24	0.23	0.23	1.00

The last six columns show covariance and correlation matrices for the least squares means.

Use of the DIFF, PDIFF and ADJUST options

Differences of Least Squares Means

Effect	TREAT	_TREAT	Difference	Std Error	DF	t	Pr>|t|
TREAT	A	B	1.28592019	1.39135076	23.8	0.92	0.3646
TREAT	A	C	2.92739048	1.37270937	25.6	2.13	0.0427
TREAT	B	C	1.64147028	1.40487730	25.7	1.17	0.2534

Differences of Least Squares Means

Adjustment	Adj P	Alpha	Lower	Upper	Adj Low	Adj Upp
Bonferroni	1.0000	0.05	-1.5867	4.1585	-2.2844	4.8562
Bonferroni	0.1289	0.05	0.1036	5.7512	-0.5951	6.4498
Bonferroni	0.7610	0.05	-1.2482	4.5312	-1.9635	5.2465

The Bonferroni adjustment works by multiplying the *p*-value by the number of comparisons made. The denominator DF for the least squares means and their differences is calculated using Satterthwaite's approximation, because this was specified in the MODEL statement. The DF differ markedly between the least squares means and their differences, because they are estimated using information from different combinations of error strata.

The ESTIMATE statement

This statement calculates a linear function of fixed and/or random effects estimates ($\hat{\alpha}$ and $\hat{\beta}$).

```
ESTIMATE '<label>' <fixed effect1> <values>
                   <fixed effect2><values>
                   ... |
                   <random effect 1> <values>
                   <random effect 2> <values>
                   ...                      / options;
```

Options

CL   Requests *t*-type confidence intervals for estimate.

ALPHA=<p>   Specifies size of confidence interval (default is 0.05).

DF=<n>   DF for *t* test.

DIVISOR=<n> Value by which to divide all coefficients so that fractional coefficients can be entered as integer numerators.

E   Shows the effect coefficients used in the estimate (useful to check ordering if interaction effects are used).

When the effect of interest is contained within another fixed interaction effect, the coefficients for this effect are automatically included in the estimate by SAS. For example, the following code for a fixed effects analysis of the multi-centre hypertension data,

```
PROC MIXED; CLASS centre treat;
MODEL dbp = dbp1 treat centre centre*treat/ DDFM=SATTERTH;
ESTIMATE 'A-B' treat 1 -1 0/ E CL;
```

would cause the coefficients below to be used by SAS for constructing the estimate.

```
                  Coefficients for A-B

    Effect        CENTRE      TREAT       Row 1
    INTERCEPT                               0
    DBP1                                    0
    TREAT                       A           1
    TREAT                       B          -1
    TREAT                       C           0
    CENTRE          1                       0
    CENTRE          2                       0
    CENTRE          3                       0
    .
    .
    .
    CENTRE*TREAT    1           A       0.0384615385
    CENTRE*TREAT    1           B          -0.04
```

```
CENTRE*TREAT    1           C                   0
CENTRE*TREAT    2           A          0.0384615385
CENTRE*TREAT    2           B                -0.04
CENTRE*TREAT    2           C                   0
CENTRE*TREAT    3           A          0.0384615385
CENTRE*TREAT    3           B                -0.04
CENTRE*TREAT    3           C                   0
CENTRE*TREAT    4           A          0.0384615385
        .
        .
        .
```

The centre·treatment coefficients are equal to $1/n_i$, where n_i is the number of centres at which treatment i is received (26 for treatment A and 25 for treatment B). Because treatments A and B are not received at every centre this estimate is in fact 'non-estimable' and SAS gives the following output.

```
                     ESTIMATE Statement Results
     Parameter    Estimate    Std Error   DF    t    Pr>|t|  Alpha
     A-B             .            .    .    .         .      .

                     ESTIMATE Statement Results
                          Lower      Upper
                            .          .
```

However, if the containing effect is fitted as random, SAS does not include any coefficients for this containing effect in the estimate. For example, the following code for a random effects analysis of the multicentre hypertension data:

```
PROC MIXED; CLASS centre treat;
MODEL dbp = dbp1 treat/DDFM=SATTERTH;
RANDOM centre centre*treat;
ESTIMATE 'A-B' treat 1 -1 0/ E CL;
```

would cause the following coefficients to be used by SAS for constructing the estimate:

```
            Coefficients for A-B
     Effect         TREAT        Row 1
     INTERCEPT                      0
     DBP1                           0
     TREAT          A               1
     TREAT          B              -1
     TREAT          C               0
```

and would lead to the estimate below.

```
                     ESTIMATE Statement Results
     Parameter Estimate   Std Error    DF     t    Pr>|t|   Alpha
     A-B      1.28592019 1.39135076   23.8   0.92  0.3646    0.05
```

The ESTIMATE statement can also be used to estimate shrunken random effects by specifying the coefficients of the random effects. For example, to estimate the shrunken difference between the first two treatments within the first two centres the following ESTIMATE statements can be added to the code above:

```
ESTIMATE 'C1, A-B' treat 1 -1 0| centre*treat 1 -1
                                            0/ E CL;
ESTIMATE 'C2, A-B' treat 1 -1 0| centre*treat 0 0 0 1 -1
                                            0/ CL;
```

These cause the following coefficients to be used for the first estimate,

```
                  Coefficients for C1, A-B
      Effect           CENTRE   TREAT        Row 1
      INTERCEPT                                 0
      DBP1                                      0
      TREAT                       A             1
      TREAT                       B            -1
      TREAT                       C             0
      CENTRE            1                       0
      CENTRE            2                       0
      CENTRE            3                       0
      .
      .

      .
      CENTRE           41                       0
      CENTRE*TREAT      1         A             1
      CENTRE*TREAT      1         B            -1
      CENTRE*TREAT      1         C             0
      CENTRE*TREAT      2         A             0
      CENTRE*TREAT      2         B             0
      CENTRE*TREAT      2         C             0
      CENTRE*TREAT      3         A             0
      CENTRE*TREAT      3         B             0
      CENTRE*TREAT      3         C             0
      .
      .

      .
      CENTRE*TREAT     41         A             0
      CENTRE*TREAT     41         B             0
      CENTRE*TREAT     41         C             0
```

and give the estimates below.

```
                  ESTIMATE Statement Results
      Parameter    Estimate    Std Error    DF      t  Pr>|t|  Alpha
      C1, A-B    -0.86460064  2.26674144  7.35  -0.38  0.7137  0.05
      C2, A-B     1.64050148  2.84725614  2.65   0.58  0.6097  0.05
```

The CONTRAST statement

This statement can be used to carry out F tests on the fixed effects. A single or multiple contrast, $\mathbf{C}$, can be defined by

$$\mathbf{C} = \mathbf{L}' \begin{pmatrix} \hat{\alpha} \\ \hat{\beta} \end{pmatrix}.$$

The test may either involve a single contrast (**L** has a single column), e.g. to compare two treatments, or several contrasts (**L** has several columns), e.g. to test the equality of several treatments. Details of how the *F* statistic is obtained from a contrast are given in Section 2.4.4. When only one row is specified, the *F* test results will give the same *p*-value as the *t* test in an equivalent ESTIMATE statement.

```
CONTRAST '<label>'   <effect1> <effect1 values>
                     <effect2> <effect2 values>
                     ... |
                     <random effect 1> <values>
                     <random effect 2> <values>
                     ...
                     second row of L,
                     ...
                     / options;
```

As for the ESTIMATE statement, when the effect of interest is contained within another fixed interaction effect, the coefficients for this effect are automatically included in the estimate by SAS.

Options

CHISQ   Requests Wald chi-squared test (see Section 2.4.4) in addition to the *F* test. This is an asymptotic test which makes no adjustment for the denominator DF and therefore could be inaccurate for small samples.

DF=<n> Specifies denominator DF for the *F* test.

E   Prints the **L** matrix.

Example

```
PROC MIXED; CLASS centre treat;
MODEL dbp = dbp1 treat/ DDFM=SATTERTH;
RANDOM centre centre*treat;
CONTRAST 'TREAT' treat 1 -1 0, treat 1 0 -1/ CHISQ E;
```

```
Tests of Fixed Effects
Source    NDF    DDF    Type III F    Pr>F
DBP1       1     284         6.31    0.0125
TREAT      2      25         2.28    0.1230
```

Use of the *E* option

```
               Coefficients for TREAT
    Effect       TREAT      Row 1      Row 2
    INTERCEPT                   0          0
    DBP1                        0          0
    TREAT        A              1          1
    TREAT        B             -1          0
    TREAT        C              0         -1
```

Use of the CHISQ *option*   *(gives additional columns in the following output)*

```
         CONTRAST Statement Results
  Source NDF DDF ChiSq    F Pr>ChiSq   Pr>F
  TREAT    2  25 4.56 2.28   0.1021 0.1230
```

This contrast gives an identical F test result to that given under 'Tests of Fixed Effects' above. Printing the **L** matrix coefficients is particularly helpful when interactions are involved to check the ordering of effects used by SAS. This option may be most useful if identity of a subgroup of treatments requires testing.

The REPEATED statement

The REPEATED statement is used to specify a covariance pattern for the residual matrix, **R**. The repeated measurements should appear in a single variable with the time points specified by another variable. The blocking effect is specified by the SUBJECT variable and the repeated effect (e.g. time) is the effect used to structure the **R** matrix.

REPEATED <repeated effect>/ SUBJECT=<blocking effect>

                                                                 options;

Note that the blocking effect should have only one observation per repeated effect (otherwise an infinite covariance matrix will occur). A repeated effect should always be used unless the covariance pattern does not depend on order (e.g. the compound symmetry structure).

Options relating to the R matrix structure

TYPE=<pattern> This specifies the covariance pattern for the **R** matrix. A wide range of patterns are available, some of which are listed below. These patterns are defined in Section 6.2.

UN	General,
AR(1)	First-order autoregressive,
CS	Compound symmetry,
TOEP	Toeplitz,
UN(1)	Heterogeneous uncorrelated,
CSH	Heterogeneous compound symmetry,
ARH(1)	Heterogeneous first-order autoregressive,
TOEP(H)	Heterogeneous Toeplitz,
UN(n)	Banded general, n bands,
TOEP(n)	Banded Toeplitz, n bands.

Other covariance patterns available are listed in the SAS manual.

GROUP=<effect>   This causes a separate covariance pattern to be estimated for each category of the group effect. This option should be used cautiously if there are a large number of group categories since it can lead to many extra parameters, particularly if a complex pattern (e.g. general) is used.

Options to print the variance parameters

R=<values>   Prints **R** matrix for subjects denoted by values (first subject is listed if <values> is omitted).

RI=<values>   Prints inverse of **R** matrix for subjects denoted by values.

RCORR=<values>   Prints correlation matrix corresponding to **R** matrix for subjects denoted by values.

Other options

LOCAL   Expresses residual variance as $\mathbf{R} + \sigma_r^2 \mathbf{I}$, allowing σ_r^2 to be fitted separately from the other parameters of **R**.

Example

Examples of using the REPEATED statement are given in Sections 6.3, 7.7, 8.1.2 and 8.2.

The PARMS statement

This statement can be used to

- Fix the values of variance components. This can be useful when the covariances are known with a greater accuracy from previous studies than that likely to be obtained in the data analysed.

- Supply initial values for variance components for the iterative procedure.

- Carry out a grid search over a range of variance component values and take those with the highest likelihood as initial values for the iterative procedure.

- Carry out a grid search over a range of variance component values and take those with the highest likelihood as the final estimates.

The latter two options can be helpful in a situation where there is the possibility of local (rather than global) solutions. These are most likely to occur when a large number of variance parameters are fitted.

Since a PARMS statement is not needed for most types of mixed model analysis we have not included details of its use here.

The PRIOR statement

A random effects model can be fitted using the Bayesian approach with the PRIOR statement. It works by considering the model in a Bayesian framework and assigning non-informative priors for the parameters. The fixed effects are analytically integrated out of the posterior distribution to give a posterior density function for the variance components. This density function is proportional (or almost proportional, depending on the choice of prior distribution) to the REML likelihood.

The joint posterior density function for the variance parameters is evaluated using rejection sampling with the base distribution, $g(\theta; \mathbf{y})$, taken as the product of inverse gamma distributions for the variance parameters. The inverse gamma distributions are approximations to the marginal densities for the variance parameters. Fixed and random effects samples are produced when the SOLUTION option is used in the MODEL and RANDOM statements, respectively. These samples are taken from the posterior density conditioned on the most recent variance components sampled (see Section 2.3.5). This conditional posterior density is multivariate normal. Note that negative variance component samples are not possible from inverse gamma distributions and this truncates the simulated density function at zero for the variance components. This can sometimes cause parameter estimates to differ from those obtained using REML.

For each sample of parameters the probabilities are output for the joint posterior density, $p(\theta; \mathbf{y})$, and for the base density, $g(\theta; \mathbf{y})$, and their ratio, $p(\theta; \mathbf{y})/g(\theta; \mathbf{y})$. F statistics for tests of the equality of each of the fixed effects are also given at each sample, although it is not clear exactly how these values should be used.

Syntax

PRIOR <prior distribution>/ options;

If no prior distribution is specified in this statement, then the default Jeffreys' prior is used (see Section 2.3.4). The only other option available is a flat prior, which is specified by

PRIOR FLAT;

Options

NSAMPLE=<n> Specifies number of samples, default 1000.
Note that a MAKE statement with the NOPRINT option requesting a SAMPLE table (see below) is almost essential when the PRIOR statement is used, because it prevents the large number of sampled values being output. This statement will output the sampled values to a new SAS dataset so that parameter estimates can be obtained.

Example

Use of the PRIOR statement is illustrated in Section 2.5.

The MAKE statement

This statement reads selected parts of PROC MIXED output into a SAS dataset, thus allowing it to be manipulated by other SAS procedures. Any number of MAKE statements can be used in one PROC MIXED analysis.

```
MAKE '<table>' OUT=<dataset> option;
```

Tables are available corresponding to all results produced by PROC MIXED. Those most likely to be useful are

Table	Statement/option
PREDICTED	MODEL/PREDICTED
PREDMEANS	MODEL/PM
SOLUTIONF	MODEL/SOLUTION
SOLUTIONR	RANDOM/SOLUTION
LSMEANS	LSMEANS
ESTIMATE	ESTIMATE
CONTRAST	CONTRAST
R<n>	REPEATED/R=<n>
RCORR<n>	REPEATED/RCORR=<n>
PARMS	PARMS
SAMPLE	PRIOR
POSTERIOR	PRIOR

Other tables available are listed in the SAS manual. We suggest that a procedure such as PROC CONTENTS is used to determine the names of the variables produced by each table.

Option

NOPRINT Suppresses printing of output from the corresponding statement/ option. This is particularly useful for statements or options that can give lengthly output such as PRIOR and MODEL/PREDICTED.

Example

```
PROC MIXED; CLASS centre treat;
MODEL dbp = dbp1 treat/ DDFM=SATTERTH;
RANDOM centre centre*treat;
LSMEANS treat/ CL;
ESTIMATE 'A-B' treat 1 -1 0/ CL;
ESTIMATE 'A-C' treat 1 0 -1/ CL;
ESTIMATE 'B-C' treat 0 1 -1/ CL;
CONTRAST 'TREAT' treat 1 -1 0, treat 1 0 -1;
MAKE 'LSMEANS' OUT=lsmeans NOPRINT;
MAKE 'ESTIMATE' OUT=estimate NOPRINT;
MAKE 'CONTRAST' OUT=contrast NOPRINT;
PROC PRINT DATA=lsmeans;
PROC PRINT DATA=estimate;
PROC PRINT DATA=contrast;
```

OBS	_EFFECT_	TREAT	_LSMEAN_	_SE_	_DF_
1	TREAT	A	92.34913293	1.10348051	52.0350
2	TREAT	B	91.06321274	1.14545866	51.3558
3	TREAT	C	89.42174246	1.11231265	58.6713

```
OBS    _T_    _PT_ _ALPHA_   _LOWER_  _UPPER_
 1    83.69   0.0001   0.05   90.1349  94.5634
 2    79.50   0.0001   0.05   88.7640  93.3624
 3    80.39   0.0001   0.05   87.1957  91.6477
```

```
                                              A      L      U
       P                                      L      O      P
 O     A      E                        P      P      W      P
 B     R      S          S       D     _      H      E      E
 S     M      T          E       F     T      T      A      R      R
 1    A-B  1.28592019  1.39135076  23.8401  0.92  0.3646  0.05 -1.5867 4.1585
 2    A-C  2.92739048  1.37270937  25.5999  2.13  0.0427  0.05  0.1036 5.7512
 3    B-C  1.64147028  1.40487730  25.6505  1.17  0.2534  0.05 -1.2482 4.5312
       OBS  SOURCE     NDF      DDF       F     P_F
        1   TREAT       2     25.0031   2.28    0.1230
```

The ID statement

Includes specified variables in output from PREDICTED and PM options in MODEL statement. An example using this statement is given in Section 6.3.4.

```
ID <ID variables>;
```

The WEIGHT statement

Fits a weighted mixed model with weights given by specified variable.

```
WEIGHT <weight variable>;
```

9.3 BASIC USE OF PROC GENMOD AND THE GLIMMIX MACRO

There is not yet a SAS procedure available to perform all types of GLMM analysis. However, a limited number of covariance pattern models can be fitted using the REPEATED statement in PROC GENMOD (Release 6.12 onwards). For other types of mixed model a SAS macro known as GLIMMIX can be used. We now give basic details on PROC GENMOD and GLIMMIX.

9.3.1 PROC GENMOD

PROC GENMOD is primarily a procedure for fitting fixed effects GLMs. However, since Release 6.12 it has been possible to fit GLMMs with covariance patterns using a REPEATED statement. A method known as generalised estimating equations (see Liang and Zeger, 1986) is used to fit the GLMM. The empirical estimator is produced by default for estimating fixed effects standard errors (see Section 2.4.3) but the model-based estimator can additionally be requested.

We give basic details of the SAS code required to fit covariance pattern models to binomial and Poisson data. More detail can be found in the SAS manual: *SAS/STAT Software: Changes and Enhancements through Release 6.12*. Use of the procedure is illustrated in the examples analysed in Sections 6.4, 8.5 and 8.13, and further examples will not be given here.

Basic syntax

```
PROC GENMOD; CLASS <class effects>;
MODEL <y variable(s)> = <fixed effects>/ <options>;
REPEATED SUBJECT=<subject effect>/ <options>;
```

The y variable is specified as a single value for data assumed to have Poisson distributions, or as a numerator with denominator for binary data (e.g. r/n). The subject effect is the blocking effect for the covariance pattern (see Section 6.2) and has the same role as the subject effect in the REPEATED statement of PROC MIXED.

MODEL statement options

DIST=<distribution>   This option specifies the distribution:
    B   Binomial,
    P   Poisson.

LINK=<link function>   This option specifies the link function. When omitted the canonical link function is taken:
    LOGIT   Logit,
    LOG   Log.

TYPE3 $\Big\}$ Used together these options produce composite
WALD   chi-squared tests of fixed effects.

REPEATED statement options

TYPE=<covariance pattern>   This option specifies the covariance pattern:
    CS   Compound symmetry,
    AR(1)   First-order autoregressive,
    TOEP(n)   Toeplitz (notice that n, the number of bands of parameters, needs to be specified),
    UN   General.

WITHIN=<fixed effect>	This defines the effect to be used for structuring the covariance pattern. It has the same role as the main effect used in the REPEATED statement in PROC MIXED. In repeated measures analyses it is usually time or visit.
COVB	Prints the covariance matrix for the fixed effects.
CORRW	Prints the correlation parameters in the **P** matrix ($\mathbf{R} = \phi\mathbf{A}^{1/2}\mathbf{B}^{1/2}\mathbf{PB}^{1/2}\mathbf{A}^{1/2}$, see Section 3.2.1).
MODELSE	Prints fixed effects estimates with model-based standard errors in addition to the (default) empirical estimators.

9.3.2   The GLIMMIX macro

This SAS macro can be used to fit all types of GLMMs. It was written by Russ Wolfinger (from SAS) and fits the GLMM using the pseudo-likelihood method described in Section 3.2. At each iteration it computes a linearised pseudo variable and fits a weighted mixed model using PROC MIXED. A dispersion parameter is fitted by default by the macro, but this can be avoided by using a suitable PARMS statement (see below). Many of the options from the GENMOD and MIXED procedures can be used. While this macro appears to work well, it should be borne in mind that it has not been fully tested and therefore is not guaranteed to be as reliable as established SAS procedures. The macro can be obtained from SAS's Web site at www.sas.com, or by anonymous ftp to ftp.sas.com or from www.med.ed.ac.uk/phs/mixed. Examples of using GLIMMIX can be found in Sections 3.4, 5.7, 8.10, 8.11 and 8.12.

Basic syntax

The macro can be used very much like a SAS procedure except that macro parameters are used in place of procedure statements. The most important parameters are as follows.

```
%GLIMMIX( STMTS=%STR(<PROC MIXED statements>),
          ERROR=<distribution>,
          LINK=<link function>,
          FREQ=<frequency parameter>);
```

The STMTS parameter can include CLASS, MODEL, RANDOM, LSMEAN, ESTI-MATE and PARMS statements specified in the same way as in PROC MIXED. However, a numerator and denominator are now needed for binomial data in the MODEL statement.

The ERROR and LINK statements are used in the same way as in PROC GENMOD. As in PROC GENMOD, a canonical link is assumed whenever the LINK parameter is not used. The FREQ parameter specifies how many times each observation is repeated. However, it is not usually necessary in clinical trials where observations are normally recorded individually. Many other options are available and these are described in the macro documentation (provided with the macro).

We have found that the algorithm can sometimes be unstable, particularly when class frequencies are small. It may be possible to overcome this problem by combining categories or, alternatively, by choosing a simpler model.

An example of using the GLIMMIX code (taken from Section 3.4) is

```
%GLIMMIX(STMTS=%STR(CLASS centre treat;
MODEL cf/one=cf1 treat/ DDFM=SATTERTH;
LSMEANS treat/ diff pdiff;
RANDOM centre;),
ERROR=B);
```

In models where a variance component is modelled at the residual level it is often desirable to fix the dispersion at 1.0. This can be done using the PARMS statement as shown in the code below (taken from Section 8.12).

```
%GLIMMIX(STMTS=%STR(
    CLASS surg;
    MODEL outcome/n=/ SOLUTION DDFM=SATTHERTH;
    RANDOM surg/ SOLUTION;
    PARMS (0) (1.0)/ EQCONS=2;),
    ERROR=B
    );
```

Glossary

Terms referred to frequently within the book are defined below. Because mixed models have been developed for use in several areas there is sometimes ambiguity in the meanings of the same terms. Thus, the definitions given here may not always agree with those in other sources. They will, however, be adhered to within this book.

Balance. See Section 1.6.

Blocking effect. An effect used to block the variance matrix. A covariance pattern is specified for data within the categories of the blocking effect. For example, in a repeated measures trial the covariance matrix is blocked by patient effects (see Section 6.2).

Bernoulli form. Binary data specified as observations of zero and one. This form must be used if covariates at the residual level are modelled; for example, if a 0/1 variable is recorded pre-and post-treatment.

Binomial form. Binary data specified as frequencies with denominators.

Containment. An effect (A) is described as contained within another effect (B) either if (B) is nested within (A), or if (B) is a random interaction term containing (A) (e.g. B = A·C) (see Section 1.6).

Containment level. This is defined as the residual error level unless the fixed effect is contained within another effect (see Section 1.6.)

Contrast. A contrast defines a specific linear comparison of fixed effect catagories. The objective is usually to determine whether the categories differ significantly by defining an appropriate test statistic from the contrast. Pairwise differences between two categories (e.g. treatments) are a common type of contrast. Multiple contrasts can be used to test the overall equality of a set of contrasts; for example, to test the overall equality of a group of treatments. Contrasts are defined in more detail in Section 2.4.4.

Crossed effects. An effect (A) is described as crossed with another effect (B) when different categories of (A) occur within each category of (B). For example in a cross-over trial, different treatment categories occur within patients, so treatment effects are crossed with patient effects.

Effect1·effect2. Interaction effect, e.g. centre·treatment denotes the interaction effect between centres and treatments.

Error stratum. See Section 1.6.

Fixed effects model. A model fitting only fixed effects.

Fixed or random effect category. An effect level; for example, if three treatments are fitted, then the treatment effect has three categories.

Full residuals. Residuals calculated by deducting only the fixed effects, $\mathbf{y} - \mathbf{X}\hat{\alpha}$ (ordinary residuals are defined when both the fixed and random effects are deducted, $\mathbf{y} - \mathbf{X}\hat{\alpha} - \mathbf{Z}\hat{\beta}$).

Generalised linear mixed model (GLMM). A mixed model for non-normal data that assumes residuals have variances proportional to those specified by a chosen distribution from the exponential family.

Least squares mean. Mean estimate for an effect category, adjusted for other effects in the model.

Mean predicted values. Values predicted for individual observations based on fixed effects only as $\mathbf{X}\hat{\alpha}$.

Mixed model. The description 'mixed' was originally used to describe a model fitting both fixed effects and random effects. Here, we use it more widely to encompass random effects models, random coefficients models and covariance pattern models (see Section 1.1).

Nested. An effect (A) is defined as nested within another effect (B) if B takes the same value for all observations within every category of A. For example, in repeated measures trials each patient receives only one treatment and patients are nested within treatments. However, in cross-over trials treatments vary between periods and patients are not nested within treatments. Note that nesting is the reverse of containment: if A is nested within B, then B can be described as contained within A.

Normal mixed model. A mixed model for normally distributed data.

Predicted values. Values predicted for individual observations based on both fixed and random effects as $\mathbf{X}\hat{\alpha} + \mathbf{Z}\hat{\beta}$.

Residuals. Residuals from the residual error strata, equal to $\mathbf{y} - \mathbf{X}\hat{\alpha} - \mathbf{Z}\hat{\beta}$.

Random effects model. A model fitting fixed and random effects. The residuals are assumed independent.

Random effect predictions. These are effectively random effect estimates. However, since they are not formally estimated within the model they are sometimes referred to as predictions. They are shrunken compared with their fixed effect counterparts (see Section 1.3).

SAS® Statistical Analysis System. The most commonly used statistical analysis package within the pharmaceutical industry.

Uniform fixed/random effect category. See Sections (3.3.3) and (3.3.4).

Variance component. Additional variation due specifically to a random effect.

Variance matrix. Matrix of variance and covariance terms.

References

Akaike, H (1974) A new look at the statistical model identification, *IEEE Transaction on Automatic Control*, AC-19, 716–723.

Box, GEP and Tiao, GC (1973) *Bayesian Inference in Statistical Analysis*, John Wiley & Sons, New York.

Breslow, NE and Clayton, DG (1993) Approximate inference in generalized linear mixed models, *Journal of the American Statistical Association*, **88**, 9–25.

Brooke, H, Gibson, A, Tappin, D and Brown, H (1997) Case–control study of sudden infant death syndrome in Scotland, 1992–1995, *British Medical Journal*, **314**, 1516–1520.

Brown, HK and Kempton, RA (1994) The application of REML in clinical trials, *Statistics in Medicine*, **16**, 1601–1617.

Carlin, JB and Louis, TA (1996) *Bayes and Empirical Bayes Methods for Data Analysis*, Chapman & Hall, London.

Carstairs, V and Morris, R (1991) *Deprivation and Health in Scotland*, Aberdeen University Press, Aberdeen,

Clayton, D and Hills M (1993) *Statistical Models in Epidemiology*, Oxford University Press, Oxford.

Collett, D (1991) *Modelling Binary Data*, Chapman & Hall London.

Diggle, PJ (1989) Testing for random dropouts in repeated measurement data, *Biometrics*, **43**, 1255–1258.

Diggle, P and Kenward, MG (1994) Informative drop-out in longitudinal data analysis, *Journal of the Royal Statistical Society, Series C*, **43**, 49–93.

Diggle, PJ, Liang, K-Y and Zeger, SL (1994) *Analysis of Longitudinal Data*, Oxford university Press, Oxford.

Dixon, JM, Ravisekar, O, Cunningham, M, Anderson, EDC, Anderson, TJ and Brown, HK (1996) Factors affecting outcome of patients with impalpable breast cancers detected by breast screening', *British Journal of Surgery*, **83**, 997–1001.

Elston, DA (1998) Estimation of denominator degrees of freedom of F-distributions for assessing Wald statistics for fixed-effect factors in unbalanced mixed models, *Biometrics*, **54**, 1085–1096.

Fisher, RA (1925) *Statistical Methods for Research Workers*, Oliver & Boyd, Edinburgh.

Giesbrecht, FG and Burns, JC (1985) Two-stage analysis based on a mixed model: large-sample asymptotic theory and small-sample simulation results, *Biometrics*, **41**, 477–486.

Gilks, WR, Richardson, S and Spiegelhalter, DJ (1995) *Markov Chain Monte Carlo in Practice*, Chapman & Hall, London.

Goldstein, H (1989) Restricted unbiased iterative generalized least-squares, *Biometrika*, **76**(3), 622–623.

Goldstein, H (1995) *Multilevel Statistical Models*, (second edition) Edward Arnold, London.

Goldstein, H (1996) Improved approximations for multilevel models with binary responses, *Journal of the Royal Statistical Society, Series A*, **159**, 505–513.

Greenhouse, SW and Geisser, S (1959) On methods in analysis of profile data, *Psychometrika*, **32**, 95–112.

Hall, S, Prescott, RJ, Hallam, RJ, Dixon, S, Harvey, RE and Ball, SG (1991) A comparative study of Carvedilol, slow release Nifedipine and Atenolol in the management of essential hypertension', *Journal of Pharmacology*, **18**(4) S36–S38.

Hartley, HO and Rao, JNK (1967) Maximum likelihood estimation for the mixed analysis of variance model *Biometrika*, **54**, 93–108.

Hay, CRM, Ludlam, CA, Lowe, GDO, Mayne, EE, Lee, RJ, Prescott, RJ, Lee, CA (1998) The effect of monoclonal or ion-exchange purified factor VIII concentrate on HIV disease progression: a prospective cohort comparison, *British Journal of Haematology*, **101**, 632–637.

Healy MJR (1986) *Matrices for Statistics*, Oxford University Press, Oxford.

Hedeker, D and Gibbons, RD (1994) A random effects ordinal regression model for multi-level analysis, *Biometrics*, **50**, 933–944.

Henderson, CR (1953) Estimation of variance and covariance components, *Biometrics*, **9**, 226–252.

Huber, P (1981) *Robust Statistics*, John Wiley & Sons, New York.

Hunter, KR, Stern, GM, Laurence, DT and Armitage, P (1970) Amantadine in Parkinsonism, *Lancet*, **i**, 1127–1129.

Jeffreys, H (1961) *Theory of Probability*, (third edition), Claredon Press, Oxford.

Jennrich, RI and Schluchter, MD (1986) Unbalanced repeated-measures models with structured covariance matrices, *Biometrics*, **42**, 805–820.

Jones, B and Kenward, MG (1989) *The Analysis of Crossover Trials*, Chapman & Hall, London.

Kacker, RN and Harville, DA (1984) Approximations for standard errors of estimators of fixed and random effects in mixed linear models, *Journal of the American Statistical Association*, **79**, 388, 853–862.

Kempthorne, O (1952) *The Design and Analysis of Experiments*, John Wiley & Sons, New York.

Kenward, MG and Roger, JH (1997) Small sample inference for fixed effects from restricted maximum likelihood, *Biometrics*, **53**, 983–997.

Kenward, MG, Lesaffre, E and Molenberghs, G (1994) An application of maximum likelihood and generalized estimating equations to the analysis of ordinal data from a longitudinal study with cases missing at random, *Biometrics*, **50**, 945–953.

Laird, NM and Ware, JH (1982) Random effects models for longitudinal data, *Biometrics*, **38**, 963–974.

Liang, KL, Zeger, SL and Qaqish, B (1992) Multivariate regression analyses for categorical data, *Journal of the Royal Statistical Society, Series B*, **54**, 3–40.

Liang, KY and Zeger, SL (1986) Longitudinal data analysis using generalized linear models, *Biometrika*, **73**, 13–22.

Lipsitz, SR, Kyungmann, K and Zhao, L (1994) Analysis of repeated categorical data using generalized estimating equations' *Statistics in Medicine*, **13**, 1149–1163.

Littell, RC, Milliken, GA, Stroup, WW, and Wolfinger, RD (1996) *SAS System for Mixed Models*, SAS Institute, Cary, NC, USA.

Little, RJ and Rubin, DB (1987) *Statistical Analysis with Missing Data*, John Wiley & Sons, New York.

Longford, NT (1993) *Random Coefficients Models*, Oxford University Press, Oxford.

McCullagh, P (1980) Regression models for ordinal data', *Journal of the Royal Statistical Society, Series B*, **42**, 109–142.

McCullagh, P and Nelder, JA (1989) *Generalised Linear Models*, Chapman and Hall, (second edition), London.

McLean, RA and Sanders, WL (1988) Approximating degrees of freedom for standard errors in mixed linear models, *Proceedings of the Statistical Computing Section, American Statistical Association*, New Orleans, 50–59.

Mead, R (1988) *The Design of Experiments*, Cambridge University Press, Cambridge.

Meyer, K (1986) Restricted maximum likelihood to estimate genetic parameters in practice. In GE Dickerson and RK Johnson) (eds), *Proceedings of the 3rd World Congress on Genetics applied to Animal Production Vol XII*, University of Nebraska, 454–459.

Nabugoomu, F and Allen, OB (1994) The estimation of fixed effects in a mixed linear model, *Proceedings of the 1993 Kansas State University Conference on Applied Statistics in Agriculture*, Kansas State University.

Patterson, HD and Thompson, R (1971) Recovery of inter-block information when block sizes are unequal, *Biometrika*, **58**, 545–554.

Prescott, RJ (1981) The comparison of success rates in cross-over trials in the presence of an order effect, *Journal of the Royal Statistical Society, Series C*, **30**, 9–15.

Rao, CR (1971) Estimation of variance and covariance components — MINQUE theory, *Journal of Multivariate Analysis*, **1**, 257–275.

Rao, CR (1972) Estimation of variance and covariance components in linear models, *Journal of the American Statistical Association*, **67**, 112–115.

Ridout, M (1991) Testing for random dropouts in repeated measurement data, *Biometrics*, **47**(4), 1617–1619.

Ripley, BD (1987) *Stochastic Simulation*, John Wiley & Sons, New York.

Rodriguez, G and Goldman, N (1995) An assessment of estimation procedures for multilevel models with binary responses, *Journal of the Royal Statistical Society, Series A*, **158**, 73–90.

Rowell, JG and Walters, RE (1976) Analysing data with repeated observations on each unit, *Journal of Agricultural Science, Cambridge*, **87**, 423–432.

Satterthwaite, FE (1946) An approximate distribution of estimates of variance components, *Biometrics Bulletin*, **2**, 110–114.

Schwarz, G (1978) Estimating the dimension of a model, *Annals of Statistics*, **6**, 461–464.

Searle, SR, Casella, G and McCulloch, CE (1992) *Variance Components*, John Wiley & Sons, New York.

Senn, S (1993) *Cross-over Trials in Clinical Research*, John Wiley & Sons, Chichester.

Smyth, JF, Bowman, A, Perren, T, Wilkinson, P, Prescott, RJ, Quinn, KJ and Tedeschi, M (1997) Glutathione reduces the toxicity and improves quality of life of women diagnosed with ovarian cancer treated with cisplatin: results of a double blind, randomised trial *Annals of Oncology*, **8**(6), 569–573.

Snedecor, GW and Cochran, WG (1989) *Statistical Methods*, (eighth edition) Iowa State University Press.

Talbot, M (1984) Yield variability of crop varieties in the UK, *Journal of Agricultural Science, Cambridge*, **102**, 315–321.

Tanner, MA (1996) *Tools for Statistical Inference: Methods for the Exploration of Posterior Distributions and Likelihood Functions*, (third edition) Springer, New York.

Thall, PF and Vail, SC (1990) Some covariance models for longitudinal count data with overdispersion, *Biometrics*, **46**, 657–671.

Thompson, R (1977) The estimation of heritability with unbalanced data. I. Observations available on parents and offspring, *Biometrics*, **33**, 485–495.

Thompson, SG and Pocock, SJ (1991) Can meta-analyses be trusted?, *Lancet*, **338**, 1127–1130.

Thrusfield, M (1995) *Veterinary Epidemiology*, Blackwell Science, Oxford.

Williams, DA (1982) Extra-binomial variation in logistic linear models, *Journal of the Royal Statistical Society, Series C*, **31**, 144–148.

Wolfinger, R and O'Connell, M (1993) Generalized linear mixed models: A pseudo-likelihood approach, *Journal of Statistical Computation and Simulation*, **48**, 233–243.

Yates, F (1940) The recovery of inter-block information in balanced incomplete block designs, *Annals of Eugenics*, **10**, 317–325.

Contacts

BMDP Statistical Software Inc., 1440 Sepulveda Boulevard, Suite 316, Los Angeles, CA 90025.

BUGS: Bayesian inference using Gibbs sampling, MRC Biostatistics Unit, Institute of Public Health, Robinson Way, Cambridge, CB2 2SR. Web page: `www.mrc-bsu.cam.ac.uk/bugs`.

Genstat Numerical Algorithms Group Ltd., Mayfield House, 256 Banbury Road, Oxford.

LOGEXACT CYTEL Software Corporation, 137 Erie St, Cambridge, MA 02139.

MLn Multi-level Models Project, 11 Woburn Square, London, WC1H 0NS. Web page: `www.ioe.ac.uk/multilevel`.

SAS Institute Inc., SAS Campus Drive, Cary, NC 27513. Web page: `www.sas.com`.

Index